Springer Series in

MATERIALS SCIENCE 107

Springer Series in

MATERIALS SCIENCE

The Springer Series in Materials Science covers the complete spectrum of materials physics, including fundamental principles, physical properties, materials theory and design. Recognizing the increasing importance of materials science in future device technologies, the book titles in this series reflect the state-of-the-art in understanding and controlling the structure and properties of all important classes of materials.

Volumes 50–98 are listed at the end of the book.

D.A. Bernards R.M. Owens
G.G. Malliaras

Editors

Organic Semiconductors in Sensor Applications

With 183 Figures

Springer

Dr. Daniel A. Bernards
Professor George G. Malliaras
Cornell University, Department of Materials Science and Engineering
327 Bard Hall, Ithaca, NY 14853, USA
E-mail: dab68@cornell.edu, ggm1@cornell.edu

Dr. Róisín M. Owens
Cornell University, Department of Biomedical Engineering
The Baker Institute, Hungerford Hill Road, Ithaca, NY 14853, USA
E-mail: ro27@cornell.edu

ISSN 0933-033X

ISBN 978-3-642-09517-7 e-ISBN 978-3-540-76314-7

Springer is a part of Springer Science+Business Media.
springer.com

Softcover reprint of the hardcover 1st edition 2008

Cover concept: eStudio Calamar Steinen
Cover production: WMX Design GmbH, Heidelberg

Preface

One of the most exciting developments of the past two decades has been the emergence of organic electronics. This term refers to the use of organic materials as active layers in a variety of semiconductor devices. One example is organic light emitting diodes, devices which emit visible light upon the passage of current through an organic semiconductor film. These devices are now commercially available in small format flat panel displays and are being intensely developed for solid-state lighting. A second example is organic thin film transistors, which control the flow of electricity in circuits and are being developed for applications in smart tags and flat panel displays.

Concurrently, our need for sensors is ever increasing. Food safety, environmental monitoring, medical diagnostics, and homeland security are all areas that would benefit from the deployment of sensors and sensor networks. Improvements would be useful or even required on multiple fronts, including sensitivity and specificity, power consumption, portability, and cost. Organic semiconductors offer many advantages in comparison with their inorganic counterparts, which make them particularly attractive for sensor applications. First, they can be deposited at or near room temperature on large area surfaces and are compatible with mechanically flexible supports such as paper and plastic. This enables their use in roll-to-roll fabrication techniques, which can dramatically decrease manufacturing costs, an important attribute for disposable sensors. Second, their properties can be tailored by means of chemical synthesis. This includes electronic properties (such as energy gap and electron affinity) but also properties such as surface energy. Of particular interest for sensors is the ability to covalently attach biologically relevant moieties to organic semiconductor molecules. Such hybrid materials have the potential to lead to the fabrication of sensors with high sensitivity and specificity.

It is important to note that the main drawbacks of organic semiconductor devices are not detrimental to their application in sensors: Low-end performance, for example in terms of device speed, prohibits organics semiconductor devices from competing with silicon in high-end computing applications. This does not, however, constitute a limitation, as sensors can tolerate considerably slower detection speeds. Moreover, long-term stability issues, which often plague organic semiconductor devices, are not relevant for disposable sensors. Therefore, sensors represent an application that can benefit from all the advantages of organic semiconductors and in principle suffers from none of the limitations. Quite naturally, the application of organic semiconductors and their devices in sensors has been attracting increased attention in the past few years.

This book covers central research directions in this rapidly emerging field. The first two chapters discuss fluorescence-based sensors and show how one can tailor organic semiconductors to yield large changes in their emission properties upon interaction with an analyte. The next two chapters deal with the application of organic light emitting diodes and photodetectors in sensors and their integration with lab-on-a-chip concepts. A variety of solid state devices are analyzed in the fifth chapter and the applications of lasing and photoconducting organic devices in sensors are proposed. The emphasis then shifts to electrical detection, first with field-effect transistors and then with electrochemical ones. In the remaining four chapters the mechanism of operation, the merits, and the potential applications of these devices in signal transduction are discussed.

We believe that the application of organic semiconductors and their devices in sensors will experience significant growth in the years to come. We hope that this book will serve as a useful text and reference for this emerging field.

Ithaca, New York, 2007

Dan Bernards
Róisín Owens
George Malliaras

Contents

Contributors

K.E. Achyuthan
Department of Biosensors and Nanomaterials, Sandia National Laboratories
Albuquerque, NM 87123, USA

G.C. Bazan
Departments of Materials and Chemistry & Biochemistry, University of California
Santa Barbara, CA 93106, USA

A. Becknell
Applied Physics Laboratory, Johns Hopkins University
Laurel, MD 20723, USA

M. Berggren
Organic Electronics, ITN, Linköpings Universitet
SE-601 74 Norrköping, Sweden

J. Bobacka
Laboratory of Analytical Chemistry, Process Chemistry Centre, Åbo Akademi
Biskopsgatan 8
FI-20500 Åbo-Turku, Finland

A. Bonfiglio
Department of Electrical and Electronic Engineering, University of Cagliari, and INFM-CNR S3 Centre
Modena, Italy

D.D.C. Bradley
Department of Physics, Imperial College London
South Kensington
London SW7 2AZ, UK

V. Bulovič
Department of Electrical Engineering and Computer Science, Massachusetts Institute of Technology
Cambridge, MA 02139, USA

S. Chemburu
Department of Chemical and Nuclear Engineering, Center for Biomedical Engineering, University of New Mexico
Albuquerque, NM 87131, USA

P. Cosseddu
Department of Electrical and Electronic Engineering, University of Cagliari, and INFM-CNR S3 Centre
Modena, Italy

A.J. deMello
Department of Chemistry
Imperial College London
South Kensington
London SW7 2AZ, UK

J.C. deMello
Department of Chemistry
Imperial College London
South Kensington
London SW7 2AZ, UK

R. Forchheimer
Image Coding, ISY
Linköpings Universitet
SE-581 83 Linköping, Sweden

J. Ho
Department of Electrical Engineering
and Computer Science,
Massachusetts Institute
of Technology
Cambridge, MA 02139, USA

O. Hofmann
Molecular Vision Ltd, BioIncubator
Unit, Bessemer Building (RSM)
Level 1, Prince Consort Road
London SW7 2BP, UK

J. Huang
Department of Materials Science
and Engineering
Johns Hopkins University
Baltimore, MD 21218, USA

A. Ivaska
Laboratory of Analytical Chemistry,
Process Chemistry Centre,
Åbo Akademi
Biskopsgatan 8
FI-20500 Åbo-Turku, Finland

E. Ji
Department of Chemistry,
University of Florida
Gainesville, FL 32611, USA

H. Katz
Department of Materials Science
and Engineering
Johns Hopkins University,
Baltimore, MD 21218, USA

A. Kumar
Department of Chemistry, Indian
Institute of Technology Bombay
Mumbai 400076, India

O. Larsson
Organic Electronics, ITN,
Linköpings Universitet
SE-601 74 Norrköping, Sweden

Y. Liu
Department of Chemistry,
University of Florida
Gainesville, FL 32611, USA

G.P. Lopez
Department of Chemical
and Nuclear Engineering,
Center for Biomedical Engineering,
University of New Mexico
Albuquerque, NM 87131, USA

I. Manunza
Department of Electrical
and Electronic Engineering,
University of Cagliari,
and INFM-CNR S3 Centre
Modena, Italy

D. Nilsson
Acreo AB, Bredgatan 34
SE-602 21 Norrköping, Sweden

K. Ogawa
Department of Chemistry,
University of Florida
Gainesville, FL 32611, USA

E. Orgiu
Department of Electrical
and Electronic Engineering,
University of Cagliari,
and INFM-CNR S3 Centre
Modena, Italy

A. Rose
Department of Chemistry,
Massachusetts Institute of
Technology
Cambridge, MA 02139, USA

K.S. Schanze
Department of Chemistry,
University of Florida
Gainesville, FL 32611, USA

K. See
Department of Materials Science
and Engineering,
Johns Hopkins University
Baltimore, MD 21218, USA

J. Shinar
Ames Laboratory-USDOE
and Department of Physics and
Astronomy, Iowa State University
Ames, IA 50011, USA

R. Shinar
Microelectronics Research Center,
Iowa State University
Ames, IA 50011, USA

J. Sinha
Department of Chemistry, Indian
Institute of Technology Bombay
Mumbai 400076, India

P.-O. Svensson
Organic Electronics, ITN,
Linköpings Universitet
SE-601 74 Norrköping, Sweden

T. Swager
Department of Chemistry,
Massachusetts Institute of
Technology
Cambridge, MA 02139, USA

S. Wang
Key Laboratory of Organic Solids,
Institute of Chemistry,
Chinese Academy of Sciences
Beijing 100080, P.R. China

D.G. Whitten
Department of Chemical
and Nuclear Engineering,
Center for Biomedical Engineering,
University of New Mexico
Albuquerque, NM 87131, USA

1

Water Soluble Poly(fluorene) Homopolymers and Copolymers for Chemical and Biological Sensors

G.C. Bazan and S. Wang

1.1 Introduction

Conjugated polymers (CPs) are characterized by a delocalized electronic structure. These materials have established themselves as useful components in optoelectronic devices such as light-emitting diodes (LEDs), field effect transistors (FETs), and photovoltaic devices [1–3]. Less explored is their use in chemical or biological sensor applications, particularly in homogeneous formats. The chemical structures of CPs offer several advantages as the responsive element in optical, chemical, and biological-detection schemes. As a result of their electronic structure, one often observes very efficient coupling between optoelectronic segments [4–14]. Excitations can be efficiently transferred to lower energy electron/energy acceptor sites over long distances in ways not easily accessible for assemblies of weakly interacting chromophores. Charge transport [15], conductivity [16–19], emission efficiency [20], and exciton migration [2] are easily perturbed by external agents, leading to large changes in measurable signals [21]. Trace detection of analytes has been successfully accomplished by making use of these amplification mechanisms [22–40].

Solubility in aqueous media is essential for interfacing with biological analytes. Water-soluble conjugated polymers typically incorporate charged functionalities as pendant groups on the conjugated backbone [41]. This class of materials, often referred to as conjugated polyelectrolytes, embody the semiconducting and optical properties of conjugated polymers and the complex charge-mediated behavior of polyelectrolytes. From the perspective of developing biosensors, the charged nature of the polymers provides for a convenient tool to control the average distance between optical partners. A widely used approach involves coordinating electrostatic interactions upon target recognition by a probe structure.

Of the various primary conjugated polymer backbones, such as poly (*p*-phenylenevinylene)s, poly(thiophene)s, poly(phenyleneethynylene), etc., [42] poly(fluorene) and related structures have been widely used in biological

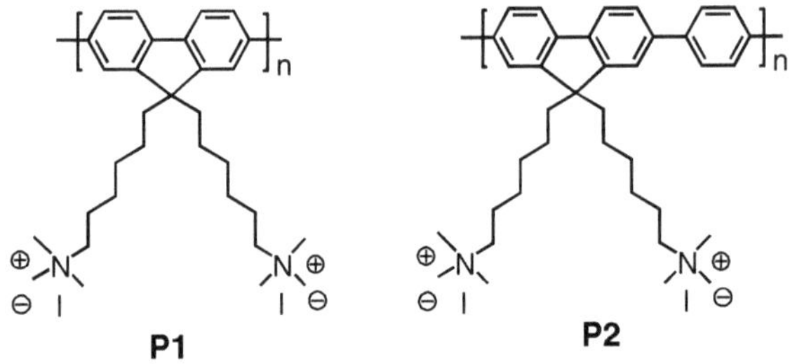

Scheme 1.1. Molecular structures of a water-soluble poly(fluorene) (**P1**) and a poly(fluorene-*alt*-phenylene) (**P2**)

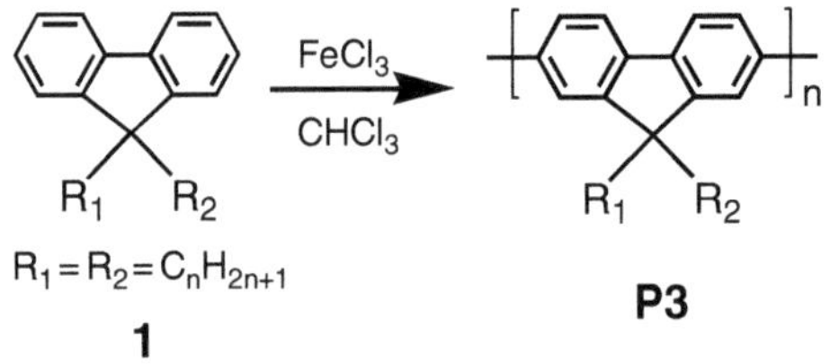

Scheme 1.2. The synthetic scheme of the soluble poly(fluorene) **P3**

and chemical sensing applications because of their facile substitution at the fluorene C9 position, good chemical and thermal stability, and high fluorescence quantum yields in water [43]. Typical molecular structures include the homopolymer **P1** and the copolymer **P2**, which are shown in Scheme 1.1. The purpose of this review is to highlight some of the fundamental properties of charged poly(fluorene)s and related copolymers and their use in optical sensory processes. Preparation methods are provided, which serve to emphasize typical synthetic approaches but do not constitute an exhaustive list of published procedures. The function and properties of conjugated polyelectrolytes with repeat units other than fluorene can be found in the literature [20–40].

1.2 General Structures and Properties

1.2.1 Design, Synthesis, and Structural Properties

The first synthesis of a soluble and processible poly(2,7-fluorene) was reported by Yoshino and coworkers in 1989 [44, 45]. Their approach involved coupling fluorene monomers with substituents at the C9 position by chemical oxidation with $FeCl_3$ (see **P3** in Scheme 1.2). However, the polymerization process was not regiospecific and the polymer backbone contained structural defects that influence the electronic delocalization. Thus, the synthesis of defect-free poly(fluorene)s became a major challenge and advances in carbon–carbon bond forming reactions promoted by organometallic catalysts provided the

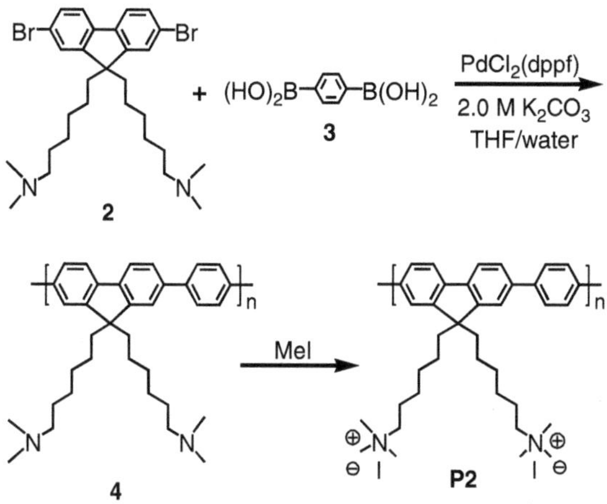

Scheme 1.3. The synthetic scheme of the water-soluble copolymer **P2**

solution. The most popular polymerization methods include Suzuki, Stille, and Yamamoto coupling reactions [46, 47].

Water-soluble copolymers containing fluorene repeat units were initially prepared by quaternization with methyl iodide of pendant amine groups on a neutral precursor, as shown in Scheme 1.3 [48]. Neutral precursor materials are particularly useful for proper characterization because of their solubility in common organic solvents. Special care must be taking during the "charging up" of the polymer structure by way of reactions such as quaternization reactions (cationic backbones) or deprotonation steps (anionic backbones). These reactions transform a material that is soluble in nonpolar solvents into one that is soluble in polar media. Precipitation can take place during the transition, which limits the degree of quaternization and the final solubility in water. To circumvent this problem, the solvent is changed during the course of the reaction to avoid formation of polymer particles or deposition of the product onto the reaction vessel surface. In the case of Scheme 1.3, the reaction begins in THF and water is added after a few minutes to redissolve the mixture. Purification of the final products is typically accomplished by precipitating into a poor solvent.

The water-soluble homopolymer **P1** was reported via the set of reactions shown in Scheme 1.4 [49]. Quaternization of neutral polymer precursor **7** with methyl iodide in THF/DMF/water gave the desired **P1** target. Because of the difficulty in purifying the monomers containing amine groups, and their propensity to strongly adsorb onto chromatographic supports, an improved synthetic approach was developed, as shown in Scheme 1.5 [15]. In Scheme 1.5, the monomers and the neutral polymer precursor are obtained with high purity. Subsequent quaternization proceeds with excellent yield via the reaction of the precursor polymer **10** with trimethylamine in THF/water, followed by precipitation from acetone.

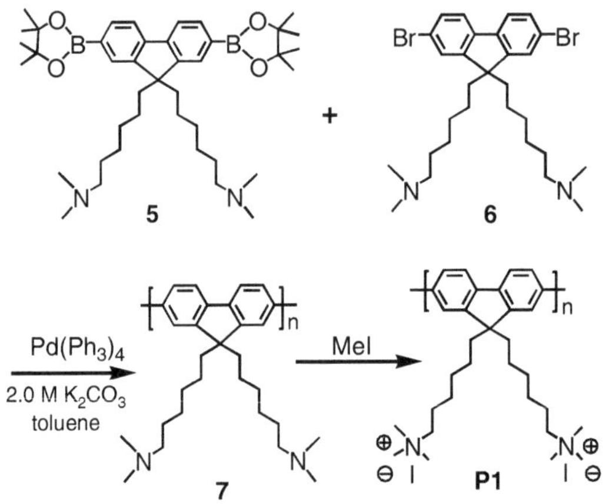

Scheme 1.4. Synthetic entry to the water-soluble poly(fluorene) **P1**

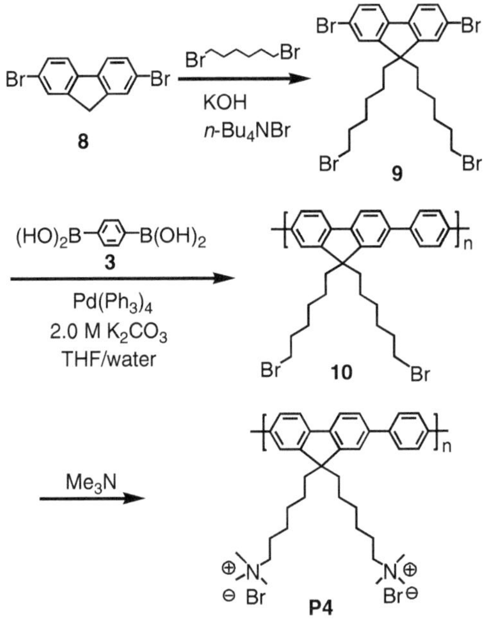

Scheme 1.5. Improved synthetic scheme for the synthesis of the water-soluble copolymer **P4**

The cationic species **P5** in Scheme 1.6 bears charged groups on the phenylene repeat units and was prepared by applying the neutral polymer precursor approach [50]. Specifically, the neutral polymer precursor **13** was obtained by the Suzuki copolymerization using **11** and **12**. Addition of EtBr to **13** provides

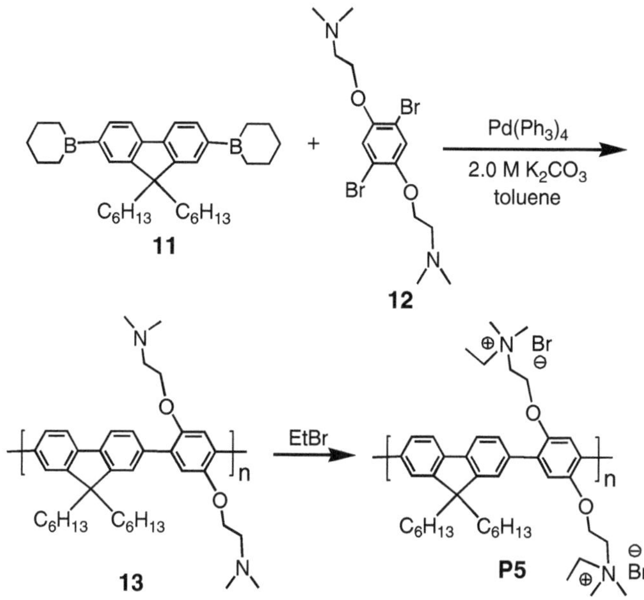

Scheme 1.6. The synthesis of the water-soluble cationic poly(fluorene) **P5**

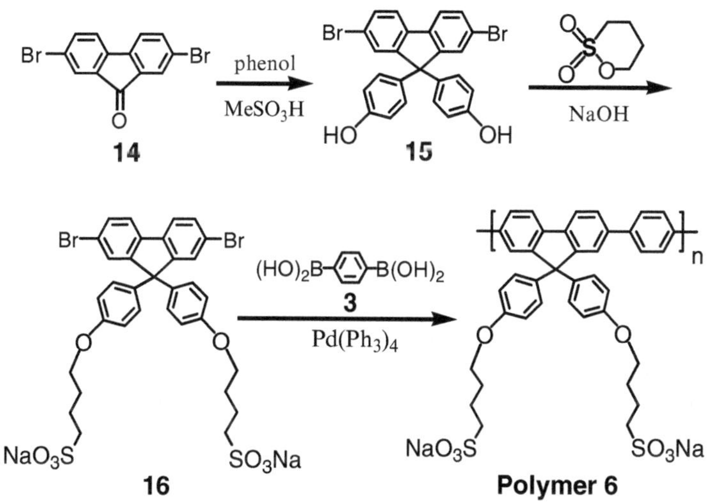

Scheme 1.7. The synthesis of the water-soluble anionic poly(fluorene) **P6**

the target material, which is soluble in DMSO, methanol, and water, and displays negligible solubility in less polar solvents such as THF and $CHCl_3$.

The synthesis of the anionic **P6** is shown in Scheme 1.7. Copolymerization of 2,7-dibromo-9,9-bis(4-sulfonylbutoxyphenyl)fluorene (**16**) with 1,4-phenylenediboronic acid (**3**) using $Pd(PPh_3)_4$ produces a polymer that is

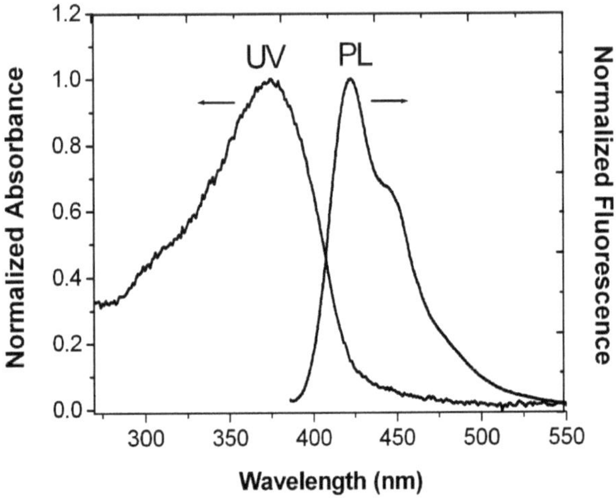

Fig. 1.1. Absorption and PL spectra of **P1** in aqueous solution. The excitation wavelength for PL spectra is 380 nm

soluble in DMSO, methanol, and water [51]. The polymer had a molecular weight of 0.65 kDa, as determined by gel permeation chromatography (GPC).

1.2.2 Optical Properties

Aqueous solutions of charged poly(fluorene)s and their derivatives typically display unstructured absorption spectra with maxima centered at approximately 380 nm. Their photoluminescence (PL) spectra exhibit vibronically well-resolved bands with maxima centered at 420 nm (Fig. 1.1). PL quantum efficiencies in aqueous media are in the range of 20–40%, depending on factors such as the substitution pattern, the degree of aggregation, the nature of counter ions, and ionic content of the medium [48,49,52]. Aggregation in water is a common feature of conjugated polyelectrolytes. This process is driven by the highly hydrophobic backbone and leads to proximity of the optically active units. Increased interchain contact often, but not always, translates to self-quenching and increased rates of energy transfer and photo-induced charge transfer. The number of charges per repeat unit and the nature of the linkers between the backbone and the charged groups influence solubility and thus aggregation tendency. Less soluble polymers exhibit a greater tendency to aggregate [53, 54].

A recent study on polymer **P1-BT**$_x$ provides insight into the effect of polymer charge on optical performance and aggregation in aqueous media [55]. The molecular structure of **P1-BT**$_x$ is given in Scheme 1.8. The main backbone is composed of alternating substituted fluorene and phenylene units together with a fractional substitution of the phenylene fragments with 2,1,3-benzothiadiazole (BT) chromophores. The subscript "x" refers to the

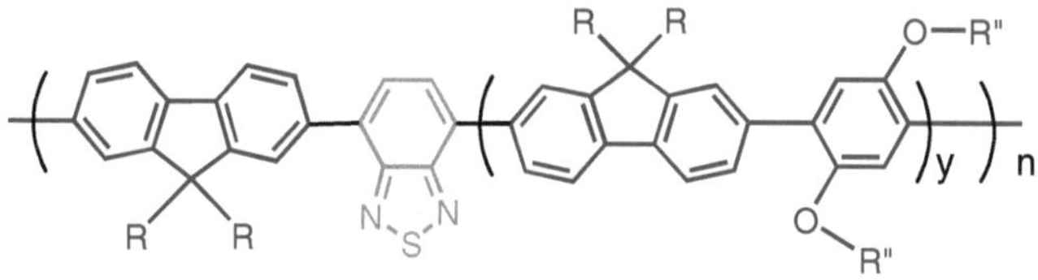

R" = $(CH_2CH_2O)_3CH_2CH_2COONa$

P1-BT$_x$

x = % BT = 100/(y+1)

Scheme 1.8. Molecular structure of **P1-BT$_x$**. Reprinted with permission from [55]. Copyright 2006 American Chemical Society

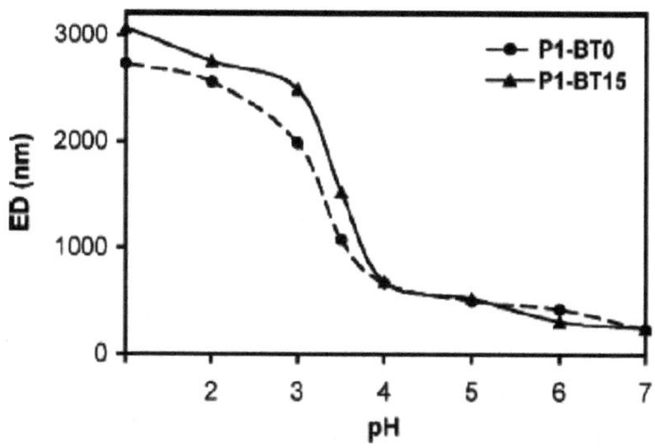

Fig. 1.2. Effective diameter (ED) determined by dynamic light scattering of **P$_1$-BT$_0$** and **P$_1$-BT$_{15}$**($[RU] = 3.8 \times 10^{-5}$ M) in water as a function of pH. Reprinted with permission from [55]. Copyright 2006 American Chemical Society

percentage of BT sites relative to the phenylene units. Two specific examples were used in these studies, namely **P1-BT$_0$** (i.e., no BT units) and **P1-BT$_{15}$**. Because of the high content of ethylene oxide groups in the backbone substituents, both neutral and anionic versions of the polymer are water-soluble.

Protonation of the carboxylic sites at low pH renders the **P1-BT$_x$** structures neutral and leads to a decrease in the electrostatic repulsion between chains. Dynamic light scattering analysis (DLS), shown in Fig. 1.2, reveals that both **P1-BT$_0$** and **P1-BT$_{15}$** show a sudden increase in particle size when the solution pH is lower than ~3.5. The concentrations of the polymers throughout this chapter are given in terms of repeat units (RUs). These data are in good agreement with substantial interchain aggregation upon protonation of the pendant carboxylic groups. The data in Fig. 1.2 should be interpreted with care since the effective diameters (EDs) are larger than would be anticipated on the basis of single chain molecular dimensions. Several assumptions are made in the treatment of DLS data. Discrepancies between calculated and

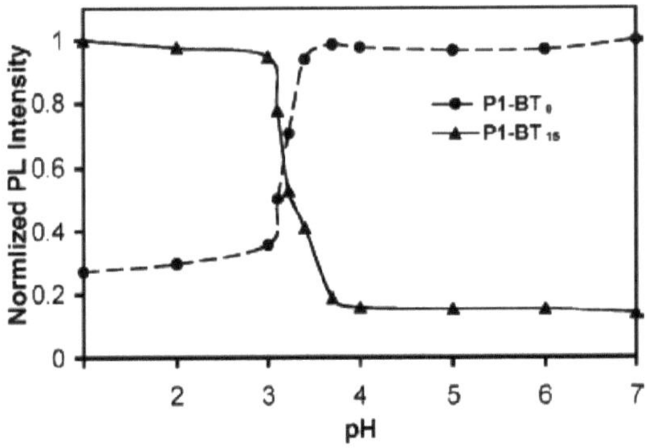

Fig. 1.3. PL response of **P_1-BT$_0$** ([RU] $= 3.8 \times 10^{-5}$ M) and **P_1-BT$_{15}$** ([RU] $=$ 4.3×10^{-5} M) as a function of pH. Data were normalized relative to the maximum PL intensity of each polymer. Reprinted with permission from [55]. Copyright 2006 American Chemical Society

actual values may arise from nonspherical aggregate shapes and from the stiff, rod-like, aspect ratio of the individual chains. Nonetheless, there is a transition to higher aggregates after the pH of the solution becomes acidic enough to neutralize the polymer charge.

Several significant observations were made in the two regimes of Fig. 1.3. Upon aggregation, the PL quantum yield of **P1-BT$_0$** decreases, possibly as a result of self-quenching via interchain contacts. For **P1-BT$_{15}$** one observes a change in the emission spectra, from blue to green, as a result of increased fluorescence resonance energy transfer (FRET) from the blue-emitting phenylene-fluorene segments to the BT sites. Additionally, the PL efficiency of the BT chromophores at low pH is considerably higher relative to conditions at high pH. Water induces quenching in the BT emission and aggregation appears to reduce BT/water contacts, ultimately decreasing energy wasting nonradiative relaxation processes, and thereby increasing optical output. Changes in optical properties are reversible when the solution pH is cycled between low and high conditions. Furthermore, as shown in Fig. 1.3, these changes occur at approximately the same pH at which aggregation was detected by DLS experiments. The collective set of observations provides insight into how interchain contacts in these aggregates influences properties that one would like to take advantage in the design of sensory schemes. A major and yet unanswered challenge rests in determining the exact organization of chains in these aggregates, particularly at low concentrations.

Recent work demonstrated that increasing the counter anion (CA) size can decrease the interchain contacts of copolymer **PFBT-X** (Scheme 1.9), which contains alternating fluorene and BT units, and leads to a substantial increase of quantum yield in the bulk [56]. Size analysis of polymers containing Br^- and $B[(3,5-(CF_3)_2C_6H_3)]_4^-$ ($BAr^F{}_4^-$) in water by DLS techniques indicates suppression of aggregation by the large and hydrophobic $BAr^F{}_4^-$. The

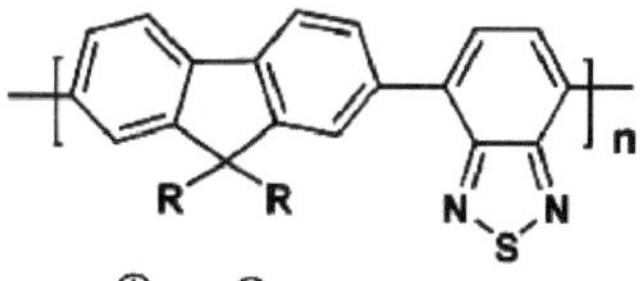

R = $(CH_2)_6\overset{\oplus}{N}Me_3\ \overset{\ominus}{X}$

X = Br, BF_4, CF_3SO_3, PF_6,
$B(C_6H_5)_4$ (BPh_4), $B(3,5\text{-}(CF_3)_2C_6H_3)_4$

PFBT-X

Scheme 1.9. Molecular structure of **PFBT-X**. Reprinted with permission from [56]. Copyright 2006 American Chemical Society

charge compensating ions are, therefore, important structural components in these materials and can be used to regulate optoelectronic properties for specific applications. Starting with a parent structure, simple ion exchange methods can be taken advantage of to generate a range of materials with substantially different performances.

The amphiphilic characteristics of charged poly(fluorene)s lead to different aggregation structures in different solvents as well as optical coupling of the optical partners [57]. In the case of **P1**, there exist two different types of aggregates depending on the solvent medium (Scheme 1.10). Single chain behavior or minor aggregation occurs when the THF content is in the range from 30% to 80% (Scheme 1.10a). These solvent mixtures allow for solvation of both components of the polymer structure. In pure water, the polymer likely forms tight aggregates, with chains coming together and forming the inner core (Scheme 1.10b). Interchain aggregation in water is dominated by hydrophobic interactions, and leads to lower emission intensities due to π-π stacking and self-quenching. Addition of THF to aqueous solutions breaks up the aggregates. Longer interchain distances reduce self-quenching and give rise to higher emission frequencies. When the THF content is higher than 80% the ionic interactions of charged groups with the nonpolar medium lead to the groups becoming buried within a new aggregate structure (Scheme 1.10c), which is dominated by the electrostatic interactions between the charged quaternary amine groups and charge compensating iodide ions [58].

1.3 Signal Transduction Mechanisms in Sensors

Biosensors are devices that transduce a biological recognition event (such as antibody–antigen binding) into measurable signals. A particular function of conjugated polymers is to amplify the signals so that lower concentrations of analyte can be interrogated. Within the context of water-soluble poly(fluorene)s, their action has been primarily to amplify fluorescent signatures. This amplification is the result of a higher optical cross section of

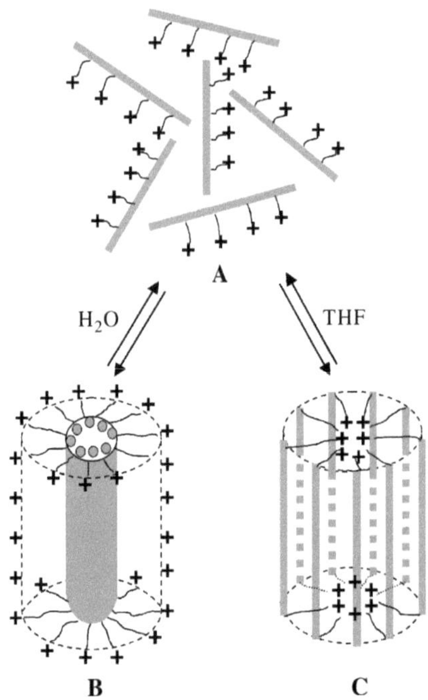

Scheme 1.10. Proposed aggregation modes of **P1** in water with different THF content [58]. Reproduced by permission of The Royal Society of Chemistry

the polymer, relative to small molecule reporters, and efficient FRET to a signaling chromophore that is triggered upon a molecular recognition event.

In comparison to intensity-based methods, techniques that rely on FRET provide large changes in emission profiles and open the opportunity for ratiometric fluorescence measurements [59]. Assays of this type are less prone to false positives from nonspecific binding events.

As shown by Förster [60], dipole–dipole interactions lead to long-range resonance energy transfer from a donor chromophore to an acceptor chromophore. Equation (1) describes how the FRET rate changes as a function of the donor–acceptor distance (r), the orientation factor (κ), and the overlap integral (J). The FRET efficiency falls off with the sixth power of distance and thus the modulation of energy-transfer processes provides a ready means for signal generation [59, 60].

$$k_{t(r)} \propto \frac{1}{r^6} \cdot \kappa^2 \cdot J(\lambda) \tag{1.1}$$

$$J(\lambda) = \int_0^\infty F_{\mathrm{D}}(\lambda) \in_{\mathrm{A}} (\lambda)\lambda^4 \, \mathrm{d}\lambda$$

In FRET-based assays, the light harvesting conjugated polymer and a fluorophore capable of introduction into a probe structure are generally designed

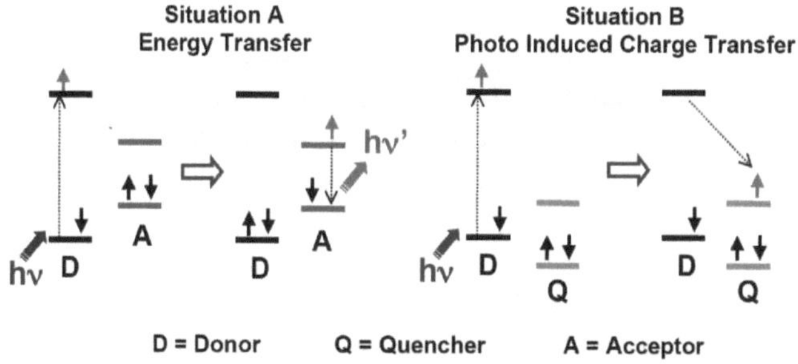

Scheme 1.11. Effect of relative orbital energy levels on FRET vs. PCT preferences

to function as the donor and the acceptor, respectively. The overlap integral expresses the spectral overlap between the emission of the donor and the absorption of the acceptor. The components of the sensor can be chosen so that their optical properties meet this requirement.

A competing mechanism to FRET is photo-induced charge transfer (PCT). Although PCT provides the basis for assays that modulate emission intensity, it constitutes an energy-wasting scheme in FRET assays, reducing the intensity of signals and the overall sensitivity. The rate for PCT shows an exponential dependence on the donor–acceptor distance (r), i.e., $k_{PCT} \alpha \exp(-\beta/r)$, where β reflects the electronic coupling. Thus, there is a more acute distance dependence relative to (1) and, as we will discuss in the following section, the chemical structure of the polymer chain makes a strong impact on the contribution of the two processes with a given fluorophore reporter.

Scheme 1.11 provides a simplified illustration of two situations that may occur upon excitation of the polymer donor [61]. Situation A corresponds to the ideal situation for FRET, where the HOMO and LUMO energy levels of the acceptor are located within the orbital energy levels of the donor. Upon excitation of the donor, energy transfer to the acceptor takes place, leading to an emissive process, provided that the emission quantum yield of the acceptor is sufficiently large. Similarly, direct excitation of the acceptor under Situation A does not quench the acceptor emission. When the energy levels of the acceptor are not contained within the orbital energies of the donor, in other words when both the electron affinity and the ionization potential are higher in one of the optical partners, as in situation B, donor excitation may lead to PCT [62–66]. As shown in Scheme 1.11, donor excitation would lead to photo-induced electron transfer to the acceptor. Excitation of the acceptor would lead to a similar charge separated state via hole transfer to the donor [61]. Although Scheme 1.11 is widely used for choosing suitable optical partners for a specific application, it fails to be accurate for intermediate cases, since it neglects contributions from the exciton binding energy, the intermolecular charge transfer state energy, and the stabilization of the charged species by the medium. The mechanism by which FRET or PCT is preferred is complex for conjugated polymer blends and may involve geminate electron-hole pairs that

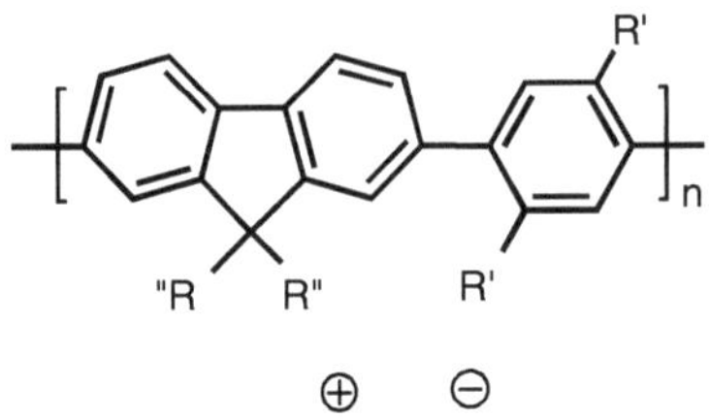

Scheme 1.12. Molecular structures of **P4**, **P7**, and **P8**

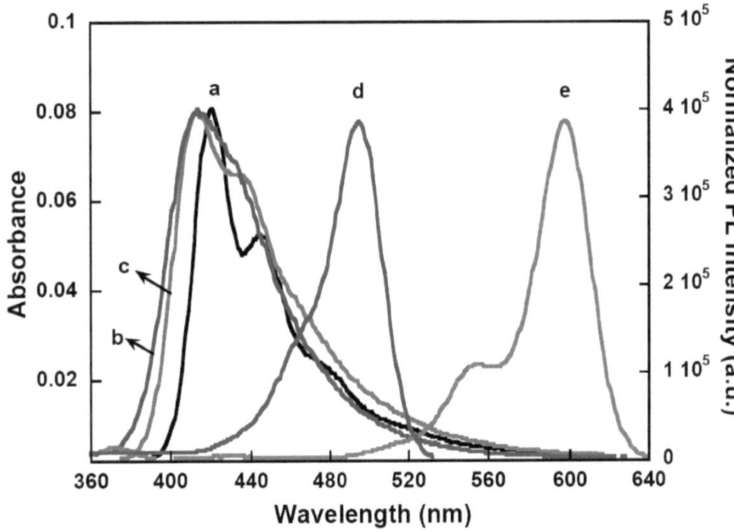

Fig. 1.4. Photoluminescence (PL) spectra of **P4** (a), **P7** (b), and **P8** (c) and absorbance of ssDNA-C* (Fl, d and TR, e) in 25 mM phosphate buffer (pH 7.4). Excitation wavelength is 385 nm for **P4**, 365 nm for **P7**, and **P8**. Reprinted with permission from [68]. Copyright 2006 American Chemical Society

may convert to exciplexes and ultimately excitons [67]. Despite these uncertainties, Scheme 1.11b provides for a necessary, but not sufficient, condition for PCT.

A series of cationic poly(fluorene-*co*-phenylene) derivatives was prepared to probe the effect of the molecular orbital energy levels on FRET efficiency [68]. As shown in Scheme 1.12, three different polymers were used in which electron donating (OMe, **P7**) or withdrawing (F, **P8**) groups were introduced into the phenylene unit. As shown in Fig. 1.4, the PL spectra of the three polymers

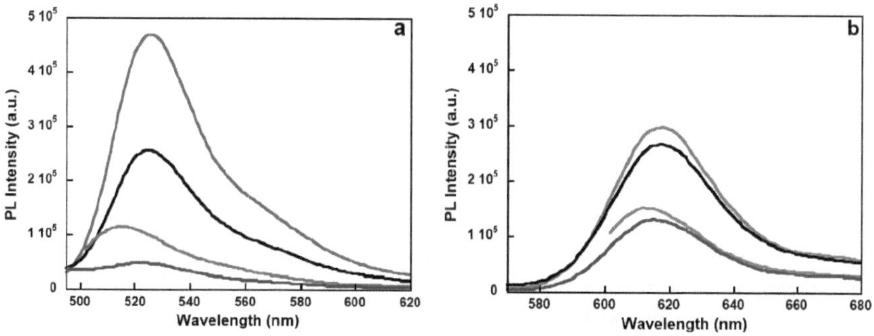

Fig. 1.5. Fluorescence spectra of ssDNA-Fl (**a**) and ssDNA-TR (**b**) in the presence of **P4** (*black*), **P7** (*blue*), and **P8** (*red*) in 25 mM phosphate buffer at [ssDNA − Fl or ssDNA − TR] $= 2 \times 10^{-8}$ M and [RU] $= 4 \times 10^{-7}$ M. The excitation wavelengths are 385 nm for **P4** and 365 nm for **P7** and **P8**. Direct excitation of ssDNA-Fl and ssDNA-TR *prior* to polymer addition are also shown in *green* and *orange*, respectively. Reprinted with permission from [68]. Copyright 2006 American Chemical Society

are similar; furthermore, they all exhibit nearly identical PL quantum yields. Figure 1.4 also shows the absorbance of two typical dyes used in labeling single-stranded DNA probes (ssDNA) namely, fluorescein (Fl) and Texas Red (TR). There is much better spectral overlap between the absorption of Fl and the emission of polymers **P4**, **P7**, and **P8**, relative to the situation with TR. By looking at (1), one would estimate nearly identical donor capabilities for the three polymers and that Fl would be a better acceptor, relative to TR.

FRET experiments were performed by monitoring dye emission upon excitation of **P4**, **P7**, and **P8** under conditions relevant for biosensor applications and the results are summarized in Fig. 1.5. The most intense Fl emission was observed for **P8**/ssDNA-Fl, which is approximately twofold more intense than that observed for **P4**/ssDNA-Fl and is over an order of magnitude larger than that for **P7**/ssDNA-Fl. For **P8**/ssDNA-Fl, the integrated Fl emission is approximately fivefold greater than that obtained by direct excitation of Fl at its absorption maximum (495 nm) in the *absence* of the polymers, while over 20-fold enhancement is observed relative to direct Fl excitation in the *presence* of **P8**. This enhancement is indicative of the signal amplification provided by the light harvesting capabilities of the poly(fluorene-*co*-phenylene) backbone. FRET experiments using ssDNA-TR as the acceptor show that TR emission intensities are similar for **P4**/ssDNA-TR and **P8**/ssDNA-TR, which are approximately twice more intense than that observed with **P7**/ssDNA-TR. For **P8**/ssDNA-TR, the integrated TR emission through FRET is approximately twice greater than that obtained by direct excitation of TR at its absorption maximum (590 nm) in the absence of the polymers, while there is a tenfold enhancement relative to direct excitation of TR in the presence of **P8**. *Of particular significance is that the TR emission* *with* **P7**/*ssDNA-TR is more intense than the Fl emission with* **P7**/*ssDNA-Fl, despite the less effective spectral overlap* ($J(\lambda)$ in (1)).

Additional experiments with **P4**, **P7**, **P8** ssDNA-Fl and ssDNA-TR revealed the following observations. First, there is self-quenching of the dyes within the conjugated polyelectrolyte/ssDNA-Fl (or ssDNA-TR) aggregates. This effect can be diminished by including unlabeled ssDNA and is less severe for TR than for Fl. Second, there is self-quenching of the polymer emission upon aggregation with unlabeled ssDNA. The exact sizes, shapes, and orientation of the components in these polyplexes are unknown at this stage but are likely to be dominated by the structural attributes of the polyelectrolyte components, namely the rigid backbones and the ssDNA. Third, the three polymers are quenched to the same extent with ssDNA-Fl. In other words, the chemical nature of the dye at the terminus of the ssDNA does not appear to influence the general arrangement of the components so vastly that different optical coupling occurs. Thus, the differences in the sensitization of Fl or TR cannot be attributed to different abilities of **P4**, **P7**, and **P8** to serve as FRET donors.

Examination of the absolute energy levels of the three polymers sheds light into the differences of dye sensitization. Cyclic voltammetry was coupled with optical measurements to estimate the HOMO and LUMO energy levels. Fluorine substitution lowers the energy levels, while the electron donating methoxy group raises the levels, relative to the unsubstituted parent structure **P4**. The Fl LUMO energy is contained within the HOMO-LUMO gap of the three polymers. However, the Fl HOMO energy (-5.9 eV) is lower than those of **P4** (-5.6 eV) and **P7** (-5.4 eV). For these two structures, it is reasonable to expect that situation B in Scheme 1.11, i.e., PCT to the LUMO of Fl, is taking place. For polymer **P8**, with a HOMO energy of -5.8 eV, the situation is less clear and, given the limitations of Scheme 1.11, both processes may be taking place to some extent. The TR HOMO level (-5.4 eV) is higher than those of **P4** and **P8** and is close in energy to the level of **P7**. In fact, the HOMO-LUMO levels of TR are well contained between the levels of **P8**, as in situation A in Scheme 1.11, which favors FRET over PCT. Additionally, the fact that the energy gap difference between TR and the three conjugated polyelectrolytes is smaller than that observed with Fl should provide additional driving force for energy transfer, relative to charge transfer.

Similar comparisons of FRET efficiencies were carried out with poly(9,9′-bis(6-*N,N,N*-trimethylammoniumhexyl)fluorene-*alt*-1,4-(2,5-bis(6-*N,N,N*-trim ethylammoniumhexyloxy))phenylene) tetrabromide (**P9**) and poly((10,10′-bis(6-*N,N,N*-trimethylammoniumhexyl)-10H-spiro(anthracene-9,9′-fluorene))-*alt*-1,4-(2,5-bis(6-*N,N,N*-trimethylammoniumhexyloxy))phenylene) tetrabromide (**P10**); structures can be found in Scheme 1.13 [69]. The 10H-spiroanthracenyl group is orthogonal to the main-conjugated backbone vector and was not anticipated to contribute to electron delocalization. These groups behave as "molecular bumpers" that effectively shield the backbone. Accordingly, both **P9** and **P10** show similar absorption and PL spectra and their HOMO and

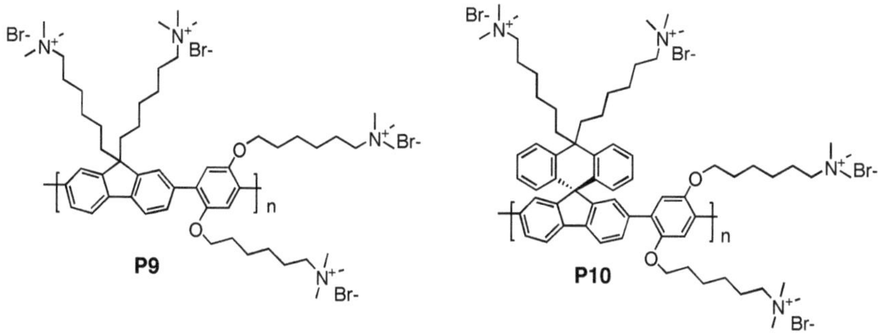

Scheme 1.13. Molecular structures of **P9** and **P10**

LUMO levels are nearly identical, as determined by electrochemical measurements. The PL quantum yields are also identical within experimental error.

These polymers served as excitation donors to ssDNA-Fl. Similar levels of polymer emission quenching are observed upon addition of ssDNA-Fl; however, the fate of the excitations is very different. In the case of **P9**, one observes negligible Fl emission, whereas for **P10**, it is possible to observe Fl emission with a FRET efficiency of 60%. Since the optical properties and orbital energy levels for **P9** and **P10** are identical, the arguments presented for polymers **P4**, **P7**, and **P8** are not applicable to rationalize the differences in optical output. However, it is important to note that FRET and PCT rates, and thereby their probabilities, vary to different extents depending on the donor–acceptor distance. PCT is essentially a contact process described by an exponential distance dependence [61, 70] and functions effectively at D-A distances considerably shorter than those probed by FRET processes (1) [71,72]. The nearly complete Fl emission quenching in **P9**/ssDNA-Fl suggests that polymer excitation results in PCT to Fl [73]. With **P10**/DNA-Fl, one observes much less Fl quenching. The introduction of the "molecular bumpers" in **P10** increases the average donor–acceptor distance. This increased separation reduces the probability of PCT, relative to the parent **P9** structure, but allows for FRET to occur with good efficiency. These results indicate that careful attention needs to be paid to molecular design strategies that fine tune distances at the molecular level to favor FRET over quenching by PCT mechanisms.

1.4 Chemo- and Biosensor Applications

1.4.1 DNA Sensors

Recent studies using **P2** have shown that it can be used to optically amplify fluorescent DNA assays [74]. The method comprises two components: (a) the light harvesting luminescent conjugated polymer **P2** and (b) a probe oligonucleotide consisting of a peptide nucleic acid (PNA) labeled with a reporter dye such as Fl (PNA-Fl). Addition to the solution of a target polynucleotide with a sequence complementary to the PNA strand yields the duplex structure. Electrostatic interactions bring the conjugated polyelectrolyte in proximity to the PNA/ssDNA-Fl duplex resulting in energy transfer from the polymer to the signaling chromophore. When a nontarget polynucleotide is added, complexation between **P2** and the probe oligonucleotide does not occur. Because the average distance between **P2** and the signaling chromophore is too large for effective energy transfer in the absence of such hybridization, there is little or no emission from the signaling chromophore (Scheme 1.14 and Fig. 1.6). The resulting emission intensity of Fl was enhanced by approximately 25-fold, compared with that obtained upon direct excitation at the absorption maximum of Fl.

By adding an S1 nuclease enzyme, it is possible to modify the strategy in Scheme 1.14 so that the overall assay is sensitive to single nucleotide

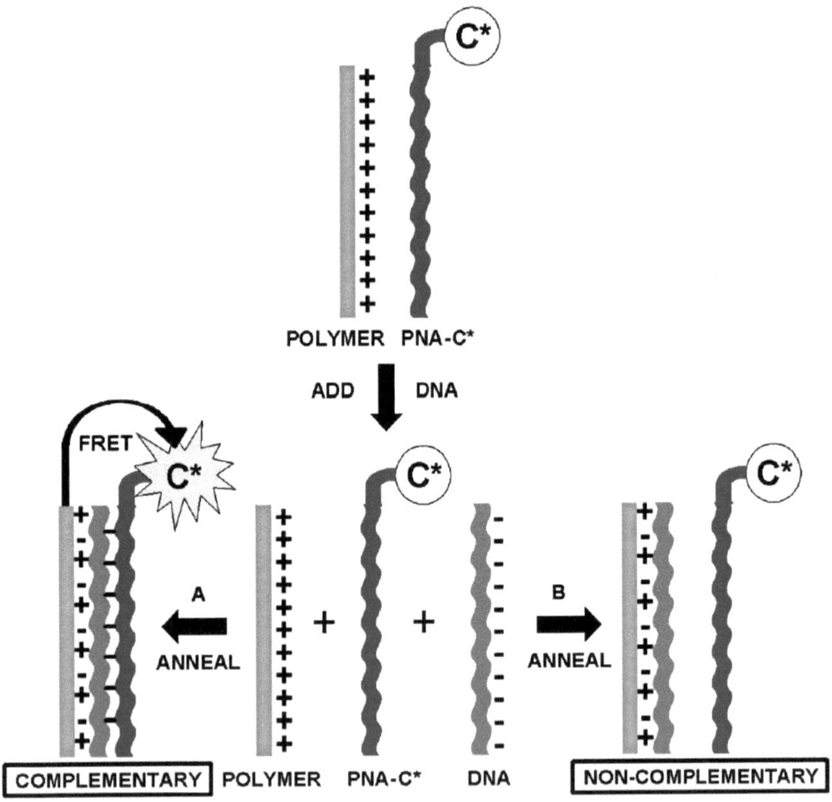

Scheme 1.14. DNA assay using **P2** (shown in *green*) and PNA-C* (shown in *blue*, where C* is Fl). The system responds depending on whether the ssDNA (shown in *red*) is complementary or noncomplementary to the PNA sequence [74]. Copyright 2002 National Academy of Sciences, U.S.A.

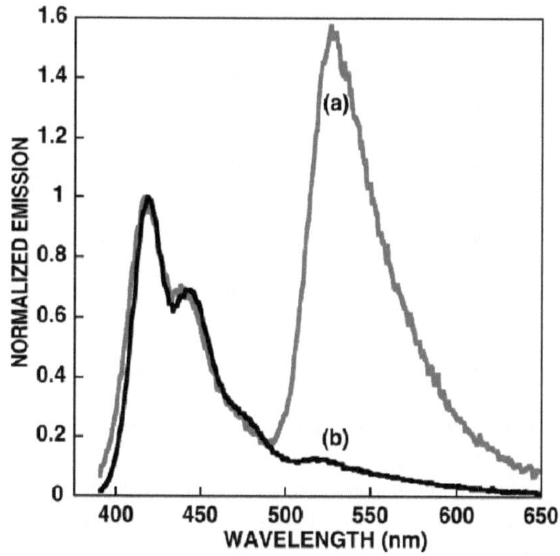

Fig. 1.6. Emission spectra of PNA-Fl in the presence of complementary (a) and noncomplementary (b) DNA by excitation of polymer **P2**. The spectra are normalized with respect to the emission of polymer **P2** [74]. Copyright 2002 National Academy of Sciences, U.S.A.

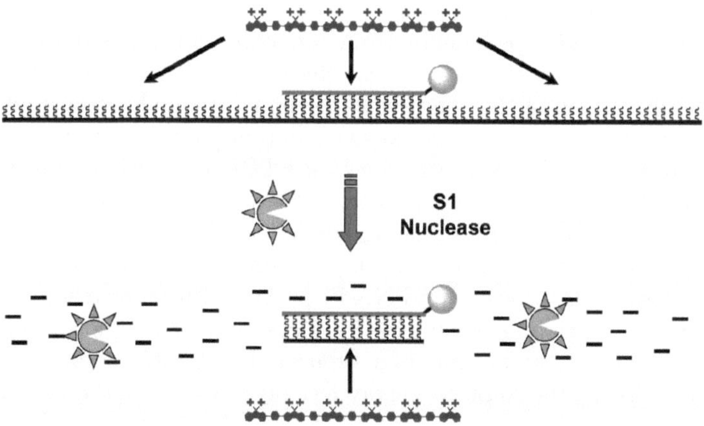

Scheme 1.15. Long ssDNA target sequences (*black*) are digested by S1 nuclease, leaving intact only those regions bound to the PNA probe (*red*). **P2** (*blue*) added directly to the resulting solutions can only associate with the remaining PNA-Fl/DNA duplex. Any PNA/DNA mismatches will result in complete DNA digestion; therefore, energy transfer from **P2** occurs only for the perfect PNA-Fl/DNA complement [75]. Copyright 2005 National Academy of Sciences, U.S.A.

polymorphisms. The overall strategy is illustrated in Scheme 1.15 [75]. The recognition is accomplished by sequence-specific hybridization between the uncharged; PNA-Fl probe and the ssDNA sequence of interest. Treatment with S1 nuclease leads to digestion of DNA, except for those regions

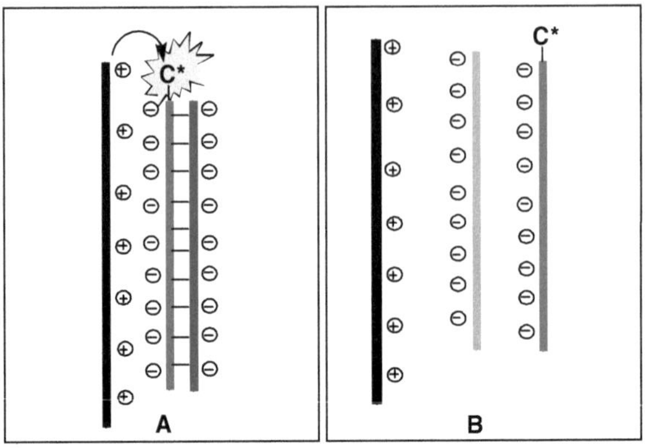

Scheme 1.16. ssDNA-C*/ssDNA sensory process involving FRET from a conjugated polyelectrolyte. The probe ssDNA-C*, where C* is the reporter fluorophore is shown in *red*. Situation A corresponds to complementary ssDNA, while B shows nonhybridized strands when noncomplementary ssDNA is present. Reprinted with permission from [76]. Copyright 2003 American Chemical Society

"protected" by a perfectly matching hybridized PNA. The polymer **P2** can function as an excitation donor to Fl only when in the PNA/ssDNA duplex. The overall method was tested by measuring the ability of the sensor system to detect wild type human DNA sequences, as opposed to sequences containing a single base mutation. Specifically, the PNA probe was complementary to a region of the gene encoding the microtubule associated protein Tau. The probe sequence covers a known point mutation implicated in a dominant neurodegenerative dementia known as FTDP-17, which has many clinical and molecular similarities to Alzheimer's disease. Using an appropriate PNA probe, unambiguous FRET signaling is achieved for only the wild type DNA and not the mutant sequence harboring the single nucleotide polymorphism.

It is possible to replace the PNA-Fl probes with more readily available ssDNA-Fl, as shown in Scheme 1.16 [76]. Because the higher local charge density of the double strand results in a stronger electrostatic attraction between the dsDNA-Fl and the conjugated polyelectrolyte relative to the situation with ssDNA, upon adding a complementary strand ssDNA, the hybridized strand will result in an efficient FRET from **P2** to Fl. In the case of a noncomplementary strand, the nonhybridized strand will interfere with the ssDNA-Fl–**P2** interactions and FRET from **P2** to Fl does not occur. The overall scheme serves as a probe for the presence of specific ssDNA sequences in solution.

In this DNA assay experiment, a $ssDNA_1$-Fl probe was annealed in the presence of an equal molar amount of its 20 base pair complementary strand, $ssDNA_2$, and in an identical fashion with a noncomplementary 20 base strand, $ssDNA_3$. The resulting solutions were mixed with **P2**. A direct comparison of the sensitized fluorescence upon excitation of the polymer (Fig. 1.7) revealed an approximately threefold higher FRET ratio for the hybridized DNA. The

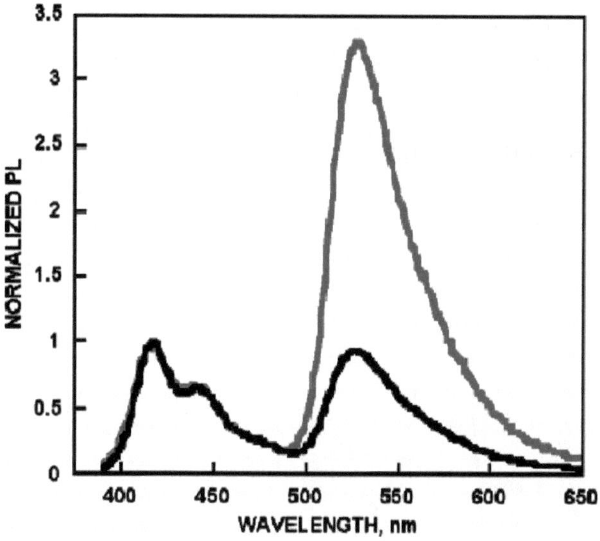

Fig. 1.7. Emission spectra of **P2** with hybridized (DNA_1-Fl/$ssDNA_2$) and nonhybridized ($ssDNA_1$-Fl + ss-DNA_3) DNA in SSC buffer. The spectra were normalized to **P2**. Reprinted with permission from [76]. Copyright 2003 American Chemical Society

data in Fig. 1.7 show that situation A in Scheme 1.16, where hybridization takes place, leads to higher FRET ratios compared with situation B, and that **P2** can be used to monitor the presence of ssDNA complementary to the $ssDNA_1$-Fl probe. However, comparison of Fig. 1.7 against the spectra shown in Fig. 1.3, shows that the polymer:Fl emission ratio is higher when using PNA-Fl probes.

Recent work demonstrates that it is possible to design DNA biosensors using aggregate-induced interchain energy transfer in polymers such as **PFPB** (see Fig. 1.8), which has 5% of the phenylene units in the typical poly(fluorene-*co*-phenylene) structure substituted with BT units chain [77,78]. The emission of **PFPB** is predominantly blue in dilute conditions. Concentrated solutions emit green light. A similar blue to green color change occurs upon addition of either ssDNA or dsDNA, and these color changes can be used to generate calibration curves that determine the concentration of DNA in solution [77]. When combined with a Cy5-labeled PNA probe (PNA-Cy5), the polymer can be used to design a three-color DNA detection assay. As shown in Fig. 1.8, **PFPB**/PNA-Cy5 emits blue color in the absence of ssDNA. When noncomplementary DNA is present ($ssDNA_{NC}$), it emits green light due to the interchain FRET characteristic of **PFPB**. When bound to the complementary $ssDNA_C$, **PFBT-Br**/PNA-Cy5 emits a red color due to the intramolecular FRET of **PFBT-Br** followed by a subsequent FRET step to Cy5.

Secondary structure modifications upon DNA folding can also be detected using water-soluble poly(fluorene)s. Single-stranded DNA with G-rich

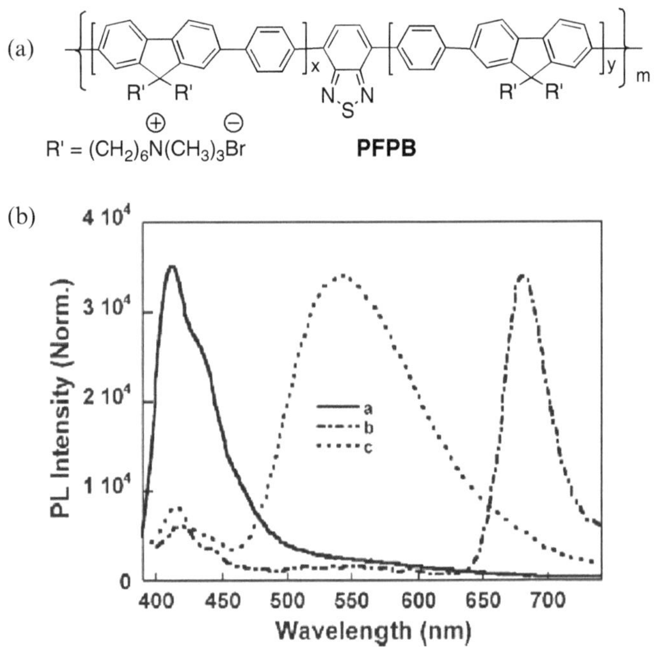

Fig. 1.8. (**a**) Molecular structure of **PFPB**. (**b**) Normalized emission spectra in water of (a) **PFPB**/PNA-Cy5, (b) **PFPB**/PNA – Cy5 + $ssDNA_{NC}$, and (c) **PFPB**/PNA – Cy5/$ssDNA_C$. [PNA – Cy5] = 2.0×10^{-8} M, [**PFBT** – **Br**] = 1.6×10^{-7} M in RUs. The excitation wavelength is 380 nm. Reprinted with permission from [78]. Copyright 2004 American Chemical Society

sequences can fold into structures named G-quadruplexes via intramolecular hydrogen bonding interactions [79]. The G-quadruplex is an unusual tetraplex conformation of telomere DNA, which has been shown to directly inhibit telomerase activity in immortalized cells as well as cancer cells [80]. Efficient recognition of the G-quadruplex structure is a key requirement for the rational design and development of telomerase inhibitors for cancer, HIV, and other diseases. This conformational change can be detected by a FRET-based homogeneous method, as shown in Scheme 1.17 [81]. A G-rich ssDNA was labeled with Fl at the 5′-end (ssDNA-Fl). The ssDNA-Fl exhibits a random coil conformation in the absence of K^+. The relatively weak electrostatic interactions of ssDNA-Fl with **P4** mean that fluorescein is spatially far away from **P4** and inefficient FRET from **P4** to fluorescein occurs (Scheme 1.17a). The formation of the more compact G-quadruplex upon addition of KCl increases the negative charge density of the DNA macromolecule and results in stronger G-quartet-DNA/**P4** complexation, relative to ssDNA-Fl/**P4**. This situation is illustrated in Scheme 1.17b, where **P4** resides in proximity to Fl. Fig. 1.9 compares the emission observed upon addition of **P4** to solutions of ssDNA-Fl in the presence of 50 mM KCl or NaCl. The Fl emission intensity obtained by excitation at 380 nm reveals a significantly higher turn-on signal

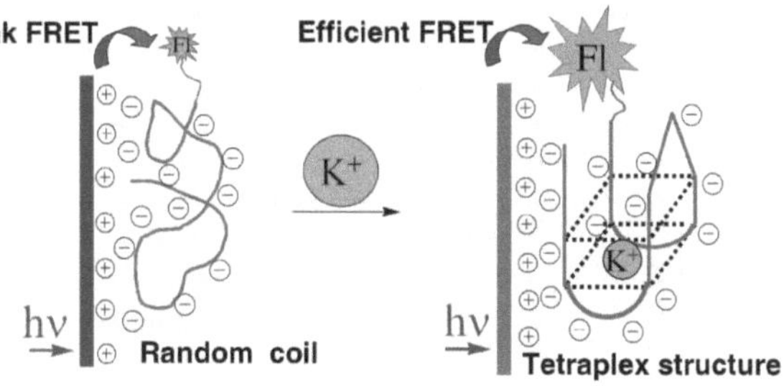

Scheme 1.17. The G-quadruplex structure assay: the G-rich ssDNA exhibits random coil conformation in the absence of K^+. The formation of a more condensed G-quadruplex upon adding KCl increases its space charge density and therefore efficient FRET from P4 to fluorescein labeled at the 5′-terminus of ssDNA is observed

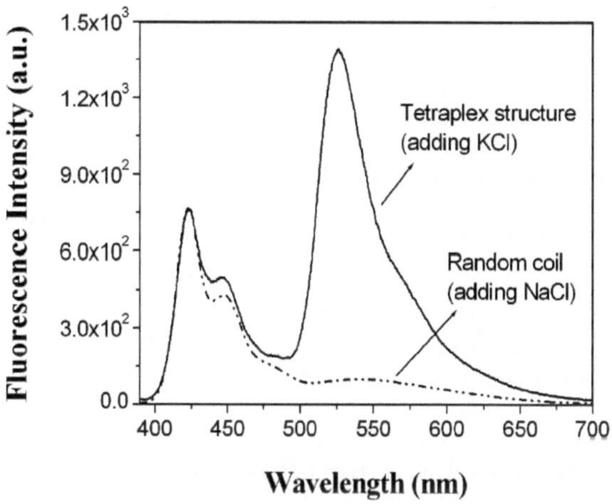

Fig. 1.9. Photoluminescence (PL) spectra from solutions containing **P4** and ssDNA-Fl/KCl (*solid line*) and ssDNA-Fl/NaCl (*dashed line*), $\lambda_{ex} = 380\,nm$, [KCl or NaCl] = 50 mM, $[ssDNA - Fl] = 1.0 \times 10^{-7}\,M$, $[\mathbf{P4}] = 2.5 \times 10^{-6}\,M$ in RUs. Measurements are in pure water. The spectra are normalized with respect to the emission of **P4**. Reprinted with permission from [81]. Copyright 2005 American Chemical Society

for the ssDNA-Fl/KCl, relative to ssDNA-Fl/NaCl. These FRET differences demonstrate the specificity of this detection method for the G-quadruplex structure. Furthermore, the Fl emission measured using 380 nm excitation is more than ten times larger than that obtained by direct excitation at the ssDNA-Fl absorption maximum, thereby demonstrating optical amplification by the conjugated polyelectrolyte.

G-quadruplex DNA can convert to the duplex form in the presence of its complementary strand [82]. Such a conformational change can also be detected using **P4** [83]. The overall strategy is illustrated in Scheme 1.18. The intercalator ethidium bromide (EB) is introduced into the assay solution. Electrostatic interactions between G-quadruplex-Fl and **P4** keep them

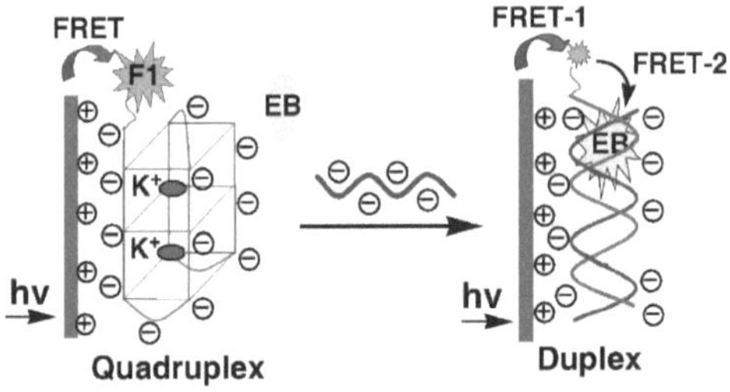

Scheme 1.18. Assay for G-quadruplex/dsDNA transitions

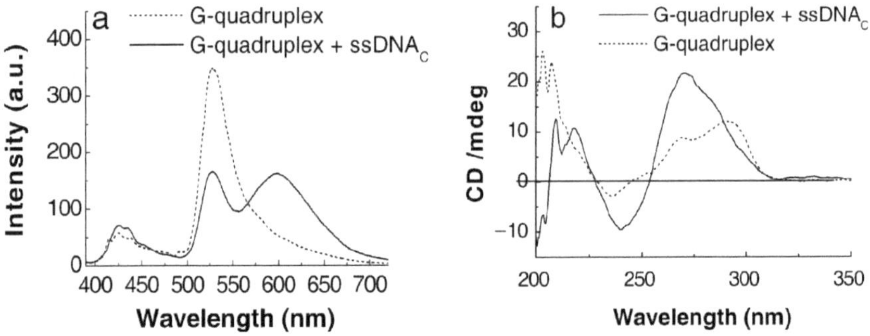

Fig. 1.10. (**a**) Photoluminescence (PL) spectra of G-quadruplex-Fl in the presence of **P4** and EB before and after addition of ssDNA$_C$, excitation wavelength is 380 nm; (**b**) CD spectra of G-quadruplex before and after addition of ssDNA$_C$. Reprinted with permission from [83]. Copyright 2006 American Chemical Society

in proximity, allowing for FRET from **P4** to Fl. EB does not intercalate into the G-quadruplex. Transition from G-quadruplex to dsDNA-Fl upon addition of complementary ssDNA allows EB to intercalate into the resulting DNA grooves. Excitation of **P4** leads to two-step energy transfer, from **P4** to dsDNA-Fl (FRET-1), followed by FRET from dsDNA-Fl to EB (FRET-2). Direct FRET from **P4** to EB is not favored because of the nonoptimized transition dipole orientations of **P4** and EB within the **P4**/dsDNA(EB) complex. By probing the EB emission intensity, the DNA conformational change can be detected. Figure 1.10a compares the emission spectra of G-quadruplex-Fl with **P4** and EB before and after addition of complementary ssDNA$_C$ ([ssDNA$_C$] $= 5.0 \times 10^{-8}$ M). The excitation wavelength was chosen at the absorbance maximum of **P4**, where no significant absorption by either Fl or EB occurs. In the absence of ssDNA$_C$, only FRET from **P4** to Fl is observed. Upon addition of ssDNA$_C$, formation of the dsDNA takes place, followed by EB intercalation, leading to detectable emission from EB. Circular dichroism (CD) spectra confirmed the quadruplex to duplex transition (Fig. 1.10b).

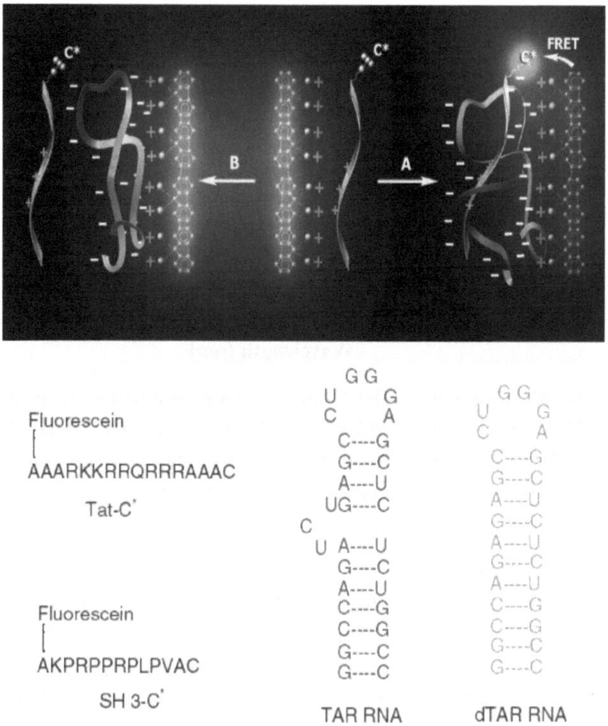

Scheme 1.19. Illustration of the HIV-1 RNA sensor operation and the chemical structures of Tat-C*, SH3-C*, TAR RNA, and dTAR RNA [49]. Copyright Wiley-VCH Verlag GmbH & Co. KGaA. Reproduced with permission

1.4.2 RNA Sensors

It is possible to extend the concepts of FRET mediation by electrostatic complexation of DNA, to detection of RNA. Detection of an HIV-related RNA using a labeled transactivator peptide (Tat) probe and **P1** has also been demonstrated [49]. As shown in Scheme 1.19, the assay contains three elements: the cationic conjugated polyelectrolyte **P1**, a protein probe (protein-C*), and the target RNA. One begins by mixing the protein-C* with the RNA in solution. Situation A corresponds to the RNA sequence that specifically binds with protein-C*. Upon binding, C* is now attached to a macromolecule that differs in charge from protein-C* by the negative charge intrinsic to the RNA. For the sensor to work, the *net* charge in the RNA–protein complex must be negative. Situation B shows that with a nonbinding RNA sequence the protein-C* remains separate, and there is no net change in charge in the vicinity of C*. Addition of **P1** results in electrostatic binding. In situation A of Scheme 1.19, **P1** resides in proximity to C*; excitation of **P1** results in energy transfer to C*. In B, the RNA and **P1** come together, but remain separated from protein-C* and there is no energy transfer to C*. Figure 1.11 compares

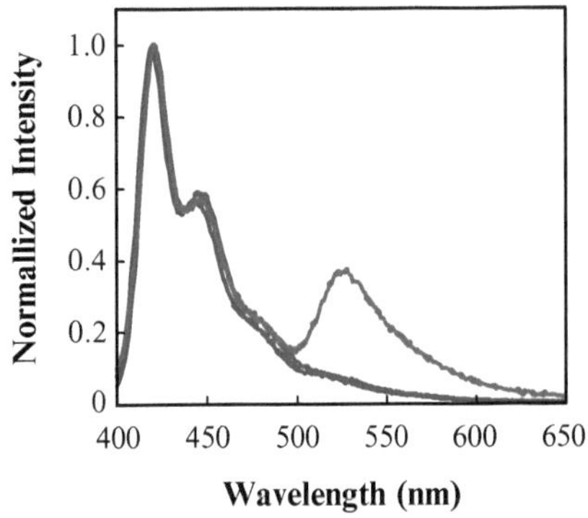

Fig. 1.11. Photoluminescence (PL) spectra from solutions containing **P1** and Tat-C*/TAR RNA (*red*), Tat-C*/dTAR RNA (*blue*), SH3-C*/TAR RNA (*green*), $\lambda_{ex} = 380$ nm, [**P1**] $= 9.6 \times 10^{-7}$ M in RUs. Measurements are in Tris-EDTA buffer solution (10 mM, pH = 7.4). The spectra are normalized with respect to the emission of **P1** [49]. Copyright Wiley-VCH Verlag GmbH & Co. KGaA. Reproduced with permission

the emission observed upon addition of **P1** to solutions of Tat-C* with TAR RNA or dTAR RNA. In these experiments, the Tat-C* probe was premixed with an equimolar amount of the RNA at room temperature. Addition of **P1** and subsequent comparison of the resulting fluorescence from Tat-C* obtained by excitation at 380 nm reveals an intensity ratio approximately 15 times higher for the Tat-C*/TAR RNA, relative to the nonspecific Tat-C*/dTAR RNA pair. These FRET differences demonstrate the specificity of the HIV-1 TAR RNA sensor to a specific RNA sequence.

Polymer **P4** can also be used as an optical platform to examine RNA–RNA interactions [84]. As shown in Scheme 1.20, the approach begins with Fl-labeled probe RNA (RNA-Fl) and treats it with a target RNA (RNA_T). Heterodimer formation, ($RNA_T/RNA - Fl$), increases the total negative charge on the Fl-bearing macromolecule and reduces the number of negatively charged molecules (relative to unbound $RNA_T + RNA - Fl$). In this case, more effective FRET from **P4** to Fl occurs in a way similar to that described in Scheme 1.17 for ssDNA. To test this approach, we investigated the assembly properties of tectoRNAs that interact via two different GNRA tetraloop/ receptor interactions (**TRI**s) (N: any nucleotide; R: purine, see Scheme 1.20): a highly specific GAAA **TRI** that was kept invariant, and a GRAA **TRI**, originally selected by in vitro selection. Specifically, RNA1 (Scheme 1.20) was tested for its ability to dimerize with three different targets, RNA**2**, RNA**3**, and RNA**4**. RNA1 contains a GAAA tetraloop (in red) that can be specifically recognized by the 11nt-motif receptor (in green) of RNA**2–4**, and a GRAA receptor (in blue) with affinity for a GGAA tetraloop and to a lesser extent GAAA and GUAA tetraloops.

Figure 1.12 shows a series of PL spectra as a function of [RNA]. At [RNA**1**-Fl] = 5 nM, the FRET ratio is equal for RNA**2–4**. At [RNA**1**-Fl] =

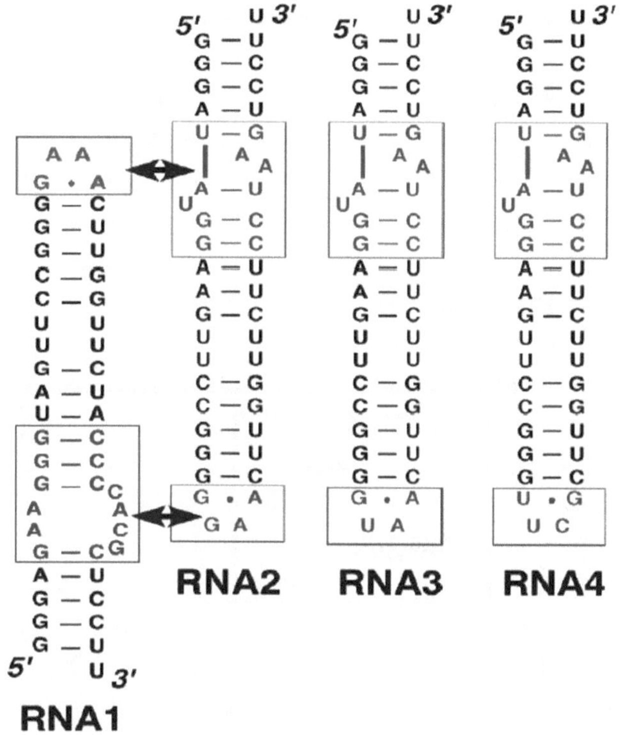

Scheme 1.20. Dimerization of tectoRNA molecules based on specific interactions between tetraloops (*red* and *orange*) and tetraloop receptors RNA molecules used in this study. RNA**1** and RNA**2** are shown as a dimer. Reprinted with permission from [84]. Copyright 2004 American Chemical Society

100 nM, RNA**1**-Fl/RNA**2** is nearly all in the dimer state; RNA**1**-Fl/RNA**3** forms ~30% of dimer and there is no association with RNA**4**. Furthermore, the integrated Fl emission from RNA-Fl/RNA**2** is ~15 times greater than by direct excitation at its absorption maximum in the absence of **P4**, indicating signal amplification by the conjugated polymer structure. When [RNA**1**-Fl] = 1,000 nM, the probe concentration is high enough to obtain RNA-Fl/RNA**2** and RNA**1**-Fl/RNA**3** as dimers, while RNA**1**-Fl + RNA**4** remain monomeric. These results demonstrate that the FRET ratio can be related to the level of tectoRNA dimerization in solution.

1.4.3 Protein Detection

A selective protein assay was demonstrated, which interfaces the FRET-based ratiometric techniques with the light-harvesting properties of conjugated polymer **P4** [85]. Polymer **P4** and a negatively charged biotinylated Fl probe (Fl-B) were used to optically detect the target protein, Streptavidin. As shown

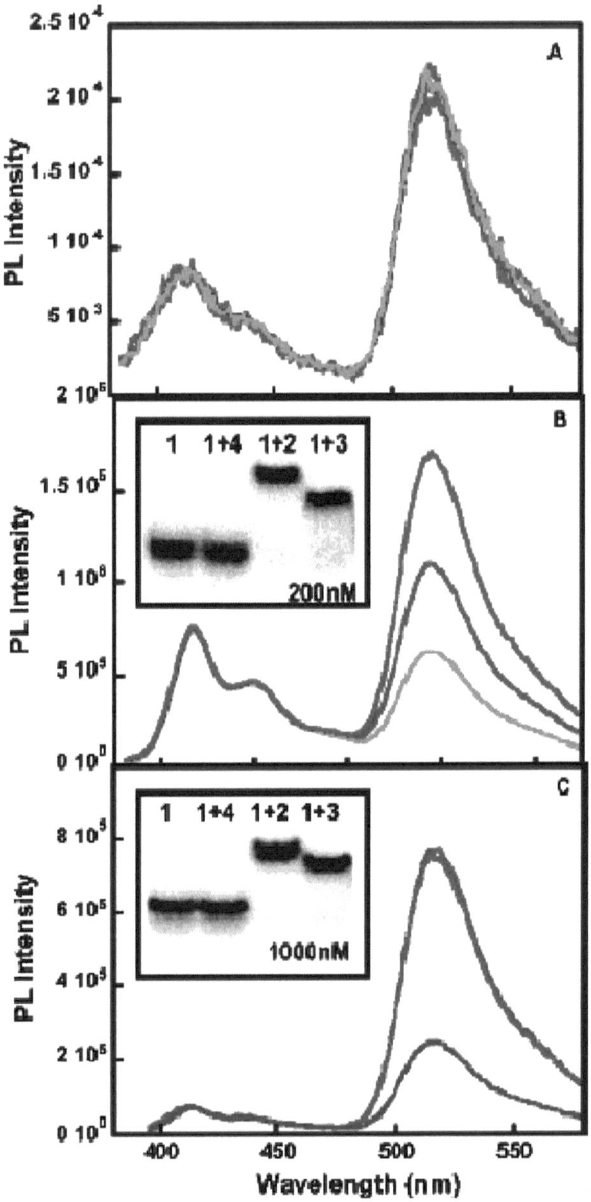

Fig. 1.12. Photoluminescence (PL) spectra for RNA**1**-Fl/RNA**2**–**4**/**P4** solutions in 25 mM HEPES buffer and $[Mg(OAc)_2] = 15$ mM. (**a**) [RNA-Fl] = 5 nM, in the presence of $[\mathbf{P4}] = 1.5 \times 10^{-7}$ M; (**b**) [RNA-Fl] = 100 nM, in the presence of $[\mathbf{P4}] = 4 \times 10^{-6}$ M; (**c**) [RNA**1**-Fl] = 1,000 nM, in the presence of $[\mathbf{P4}] = 3 \times 10^{-5}$ M. The excitation wavelength is 380 nm. Emission intensities were normalized relative to the **P4** emission. *Red*: RNA**1**-Fl + RNA**2**; *Blue*: RNA**1**-Fl + RNA**3**; *Green*: RNA**1**-Fl + RNA**4**. Reprinted with permission from [84]. Copyright 2004 American Chemical Society

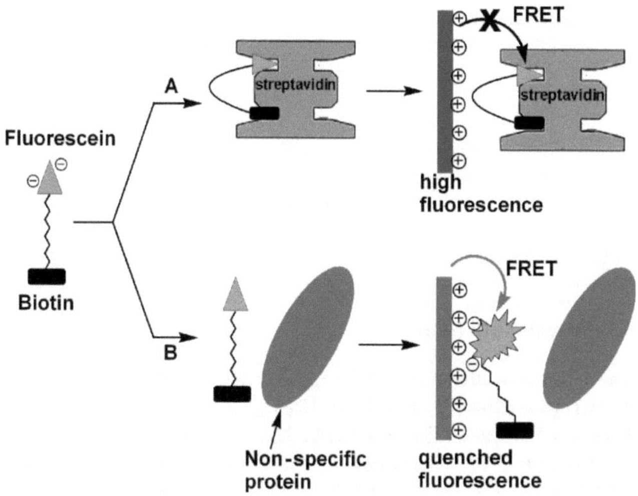

Scheme 1.21. Protein assay operation: in the presence of target Streptavidin, the biotin moiety of Fl-B specifically associates with Streptavidin and the fluorescein moiety is buried in the adjacent vacant-binding sites. This makes fluorescein far away from **P4** and inefficient FRET between them occurs (**A**). In the presence of nonspecific target, the strong electrostatic interactions between cationic **P4** and negatively charged biotinylated fluorescein (Fl-B) result in efficient FRET from **P4** to fluorescein (**B**)

in Scheme 1.21, electrostatic interactions between P4 and fluorescein result in efficient FRET from **P4** (shown in blue) to Fl. In the presence of Streptavidin, the biotin moiety of Fl-B specifically associates with Streptavidin and the fluorescein moiety is buried in the adjacent vacant binding sites [86]. The FRET efficiency from **P4** is thereby considerably reduced.

Figure 1.13a compares the PL spectra upon addition of **P4** to solutions of Fl-B with either Streptavidin or bovine serum albumin (BSA) upon excitation of the polymer. For **P4**/Fl-B, efficient FRET is observed from **P4** to fluorescein where the fluorescence of **P4** is strongly quenched, and the fluorescein emits intensely. For **P4**/Fl-B/Streptavidin, the FRET is very weak, and less fluorescein fluorescence is observed. For **P4**/Fl-B/BSA or **P4**/Fl-B/Streptavidin (biotin pre-saturated), efficient FRET is still observed from **P4** to Fl. These results confirm that the binding of biotin to Streptavidin specifically controls the efficiency of energy transfer and ultimately the emission color. Note that the BSA enhances the emission intensities of both **P4** and Fl, probably due to the surfactant-like character of BSA. Figure 1.13b compares the ratio of the intensity at 425 nm to that at 525 nm (I_{425}/I_{525}) in the presence of Streptavidin or BSA. The value of ratio (I_{425}/I_{525}) for Streptavidin is approximately 26 times higher than that for BSA.

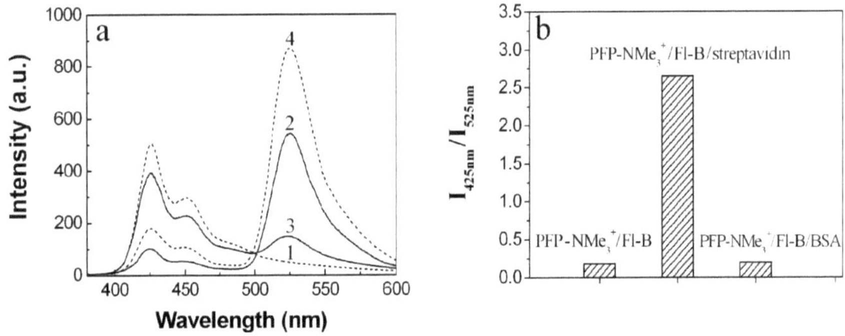

Fig. 1.13. (**a**) Photoluminescence (PL) spectra from solutions of **P4** (1), **P4**/Fl-B (2), **P4**/Fl-B/Streptavidin (3), **P4**/Fl-B/BSA (4); (**b**) the dependence of intensity ratio (I_{425}/I_{525}) on Streptavidin and BSA, $\lambda_{ex} = 380\,nm$, [**P4**] $= 1.0 \times 10^{-6}$ M in RUs, [Fl-B] $= 4.0 \times 10^{-8}$ M, [Streptavidin] = [BSA] $= 2.0 \times 10^{-8}$ M. Measurements are in phosphate buffered saline (50 mM, pH = 7.5) [85]. Copyright Wiley-VCH Verlag GmbH & Co. KGaA. Reproduced with permission

1.4.4 Glucose Sensors

There is a growing interest in the design and development of chemical sensors for hydrogen peroxide (H_2O_2) because of its importance in environmental and bioanalytical research [87]. H_2O_2 is generated by almost all oxidases, thus the activity of an enzyme, or the concentration of the enzyme substrate, for example glucose, can be quantitatively assayed by the determination of the amount of H_2O_2 produced [88]. Recently, a highly sensitive hydrogen peroxide (H_2O_2) probe has been developed by taking advantage of the amplified fluorescence quenching of conjugated polymers [89]. As shown in Scheme 1.22a, **P4** and fluorescein with boronate protecting groups (Fl-BB) can be used to optically detect H_2O_2. Without addition of H_2O_2, the absence of electrostatic interactions between cationic **P4** and the neutral Fl-BB keeps the Fl-BB far away from **P4**, and no fluorescence quenching of **P4** occurs. In the presence of H_2O_2, the formation of the anionic quencher, Fl by specific reaction of the Fl-BB with H_2O_2 results in complexation of **P4** and Fl, and therefore efficient fluorescence quenching of the **P4** emission takes place. As shown in Scheme 1.22b, since glucose oxidases (GOx) can specifically catalyze the oxidation of β-D(+)-glucose to gluconolactone, generating H_2O_2 as a byproduct, glucose detection can also be realized with the H_2O_2 probe as the signal transducer.

Figure 1.14a compares the PL spectra observed upon excitation of the polymer before and after addition of the H_2O_2 to solutions of **P4**/Fl-BB in phosphate buffer. Without addition of H_2O_2, neither fluorescence quenching of **P4** nor Fl emission is observed. The addition of the H_2O_2 leads to a significant quenching of **P4** emission, and the appearance of the Fl emission at approximately 522 nm. The plot of the fluorescence quenching of **P4** vs. [H_2O_2]

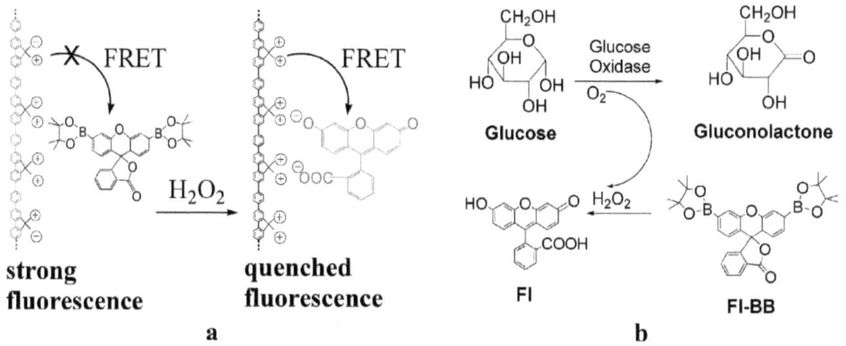

Scheme 1.22. The reaction mechanisms of (**a**) H_2O_2 and (**b**) glucose assays (reproduced from [89])

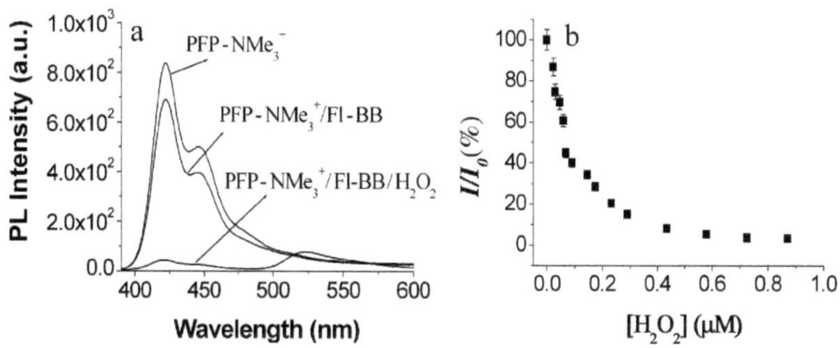

Fig. 1.14. (**a**) Photoluminescence spectra from solutions of **P4** and **P4**/Fl-BB in the presence or absence of H_2O_2 in phosphate buffer solution (6 mM, pH 7.4), [**P4**] = 2.5×10^{-6} M, [Fl-BB] = 1.0×10^{-6} M, [H_2O_2] = 8.0×10^{-5} M; (**b**) calibration plot of the H_2O_2 assay, [H_2O_2] = $0 \sim 8.7 \times 10^{-7}$ M. I_0 is the fluorescence emission intensity of **P4**/Fl-BB in the absence of H_2O_2, and I is the fluorescence emission intensity of **P4**/Fl-BB in the presence of H_2O_2. The excitation wavelength is 380 nm [89]. Copyright Wiley-VCH Verlag GmbH & Co. KGaA. Reproduced with permission

(Fig. 1.14b) shows that the H_2O_2 can be determined within a concentration range from 15 to 600 nM.

Figure 1.15 shows a typical calibration plot for evaluating glucose concentration. The emission of **P4** in the presence of the Fl-BB ([Fl-BB] = 8.0×10^{-5} M) and the GOx (1.5 mg mL^{-1}) decreases upon addition of glucose. The increasing fluorescence quenching of **P4**, as the concentration of the glucose increases, shows the glucose-dependent generation of the H_2O_2. The glucose can be detected in the concentration range from 10 μM to 1 mM with a detection limit of 5 μM. These results demonstrate that the use of cationic conjugated polyelectrolytes provides an approach to detect the H_2O_2 and glucose with excellent sensitivity.

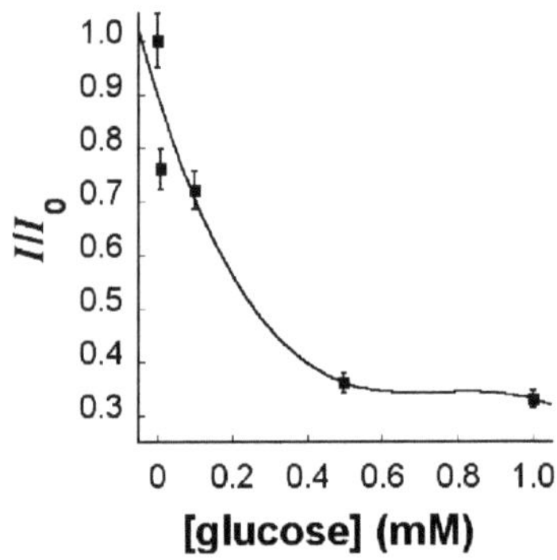

Fig. 1.15. The calibration *curve* corresponding to the fluorescence quenching of **P4** in the presence of the Fl-BB and GOx upon addition of variable concentrations of glucose in phosphate buffer solution (6 mM, pH 7.4). [**P4**] = 2.5×10^{-6} M, [Fl-BB] = 8.0×10^{-5} M, [GOx] = 1.5 mg mL^{-1}, [glucose] = $1.0 \times 10^{-3} \sim 1.0 \times 10^{-5}$ M. The excitation wavelength is 380 nm [85]. Copyright Wiley-VCH Verlag GmbH & Co. KGaA. Reproduced with permission

1.4.5 Detection of Other Small Molecules

Water-soluble poly(fluorene)s find applications in sensing other important biological small molecules, such as free radicals and antioxidants [90]. As shown in Scheme 1.23a, the anionic polyfluorene **P6** can form a complex with the cationic quencher 4-trimethylammonium-2,2-6,6-tetramethylpiperidine-1-oxyl iodide (CAT1) through electrostatic interactions. The fluorescence of **P6** is efficiently quenched by the CAT1 with a Stern–Volmer constant (K_{sv}) of 2.3×10^7 M^{-1}. Either by hydrogen abstraction or by reduction (Scheme 1.24b), the transformation of the paramagnetic nitroxide radical into diamagnetic hydroxylamine inhibits the quenching and, therefore, the fluorescence of **P6** is recovered. The fluorescence recovery can be used to probe the processes of hydrogen transfer reaction from antioxidants to radicals and the reduction of radicals by antioxidants.

As shown in Fig. 1.16a, upon addition of CAT1 the fluorescence of **P6** is efficiently quenched. The (±)-6-hydroxy-2,5,7,8-tetramethylchromane-2-carboxylic acid (trolox) is added and allowed to equilibrate for 35 min, which reverses the quenching due to the scavenging of nitroxide radicals via hydrogen transfer from trolox to CAT1. The fluorescence recovery depends on the concentration of trolox (Fig. 1.16b). The trolox can be detected in the range of $10 \sim 100$ μM. The same probe can also be used to detect the capabilities of a variety of antioxidants. Sensing for ascorbic acid can be accomplished with high sensitivity and selectivity because of its excellent antioxidant capabilities. The ascorbic acid concentration can be determined in the 50 nM to 200 μM range (Fig. 1.17a). Control experiments were also done with a nonspecific quencher, N,N'-dimethyl-4,4′-bipyridinium (MV^{2+}) for ascorbic acid. As shown in Fig. 1.17b, upon addition of MV^{2+}, the fluorescence of **P6** is

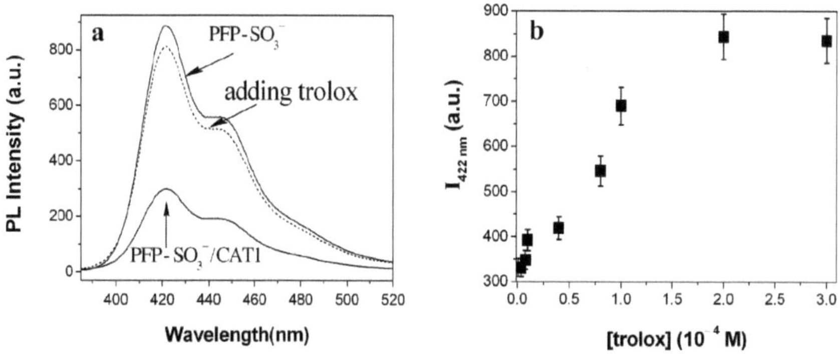

Fig. 1.16. (**a**) Photoluminescence (PL) spectra of **P6, P6**/CAT1 in the absence and presence of the trolox; (**b**) the fluorescence intensity at 422 nm of **P6**/CAT1 as a function of trolox concentrations, [trolox] = 0–3.0 × 10^{-4} M; (**c**) The fluorescence intensity at 422 nm of **P6**/CAT1 in the presence of trolox as a function of the incubating time. The excitation wavelength is 376 nm and measurements were performed in phosphate buffer (5 mM, pH 7.4), [**P6**] = 1.0 × 10^{-6} M in RUs, [CAT1] = 5.0×10^{-5} M. [trolox] = 1.0×10^{-4} M. Reprinted with permission from [90]. Copyright 2006 American Chemical Society

efficiently quenched. The ascorbic acid was added and allowed to equilibrate for 35 min, almost no recovery was observed for the fluorescence of **P6**.

1.5 Heterogeneous Platforms

Detection systems discussed thus far are homogeneous in nature. A particularly significant new technology in the arsenal of molecular biologists are DNA Chips, which consist of printed DNA microarrays that allow screening of thousands of genes on a single platform. The goal of such technology is to screen the entire genome on a single chip [91–95]. Gene chips and microarrays allow high throughput screening of hundreds to thousands of genes in a single experiment. Nucleic acid detection using such microarrays typically requires labeling of target nucleic acids with fluorophores or other reporter molecules prior to hybridization, which adds cost, complexity, and an element of uncertainty, in particular the efficiency of target labeling [91]. The synthesis of the polymer **PFBT** enabled the entry of conjugated polyelectrolytes into DNA microarrays [96]. The overall detection strategy is schematically illustrated in Scheme 1.24 [97]. One starts with a surface containing immobilized PNA (shown in yellow). Hybridization of ssDNA (shown in blue) to the PNA surface results in a negatively charged surface. Because of electrostatic attraction, the addition of the **PFBT** (shown in orange), followed by washing, results in preferential adsorption onto sites containing complementary ssDNA. After workup by washing with SSC buffer and water, polymer emission indicates that the ssDNA is complementary to the surface-bound

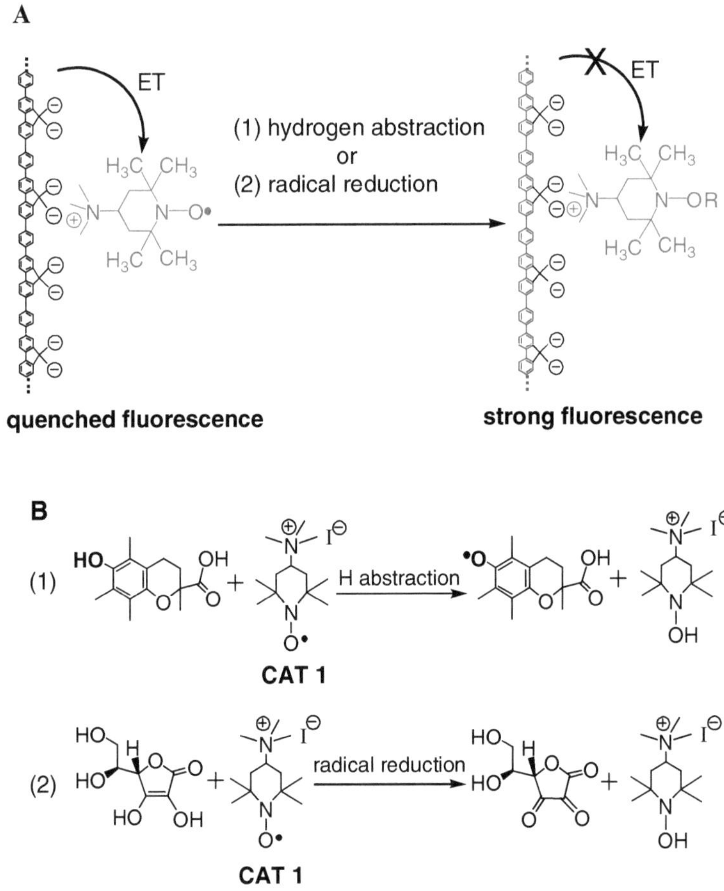

a b

Scheme 1.23. (**a**) Radical scavenging assays and (**b**) the radical scavenging reactions of CAT 1 by trolox and ascorbic acid (adapted from [90])

PNA. The overall selectivity of Scheme 1.24 relies on the successful removal of **PFBT** from nonhybridized PNA surfaces. From a practical perspective, it is important to note that this conjugated polyelectrolyte-based DNA chip approach circumvents the need to label the target oligonucleotides, thereby removing an element of uncertainty from the detection mechanism.

1.6 Summary and Outlook

In summary, it is possible to take advantage of the optical amplification of water-soluble poly(fluorene)s to design chemical and biological sensors. These assays do not require sophisticated instrumentation and should be applicable to many standard fluorescent assays. Although a great deal of progress

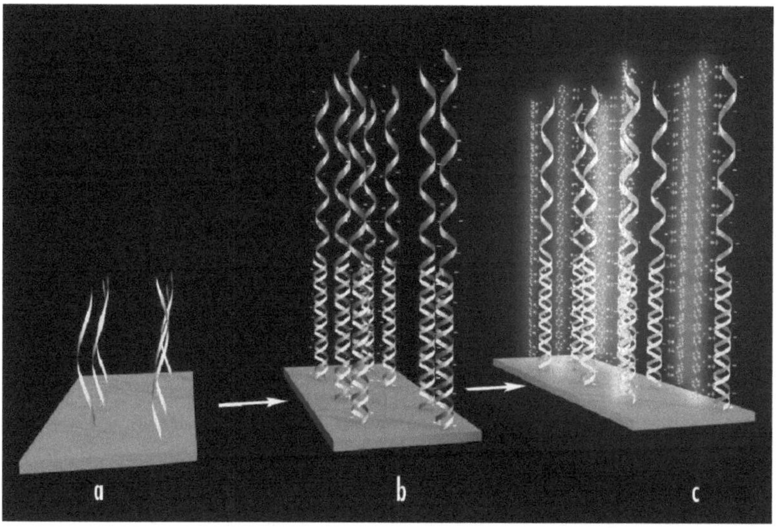

Scheme 1.24. Label-free ssDNA detection using immobilized PNA and **PFBT**. (**a**) Surface bound PNA (shown in *yellow*), (**b**) hybridization with ssDNA (shown in *blue*), (**c**) electrostatic adsorption of **PFBT** onto the PNA/ssDNA surface

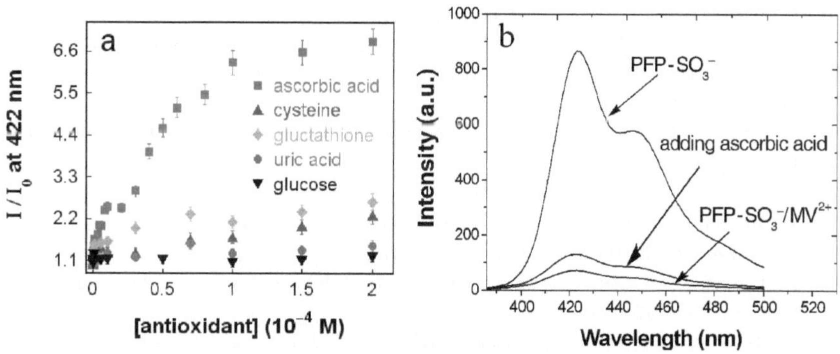

Fig. 1.17. (**a**) The fluorescence intensity at 422 nm for **5**/CAT1 as a function of concentrations of antioxidants under O_2, [antioxidants] = $0\text{–}2.0 \times 10^{-4}$ M, [**P6**] = 1.0×10^{-6} M in RUs, [CAT1] = 5.0×10^{-5} M; (**b**) the PL spectra of **P6**, **P6**/MV^{2+}, and **P6**/ MV^{2+}/ascorbic acid under O_2, [**P6**] = 1.0×10^{-6} M in RUs, [MV^{2+}] = 4.37×10^{-8} M, [ascorbic acid] = 2.0×10^{-4} M. The excitation wavelength is 376 nm and measurements are performed in phosphate buffer (5 mM, pH 7.4). Reprinted with permission from [90]. Copyright 2006 American Chemical Society

has already been made in this field, major challenges remain before these sensors can be put into real-world applications. Because these polymers tend to aggregate in solution resulting in fluorescence quenching, the synthesis of new water-soluble polyfluorenes with high quantum yield in aqueous solution is needed. Furthermore, the prevention of nonspecific interactions and the suppression of background fluorescence also need to be resolved. Despite these remaining drawbacks, there is a real possibility that the approaches described in this chapter, or related permutations, will be incorporated into routine characterization protocols in molecular biology laboratories.

References

1. A.J. Heeger, Angew. Chem. Int. Ed. **40**, 2591 (2001)
2. A.R. Brown, A. Pomp, C.M. Hart, D.M. de Leeuw, Science **270**, 972 (1995)
3. G. Yu, J. Gao, J.C. Hummelen, F. Wudl, A.J. Heeger, Science **270**, 1789 (1995)
4. J.E. Guillet, *Polymer Photophysics and Photochemistry* (Cambridge University Press, Cambridge, 1985)
5. S.E. Weber, Chem. Rev. **90**, 1469 (1990)
6. H.F. Kauffmann, *Photochemistry and Photophysics*, vol. 2, ed. by J.E. Radek (CRC, Boca Raton, 1990)
7. G.D. Scholes, K.P. Ghiggino, J. Chem. Phys. **101**, 1251 (1994)
8. S. Tasch, E.J.W. List, C. Hochfilzer, G. Leising, P. Schlichting, U. Rohr, Y. Geerts, U. Scherf, K. Mullen, Phys. Rev. B **56**, 4479 (1997)
9. T.Q. Nguyen, J.J. Wu, V. Doan, B.J. Schwartz, S.H. Tolbert, Science **288**, 652 (2000)
10. N.G. Pschirer, K. Byrd, U.H.F. Bunz, Macromolecules **34**, 8590 (2001)
11. D. Beljonne, G. Pourtois, C. Silva, E. Hennebicq, L.M. Herz, R.H. Friend, G.D. Scholes, S. Setayesh, K. Müllen, J.L. Brédas, Proc. Natl. Acad. Sci. USA **99**, 10982 (2002)
12. B. Liu, S. Wang, G.C. Bazan, A. Mikhailovsky, J. Am. Chem. Soc. **125**, 13306, (2003)
13. B. Liu, G.C. Bazan, J. Am. Chem. Soc. **126**, 1942 (2004)
14. J. Gierscher, J. Cornil, H.-J Egelhaaf, Adv. Mater. **19**, 173–191 (2007).
15. D.W. Lee, T.M. Swager, Synlett. 149 (2004)
16. Y.H. Korri, F. Garnier, P. Srivastava, P. Godillot, A. Yassar, J. Am. Chem. Soc. **119**, 7388 (1997)
17. P. Bäuerle, A. Emge, Adv. Mater. **10**, 324 (1998)
18. F. Garnier, Y.H. Korri, P. Srivastava, B. Mandrand, T. Delair, Synth. Met. **100**, 89 (1999)
19. Y.H. Korri, A. Yassar, Biomacromolecules **2**, 58 (2001)
20. Q. Zhou, T.M. Swager, J. Am. Chem. Soc. **117**, 12593 (1995)
21. T.M. Swager, Acc. Chem. Res. **31**, 201 (1998)
22. J.S. Yang, T.M. Swager, J. Am. Chem. Soc. **120**, 11864 (1998)
23. J.S. Yang, T.M. Swager, J. Am. Chem. Soc. **120**, 5321 (1998)
24. M. Leclerc, Adv. Mater. **11**, 1491 (1999)
25. D.T. McQuade, A.E. Pullen, T.M. Swager, Chem. Rev. **100**, 2537 (2000)
26. P.C. Ewbank, G. Nuding, H. Suenaga, R.D. McCullough, S. Shinkai, Tetrahedron Lett. **42**, 155 (2001)
27. H.A. Ho, M. Boissinot, M.G. Bergeron, G. Corbeil, K. Doré, D. Boudreau, M. Leclerc, Angew. Chem. Int. Ed. **41**, 1548 (2002)

28. I.B. Kim, B. Erdogan, J.N. Wilson, U.H.F. Bunz, Eur. J. Chem. **10**, 6247 (2004)
29. A. Rose, Z.G. Zhu, C.F. Madigan, T.M. Swager, V. Bulovic, Nature **434**, 876 (2005)
30. I.B. Kim, A. Dunkhorst, J. Gilbert, U.H.F. Bunz, Macromolecules **38**, 4560 (2005)
31. I.B. Kim, J.N. Wilson, U.H.F. Bunz, Chem. Commun. 1273 (2005)
32. C. Fan, S. Wang, J.W. Hong, G.C. Bazan, K.W. Plaxco, A.J. Heeger, Proc. Natl. Acad. Sci. USA **100**, 6297 (2003)
33. N. DiCesare, M.R. Pinto, K.S. Schanze, J.R. Lakowicz, Langmuir **18**, 7785 (2002)
34. B. Liu, G.C. Bazan, Chem. Mater. **16**, 4467 (2004)
35. M. Leclerc, H.A. Ho, Synlett. **2**, 380 (2004)
36. F. Le Floch, H.A. Ho, L.P. Harding, M. Bedard, P.R. Neagu, M. Leclerc, Adv. Mater. **17**, 1251 (2005)
37. H.A. Ho, M. Bers-Aberem, M. Leclerc, Eur. J. Chem. **11**, 1718 (2005)
38. K.P.R. Nilsson, O. Inganäs, Nat. Mater. **2**, 419 (2003)
39. K.P.R. Nilsson, J.D.M. Olsson, F. Stabo-Eeg, M. Lindgren, P. Konradsson, O. Inganäs, Macromolecules **38**, 6813 (2005)
40. A. Herland, K.P.R. Nilsson, J.D.M. Olsson, P. Hammarstrom, P. Konradsson, O. Inganäs, J. Am. Chem. Soc. **127**, 2317 (2005)
41. M.R. Pinto, K.S. Schanze, Synthesis 1293 (2002)
42. D.T. McQuade, A.E. Pullen, T.M. Swager, Chem. Rev. **100**, 2537 (2000)
43. U. Scherf, E.J.W. List, Adv. Mater. **14**, 477 (2002)
44. M. Fukuda, K. Sawada, K. Yoshino, J. Polym. Sci. A: Polym. Chem. **31**, 2465 (1993)
45. M. Fukuda, K. Sawada, K. Yoshino, Jpn. J. Appl. Phys. **28**, L1433 (1989)
46. A.D. Schlüter, in *Handbook of Conducting Polymers*, ed. by T.A. Skotheim, R.L. Elsenbaumer, J.R. Reynolds (Marcel Dekker, New York, 1998), p. 209
47. B. Liu, G.C. Bazan, in *Organic Electroluminescence*, ed. by Z.H. Kafafi, (Marcel Dekker, New York, 2005) Chapter 5
48. M. Stork, B.S. Gaylord, A.J. Heeger, G.C. Bazan, Adv. Mater. **14**, 361 (2002)
49. S. Wang, G.C. Bazan, Adv. Mater. **15**, 1425 (2003)
50. B. Liu, W.-L. Yu, Y.-H. Lai, W. Huang, Macromolecules **35**, 4975 (2002)
51. H.D. Burrows, V.M.M. Lobo, J. Pina, M.L. Ramos, J.S. de Melo, A.J.M. Valente, M.J. Tapia, S. Pradhan, U. Scherf, Macromolecules **37**, 7425 (2004)
52. S. Wang, B. Liu, B.S. Gaylord, G.C. Bazan, Adv. Funct. Mater. **13**, 463 (2003)
53. L. Arnt, G.N. Tew, J. Am. Chem. Soc. **124**, 7664 (2002)
54. M. Bockstaller, W. Ko¨hler, G. Wegner, D. Vlassopoulos, G. Fytas, Macromolecules **34**, 6359 (2001)
55. F. Wang, G.C. Bazan, J. Am. Chem. Soc. **128**, 15786 (2006)
56. R. Yang, A. Garcia, D. Korystov, A. Mikhailovsky, G.C. Bazan, T.-Q. Nguyen, J. Am. Chem. Soc. **128**, 16532 (2006)
57. B.S. Gaylord, S. Wang, A.J. Heeger, G.C. Bazan, J. Am. Chem. Soc. **123**, 6417 (2001)
58. S. Wang, G.C. Bazan, Chem. Commun. 2508 (2004)
59. J.R. Lakowicz (ed.), in *Principles of Fluorescence Spectroscopy* (Kluwer Academic, New York, 1999)
60. T. Förster, Ann. Phys. **2**, 55 (1948)

61. J. Cornil, V. Lemaur, M.C. Steel, H. Dupin, A. Burquel, D. Beljonne, J.L. Bredas, in *Organic Photovoltaics*, ed. by S.J. Sun, N.S. Sariciftci (Taylor and Francis, Boca Raton, 2005), p. 161
62. N.S. Sariciftci, L. Smilowitz, A.J. Heeger, F. Wudl, Science **258**, 1474 (1992)
63. R.A. Marcus, Angew. Chem. Int. Eng. **2**, 1111 (1993)
64. Q.H. Xu, D. Moses, A.J. Heeger, Phys. Rev. B **67**, 245417 (2003)
65. J.L. Brédas, D. Beljonne, V. Coropceanu, J. Cornil, Chem. Rev. **104**, 4971 (2004)
66. K.R.J. Thomas, A.L. Thompson, A.V. Sivakumar, A.J. Bardeen, S. Thayumanavan, J. Am. Chem. Soc. **127**, 373 (2005)
67. A.C. Morteani, P. Sreearunothai, L.M. Herz, R.H. Friend, C. Silva, Phys. Rev. Lett. **92**, 247402 (2004)
68. B. Liu, G.C. Bazan, J. Am. Chem. Soc. **128**, 1188 (2006)
69. J.W. Hong, W.L. Hemme, G.E. Keller, M.T. Rinke, G.C. Bazan, Adv. Mater. **18**, 878 (2006)
70. N.J. Turro, *Modern Molecular Photochemistry*, (University Science Books, 1997)
71. S. Tyagi, F.R. Kramer, Nat. Biotechnol. **14**, 303 (1996)
72. M.K. Johansson, H. Fidder, D. Dick, R.M. Cook, J. Am. Chem. Soc. **124**, 6950 (2002)
73. PCT quenching has been observed with fluorescein "locked" within a protein environment by electron transfer from either a tryptophan or tyrosine residue, see M. Götz, S. Hess, G. Beste, A. Skerra, M.E. Michel-Beyerle, Biochemistry **41**, 4156 (2002)
74. B.S. Gaylord, A.J. Heeger, G.C. Bazan, Proc. Natl. Acid. Sci. USA **99**, 10954 (2002)
75. B.S. Gaylord, M.R. Massie, S.C. Feinstein, G.C. Bazan, Proc. Natl. Acad. Sci. USA **102**, 34 (2005)
76. B.S. Gaylord, A.J. Heeger, G.C. Bazan, J. Am. Chem. Soc. **125**, 896 (2003)
77. J.W. Hong, W.L. Hemme, G.E. Keller, M.T. Rinke, G.C. Bazan, Adv. Mater. **18**, 878 (2006)
78. B. Liu, G.C. Bazan, J. Am. Chem. Soc. **126**, 1942 (2004)
79. J.T. Davis, Angew. Chem. Int. Ed. **43**, 668 (2004)
80. D.E. Gilbert, J. Feigon, Curr. Opin. Struct. Biol. **9**, 305 (1999)
81. F. He, Y. Tang, S. Wang, Y. Li, D. Zhu, J. Am. Chem. Soc. **127**, 12343 (2005)
82. A.T. Phan, J.L. Mergny, Nucleic. Acids. Res. **30**, 4618 (2002)
83. F. He, Y. Tang, M. Yu, F. Feng, L. An, H. Sun, S. Wang, Y. Li, D. Zhu, G.C. Bazan, J. Am. Chem. Soc. **128**, 6764 (2006)
84. B. Liu, S. Baudrey, L. Jaeger, G.C. Bazan, J. Am. Chem. Soc. **126**, 4076 (2004)
85. L. An, Y. Tang, S. Wang, Y. Li, D. Zhu, Macromol. Rapid. Commun. **27**, 993 (2006)
86. G. Kada, K. Kaiser, H. Falk, H.J. Gruber, Biochim. Biophys. Acta **1427**, 44 (1999)
87. D.B. Papkovsky, T.C. O'Riordan, G.G. Guilbault, Anal. Chem. **71**, 1568 (1999)
88. M. Wu, Z. Lin, M. Schäferling, A. Dürkop, O.S. Wolfbeis, Anal. Biochem. **340**, 66 (2005)
89. F. He, Y. Tang, M. Yu, S. Wang, Y. Li, D. Zhu, Adv. Funct. Mater. **16**, 91 (2006)
90. Y. Tang, F. He, M. Yu, S. Wang, Y. Li, D. Zhu, Chem. Mater. **18**, 3605 (2006)
91. R.F. Service, Science **282**, 396 (1998)
92. D.D. Schemaker, Nature **409**, 922 (2001)

93. E.M. Southern, Trends Genet. **12**, 110 (1996)
94. G. Ramsay, Nat. Biotechnol. **16**, 40 (1998)
95. C.B. Epstein, R.A. Butow, Curr. Opin. Biotechnol. **11**, 36 (2000)
96. B. Liu, G.C. Bazan, Nature Protocols **1**, 1698 (2006)
97. B. Liu, G.C. Bazan, Proc. Nat. Acad. Sci. **102**, 589 (2005)

2

Polyelectrolyte-Based Fluorescent Sensors

K. Ogawa, K.E. Achyuthan, S. Chemburu, E. Ji, Y. Liu, G.P. Lopez,
K.S. Schanze, and D.G. Whitten

2.1 General Introduction

The organic semiconductors that are discussed in this chapter are polyelectrolytes that contain a fluorescent chromophore that is present either as (1) a single large conjugated system, or as (2) an array of strongly aggregated chromophores that exhibit distinct fluorescence from the corresponding monomer by virtue of excitonic delocalization. The latter "class" of organic semiconductors consists of two different systems. In one the chromophores are covalently linked to a polymer backbone and the ensemble is thus a "permanent" molecule. In the other example to be discussed, the aggregated chromophore is associated with an oppositely charged support via adsorption or a templating polyelectrolyte scaffold via "host–guest" interactions. These different systems we will discuss are remarkable for their structural differences as well as for the similarity of their photophysical behavior and potential utility in sensing applications. Two characteristics stand out for these macromolecules and macromolecular assemblies: they are water soluble, but can be easily adsorbed onto oppositely charged supports, and they are also relatively hydrophobic and thus can associate with organic and inorganic counterions and biopolymers by nonspecific association. As we shall discuss, all three systems show amplified fluorescence quenching (super-quenching) upon association with molecules capable of acting as energy transfer or electron transfer quenchers. Structures of the three systems to be discussed are shown in Fig. 2.1.

2.1.1 Amplified Fluorescence Quenching

Amplified fluorescence quenching was first observed for the conjugated polyelectrolyte **1** by the electron acceptor methyl viologen (**5**) [1]. The Stern–Volmer quenching constant for **1** with **5** as a quencher was found to be $\sim 10^7\,M^{-1}$ in aqueous solution [1]. The quenching is "static" in nature and attributed to a ground-state "charge transfer" complex between the viologen

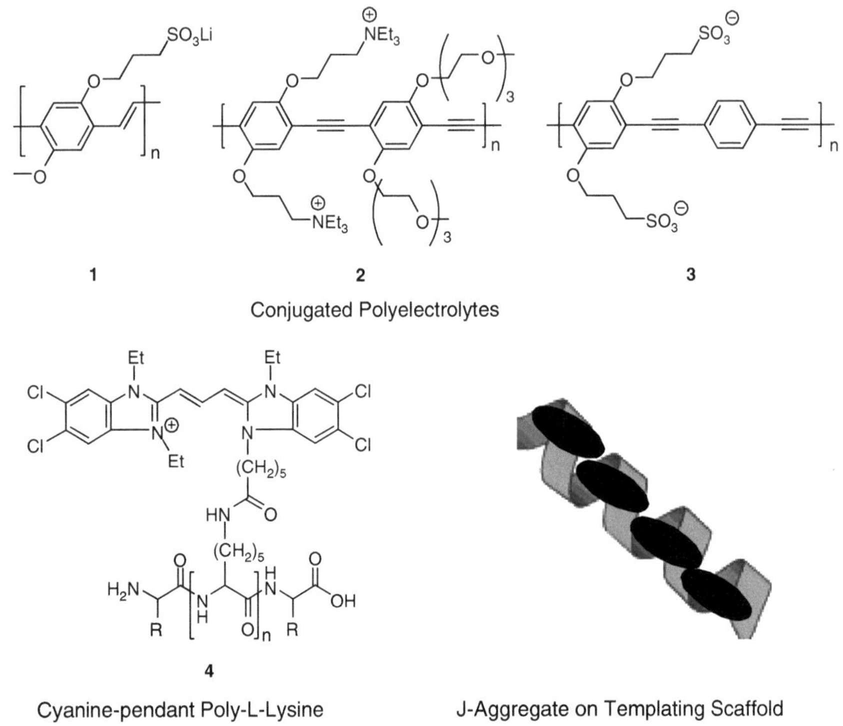

Fig. 2.1. Examples of polyelectrolyte-based fluorescent sensors

and one (or more) of the polymer repeat units. A similar ground-state charge transfer complex has been found with viologen and aromatic hydrocarbons, including the "monomer" chromophore for **1**, *trans*-stilbene; the equilibrium constant for formation of the complex in acetonitrile is $\sim 15\,\mathrm{M}^{-1}$ and this is also the Stern–Volmer constant for fluorescence quenching of *trans*-stilbene by viologen [2]. Thus the quenching constant for the polymer with **5** is enhanced by six orders of magnitude compared with the quenching of trans-stilbene or other aromatic hydrocarbons. Similar enhanced fluorescence quenching can be obtained with cationic organic dyes such as cyanines and in this case sensitized fluorescence of the dye may or may not be observed [3].

A series of cyanine-pendant poly-L-lysine polyelectrolytes (**4**) also exhibit similarly enhanced fluorescence quenching by anionic electron transfer or energy transfer quenchers (Fig. 2.2) [4]. For the cyanine-pendant poly-L-lysines it was possible to vary the number of polymer repeat units (PRU) systematically and gain an understanding of what factors influence the magnitude of the amplified quenching [5]. For the cyanine monomer (a monocation) quenching by the electron acceptor anthraquinone disulfonate (**6**) (AQS, a dianion) shows a $K_{\mathrm{SV}} = 630\,\mathrm{M}^{-1}$ (larger than the above-mentioned value for quenching of neutral *trans*-stilbene with the dicationic viologen), and quenching constants

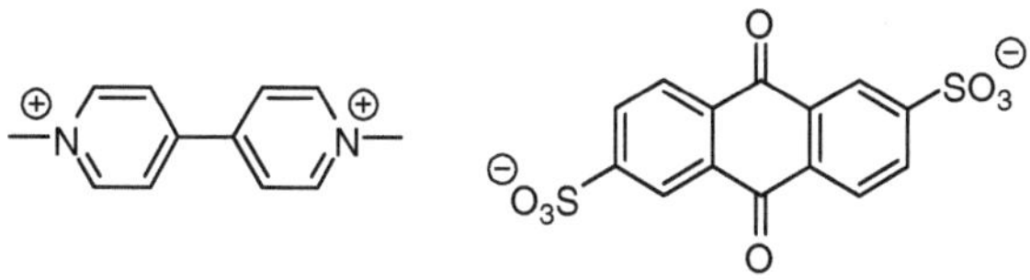

5 6

Fig. 2.2. Structures of quencher ions

for oligomers and polymers show a monotonic increase with the number of PRUs [5]. The highest value obtained is $K_{SV} = 1.2 \times 10^9\,M^{-1}$ for a polymer having ∼900 PRU [5]. The initial increase as PRU goes from 1 to about 33 can be mainly attributed to increased Coulombic attraction between the multicationic oligomer and the AQS; however, as the number of PRU increases beyond 33 the cyanine can be observed to form largely J-aggregates and the number of quenchers per polymer at 50% quenching $(Q/P)_{50}$ undergoes a further decrease; a minimum value of 2.6 for $(Q/P)_{50}$ is reached for the polymer with 250 PRU [5, 6]. Subsequent increases of polymer molecular weight lead to an increase in $(Q/P)_{50}$. More significantly the value of PRU/Q at 50% quenching levels off at ∼100 suggesting that excitonic delocalization over patches of ∼100 chromophores can be intercepted by a single quencher [5, 6]. Thus the amplified quenching (super-quenching) can be clearly shown to be a combination of enhanced Coulombic (and hydrophobic) attraction between oppositely charged polyelectrolyte and quencher, and delocalization of the excitation over an extended array of chromophores associated either by direct conjugation and/or aggregation.

When equimolar (in PRU) aqueous solutions are made from cationic-conjugated polyelectrolyte **1** and the cyanine-pendant poly-L-lysine **4** (∼250 PRU), there is clear evidence of energy transfer from the higher energy absorbing **1** to the lower energy emitting **4** [3]. Although the absorption spectrum of the mixture resembles an "addition" spectrum of the two polymers, the emission spectrum is dominated by emission from the J-aggregate of **4**. The emission is also independent of the exciting wavelength [3]. This suggests that (not surprisingly) there is strong association between the oppositely charged polyelectrolytes. Interestingly, the roughly neutral (overall) ensemble can undergo super-quenching by addition of either cationic methyl viologen or anionic AQS [3]. This indicates that the mixture must consist of regions of each individual polymer that possess sufficient residual charge to strongly bind counterions and permit quenching by either quencher. Although the two polymers, **1** and **4**, associate strongly when solutions of the polymers are mixed, there is little evidence for interaction when one of the polymers is first adsorbed onto an oppositely charged support. For example, polymer **4** (cationic) can be coated onto Laponite clay nanoparticles at submonolayer coverage such that the nanoparticles still retain a net overall negative charge. The Laponite-supported **4** exhibits J-aggregate fluorescence and can be quenched by both anionic and cationic electron acceptors (*vide infra*) [7]. Interestingly, mixing aqueous solutions of **1** with a suspension of Laponite-coated **4** results in a superposition of absorption/excitation and fluorescence from **1** and **4** suggesting that there is independent photophysical behavior for the two polymers and hence little interaction or association between them under these conditions [3]. Since both Laponite supported **4** and solution-phase **1** are anionic, it is reasonable to assume that at low concentrations there should be little interaction between them.

When polymer **4** is treated with suspensions of Laponite clay until no further uptake of polymer occurs, there is still less than a monolayer of polymer and since the clay has a higher charge density than the polymer, the overall charge on the nanoparticles is negative. Under these conditions the polymer J-aggregate fluorescence is quenched by both **5** and AQS. Interestingly, the quenching constant, K_{SV}, for AQS is 50% higher for the Laponite-supported **4** than for aqueous solutions [3]. For coating of the polymers onto other supports, a "quench reversal" may be obtained. For example, both **1** and **4** can be coated by adsorption onto commercially available cationic or anionic polystyrene microspheres. Under these conditions, the fluorescence of polymer **1** (on cationic polystyrene) is quenched by AQS, but not by **5**. Similarly, anionic polystyrene-supported **4** is quenched by **5** but not by AQS.

Not surprisingly amplified quenching of polymers adsorbed to, or otherwise bound onto, supports can be observed and also "tuned" depending on the properties (size, structure, charge density) of the support [3]. An interesting example involves the series of cyanine-pendant poly-L-lysines whose solution-phase quenching is described earlier. When this series (ranging from monomeric to small oligomers to relatively large polymers) of polymers **4** was studied in solution, there was a large increase in quenching constant with polymer molecular weight. This series of polymer **4**, including the monomer, can be coated onto silica microspheres, which at neutral pH have a negative charge, but with a charge density much lower than the Laponite clay. Under these conditions, the adsorbed monomer and range of MW polymers are all quenched by AQS. However, the magnitude of the quenching for the series of cyanines is quite different from that observed in solutions [6]. Most notable is the difference between quenching of the monomer in solution and when adsorbed on the silica. The monomer exhibits a quenching constant with AQS, $K_{SV} = 630\,M^{-1}$, which is reasonable for a small cationic monomer quenched by a dication. The quenching constant for the monomer adsorbed on silica is enhanced by almost 20,000 and approaches those of the larger polymers in solution. In fact, the quenching constants for the adsorbed series of different molecular weight cyanine polymers **4** exhibit only a small increase in quenching constant with increase in molecular weight [6]. Similar superquenching through adsorption-mediated aggregation has been observed for a series of cyanines adsorbed on Laponite and thus defines the third system indicated in the first paragraph of this chapter [8]. Although self-assembly of small molecules (monomers–oligomers) onto supports can result in amplified quenching as a general process, its utility may be limited if the "building blocks" can reversibly dissociate from the support and thus it has not been widely used in sensing applications to date.

2.1.2 General Sensor Schemes: Bioassays Based on Quench/Unquench

The high sensitivity of polymers **1** and **4** to fluorescence quenching by electron acceptors and energy transfer acceptors suggested that the quenching might be the basis for sensitive chemical or biosensing [1,4]. Initial studies with **1** confirmed that both types of sensing approaches are possible. Several studies have indicated that one of the simplest approaches, linking a small recognition molecule (or ligand) to a quencher, can lead to a molecule (bioconjugate) that quenches the fluorescence of the polymer and wherein the quenching is reversed when a larger biomacromolecule associates with the ligand portion of the bioconjugate [1,3,4,9–11]. This quenching followed by quench reversal can be observed for solution phase polymers such as **1** and similar conjugated polyelectrolytes. However, in several cases, it has been shown that although quenching by the bioconjugate occurs, quenching reversal does not readily occur. An example is the finding for the higher molecular weight cyanine polymer **4** in combination with a bioconjugate containing an AQS quencher and a biotin ligand in aqueous solution [4]. Although strong quenching is observed, there is no quench reversal when a protein (avidin) recognizing and binding very strongly to biotin is added. Interestingly, a clear quench/quench reversal is observed for the same components when the polymer **4** is supported on nanoparticle Laponite [4]. One of the problems with trying to develop fluorescence-based sensors for polymers such as **1** and **4** in solution is the occurrence of nonspecific interactions with proteins, nucleic acids, or other biopolyelectrolytes [12]. Not surprisingly in view of the results described earlier for mixtures of **1** and **4**, there is strong association between oppositely charged polyelectrolytes, which can modify fluorescence in a variety of ways. A general solution to this problem is to anchor the sensing polymer to a support such as described above, either by adsorption or by covalent attachment. In the supported format, the polymer has lower mobility but still may exhibit strong fluorescence and susceptibility to highly amplified quenching. Additionally, the choice of polymer is important in maximizing effectiveness as a sensor. Polymers **1** and **4** have very low fluorescence yields and thus relatively lower effectiveness as fluorescent sensors in quench/unquench applications. In contrast the poly(phenylene ethynylene) polyelectrolytes such as **2** or **3** (Fig. 2.1) have more "rigid rod"-like structural units and exhibit much higher fluorescence efficiencies and lower sensitivity to fluorescence modulation via nonspecific interactions [12].

Several different sensing schemes have been developed on the basis of the quench/unquench of a highly fluorescent polymer. Previously reported examples include fluorescence quenching by association of a small molecule bioconjugate with either a solution-phase or particle-supported polymer and subsequent dissociation and unquenching when a biomacromolecule binds to the ligand portion of the bioconjugate. Other formats include microsphere-bound polymer that is colocated with a biological receptor (or "capture strand"); typically these bioassays involve a competition whereby a quencher-bioconjugate competes with an unlabeled analyte for sites on the surface of the microparticle. Assays of this type have been developed for both nucleic acids and proteins [12–14].

Among the most sensitive and effective assays that have been developed are those measuring enzyme activity [10, 15, 16]. A typical case involves protease assays. Both solution phase and microsphere-based assays have been developed. In these assays, a quencher is tied to a bioconjugate peptide, which contains a recognition site and a cleavage site. Before protease catalyzed cleavage of the bioconjugate peptide, the bioconjugate associates with the fluorescent polymer and quenches its fluorescence. Subsequent to the cleavage, the quencher is released and no longer associates with the polymer [15]. Thus, these assays function as fluorescence "turn on" assays. The same principle has been used to develop assays for kinase and phosphatase enzymes [16]. In this case, differential binding of a peptide, protein, or other moiety, before or after phosphorylation, to a microsphere on which a phosphate-binding site and fluorescent polymer are colocated, can be used as the basis for fluorescence "turn-on" and "turn-off" assays. Since these assays and their development have been described in detail and reviewed elsewhere, [12, 15, 16] this chapter will focus on assay technologies that have been more recently developed. The next section of the chapter will discuss enzyme activity assays that have been developed with the fluorescent dyes and polyelectrolytes. Although some of these involve quench/unquench of fluorescent polymers, in several cases new formats and/or new approaches to assay technologies will be described.

2.2 Enzyme Activity Assays

2.2.1 Assay Formats and Types

The assays that have been described in the previous section can be carried out by measurements of fluorescence of the polymer with a conventional fluorometer or a microwell plate reader. In the latter format, they are particularly convenient to use in high throughput screening (HTS) applications and have been carried out for small samples in up to a 3,456 well plate format. More recently, we have studied amplified quenching for microsphere-supported polymers such as **2**, by flow cytometry and in microfluidic channels [17]. In addition to providing new ways of adapting the super-quenching technology, they have provided evidence that in several cases the quenchers associate very strongly with the microsphere-adsorbed polymer and are only removed by "washing" for long periods. Interestingly, the same Stern–Volmer quenching constant is obtained for **2** with AQS by direct measurement of fluorescence and by flow cytometry.

In addition to the measurements of enzyme activity through the quench-unquench procedures described earlier, we have also found a new strategy for using super-quenching unquench/quench by what we refer to as "frustrated super-quenching." A microsphere-supported polymer such as **2** (and other cationic-conjugated polymers) can be coated by a lipid bilayer to yield a layered system in which the conjugated polymer layer is separated by the bilayer from reagents present in the aqueous medium. The lipid bilayer can "protect" the polymer from quenching by electron transfer quenchers such as AQS, and for good bilayers such as dimyristoyl phosphatidyl glycerol (DMPG) the bilayer protects the polymer, such that lower than usual ($\sim$20%, reduced from 95%) quenching occurs by AQS under the same conditions [17]. The small amount of quenching that does occur may be attributed to defects in the bilayer but in general the bilayer provides strong attenuation of quenching. In the case of DMPG bilayers over the polymer, disruption of the bilayer in the presence of AQS by addition of Triton-X 100, a nonionic surfactant, results in efficient quenching of the polymer fluorescence. This has been demonstrated for microsphere supported **2** and "overlayers" of DMPG through flow cytometry and in microfluidic channels. Extensions of this principle of assay development will be discussed in more detail subsequently.

A final type of assay involves incorporating biopolymer-mediated controlled self-assembly as a means of a fluorescent switch [8, 18, 19]. In the examples to be discussed, advantage is taken of the assembly of nonfluorescent cyanine dye monomers to form highly fluorescent J-aggregates on a templating biopolymer or inorganic scaffold. For the biopolymers such self-assembly provides a fluorescence "turn-on" assay for the biopolymer as well as an enzyme activity assay for enzymes that can degrade the biopolymer to monomer or small oligomers. In the latter case, the fluorescence is quenched as the enzyme digests the biopolymer scaffold required for fluorescent J-aggregate formation.

2.2.2 Proteolytic Enzyme Assays Using Conjugated Polyelectrolytes

Proteolytic enzymes like proteases and kinases/phosphatases play a crucial role in many physiological and pathological processes, rendering the necessity for inventing high throughput screening assays for their real time analysis of crucial importance. Most of the assays developed so far are low sensitivity assays, requiring high concentrations of the enzyme or the substrate. Enzyme assays using fluorescent-conjugated polymers that are in solution as well as supported on solid surfaces like borosilicate or polystyrene microspheres accommodate real time analysis of enzyme kinetics, and have very low detection limits (nanomolar). In this section, we describe in some detail a protease sensor system that operates via a "turn on" mechanism.

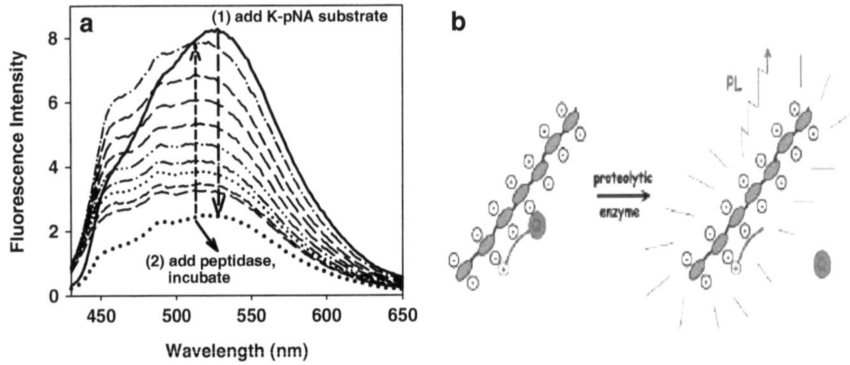

Fig. 2.3. (**a**) Fluorescence spectroscopic changes observed in a proteolytic enzyme assay using PPE-SO_3 (**3**) and K-*p*NA. Initial addition of KpNA quenches fluorescence, and then addition of peptidase gives rise to fluorescence recovery. *Solid line* Initial fluorescence, [PPE-SO_3] = 1.0 µM, phosphate buffer solution, pH 7.1; *Dotted line* fluorescence after addition of 167 nM K-*p*NA. Fluorescence intensity as a function of time (5–200 min) after addition of porcine intestinal peptidase (3.3 µg mL^{-1}). (**b**) Mechanism of the "turn-on" CPE-based sensor

Protease activity has been monitored quantitatively and in real-time using the anionic fluorescent-conjugated polyelectrolyte **3** (Fig. 2.1) [11]. The sensing mechanism is based on the electrostatic interaction between anionic **3** and a peptide substrate with a positive charge. For example, the cationic monopeptide L-lysine-*p*-nitroanilide (K-pNA) was used as a substrate-quencher for development of a prototype assay. Polyelectrolyte **3** is strongly fluorescent in aqueous buffer solution, but the fluorescence is very efficiently quenched by K-pNA ($K_{SV} \approx 10^7\,M^{-1}$). The efficient quenching is believed to arise via a charge transfer mechanism (the nitroanilide moiety is a good electron acceptor) and because the dicationic substrate-quencher ion pairs strongly to the polyelectrolyte chains. Upon addition of a peptidase enzyme (a nonspecific aminopeptidase), [20,21] the fluorescence of **3** recovers (Fig. 2.3). The fluorescence recovery arises because the enzyme cleaves the peptide bond linking the L-lysine residue to the pNA quencher moiety. As a result, the pNA is no longer ion-paired to the polyelectrolyte, so its ability to quench falls dramatically. The mechanism by which this sensor operates is illustrated schematically in Fig. 2.3.

2.2.3 Phospholipase Assays Using Conjugated Polyelectrolytes

Water-soluble conjugated polyelectrolytes interact with oppositely charged surfactants forming polymer–surfactant complexes. The polymer–surfactant interaction induces changes in the conformation [22] and optical properties [22–24] of the polymer including narrowed absorption and enhanced fluorescence quantum efficiency. In addition, the fluorescence quenching efficiency of

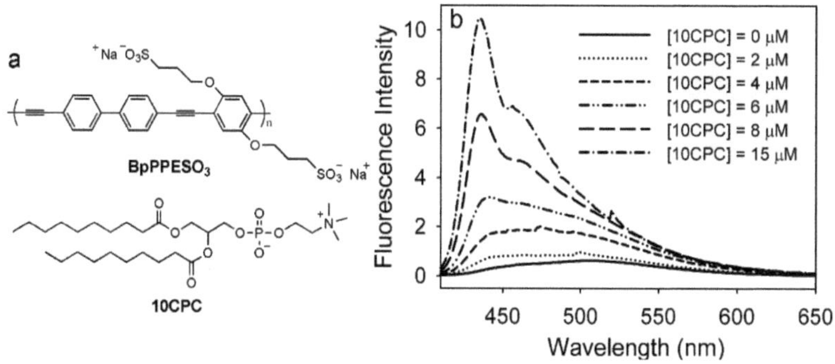

Fig. 2.4. (**a**) Structures of polymer and substrate. (**b**) Fluorescence spectroscopic changes observed upon addition of 10CPC. [BpPPESO$_3$] = 1 μM in water at ambient temperature. Reproduced from ref. [60] with permission. Copyright 2007, American Chemical Society

a conjugated polyelectrolyte by ionic and neutral quenchers can be modified significantly in the presence of surfactants [7–9]. Phospholipids are naturally occurring surfactants that feature an ionic headgroup (either cationic, anionic, or zwitterionic) and two hydrophobic tails (Fig. 2.4a). In a recent series of investigations, we have explored the effect of phospholipids on the optical properties of conjugated polyelectrolytes. As shown in Fig. 2.4b, the fluorescence intensity of an aqueous solution of the anionic polymer BpPPESO$_3$ increases dramatically and blue-shifts with increasing concentration of the added phospholipid 10CPC. Addition of 15 μM 10CPC to a BpPPESO3 solution leads to a greater than 50-fold increase in the polymer's fluorescence intensity at 435 nm. The significant change of the polymer fluorescence suggests that the phospholipid interacts with the polymer via a combination of the electrostatic attraction between the anionic units on the polyelectrolyte and the cationic head-group of the phospholipid, as well as a hydrophobic interaction between the hydrocarbon tails of the lipid and the polymer backbone. The phospholipid molecules likely bind to the polymer chains in a manner that maximizes the hydrophobic interactions, thereby inducing the polymer to take on a more extended conformation as well as disrupting the polymer–polymer interactions (aggregation).

We have taken advantage of this lipid induced change in fluorescence of the conjugated polyelectrolyte to develop a real-time fluorescence turn-off assay for the lipase enzyme phospholipase C (PLC). PLC plays a critical role in cell function and signal transduction cascades in mammalian systems [25, 26]. It catalyzes hydrolysis of the phosphate ester in a phospholipid at the glycerol side, [27] yielding as products diacylglycerol (DAG) and a phosphoryl base. Neither of these lipid hydrolysis products has a significant effect on the fluorescence of BpPPESO3 when they are present at relatively low concentrations. Figure 2.5a illustrates the mechanism of the PLC turn-off assay. The polymer aggregates in aqueous solution, but upon addition of the phospholipid

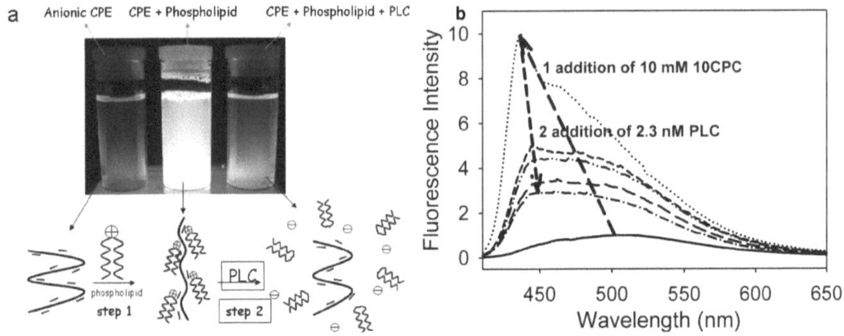

Fig. 2.5. (**a**) Mechanism and demonstration of PLC turn-off assay. (**b**) Fluorescence spectroscopic changes observed in the PLC turn-off assay in water. *Solid line* Initial fluorescence, [$BpPPESO_3$] = 1.0 µM; *Dotted line* fluorescence after step 1: addition of 10 µM 10CPC. Fluorescence intensity as a function of time after step 2: addition of 2.3 nM PLC: 1 (- - -), 5 (- ·· - ··), 20 (— —), 45 (—•—) min. Reproduced from ref. [60] with permission. Copyright 2007, American Chemical Society

a polymer–lipid complex forms, which disrupts the polymer aggregates that are present in solution (step 1). As a result, the fluorescence of the polymer is enhanced. After addition of the lipid, the solution containing the polymer–lipid complex is then incubated with the PLC enzyme (step 2). PLC hydrolyses the lipid, thereby decreasing its amphiphilic character, and disrupting the polymer–lipid complex. This effect causes the polymer to reaggregate, causing a decrease in the fluorescence intensity. As shown in Fig. 2.5, the decrease in fluorescence intensity continues as long as lipid substrate remains in the solution and the PLC-catalyzed hydrolysis reaction continues.

After optimizing the assay conditions, including ionic strength, pH, temperature, activator (Ca^{2+}) concentration, and polymer concentration, a calibration curve was developed, which allows the lipid substrate concentration to be determined from the fluorescence intensity. The calibration curve allows the enzyme catalysis kinetics parameters (e.g., K_m and V_{max}) to be measured. This PLC turn-off assay is effectively inhibited by known inhibitors (F^- and EDTA), which demonstrates that the sensor relies on the specific catalysis reaction by PLC. It has been demonstrated to be a sensitive (detection limit ~0.5 nM enzyme concentration), fast (<5 min), and selective (good specificity over phospholipase A and D, and other nonspecific proteins) PLC assay, which can be carried out at very low initial substrate concentration (in the range of micromolar to nanomolar).

This method is general and can be applied to other enzyme systems utilizing different types of substrates. For example, a similar assay to sense phosphatidylinositol phospholipase C (PI-PLC) can be constructed by utilizing the anionic phospholipid phosphatidylinositol biphosphate (PIP_2) as substrate along with a cationic-conjugated polyelectrolyte as the fluorescent signaling element. By taking the advantage of the rapid and strong response of the polymer to phospholipids, it is possible to image the interactions between polymer and liposome by means of confocal or epifluorescence microscopy. Finally, some natural or synthesized cationic lipids have been applied as DNA transfection agents, [28, 29] so this assay might open another way to detect the cationic lipids used in gene transfer.

2.2.4 Assays Based on "Frustrated Super-Quenching"

Phospholipase A_2 (PLA_2) also belongs to the family of lipase enzymes. It catalyzes the hydrolysis of the *sn*-2 fatty acid ester bond of the phospholipid substrate liberating a fatty acid and lysophospholipids. When the liberated fatty acid is arachidonic acid, a cascade of reactions promoting the formation of proinflammatory moieties is triggered. PLA_2 is also associated with circulatory low density lipoprotein and is now being exploited as a novel biomarker for coronary artery diseases. A simple fluorescence turn off assay for PLA_2 was developed by quantifying enzymatic activity of PLA_2 using frustrated super-quenching, which is achieved through the ionic interaction between a solid supported fluorescent-conjugated polyelectrolyte and its specific quencher. A cationic-conjugated polyelectrolyte (**2**) is physically adsorbed onto a solid support such as nonporous borosilicate microspheres. As illustrated in Fig. 2.6a, fluorescence of microsphere supported polymer is quenched in the presence of AQS. Using the layer by layer method of assembly, microsphere supported cationic polymer is coated with DMPG, an anionic lipid bilayer that serves the dual purpose of acting as a substrate for PLA_2 enzymatic activity and shielding the fluorescent polymer from quenching by AQS. When these lipobeads are

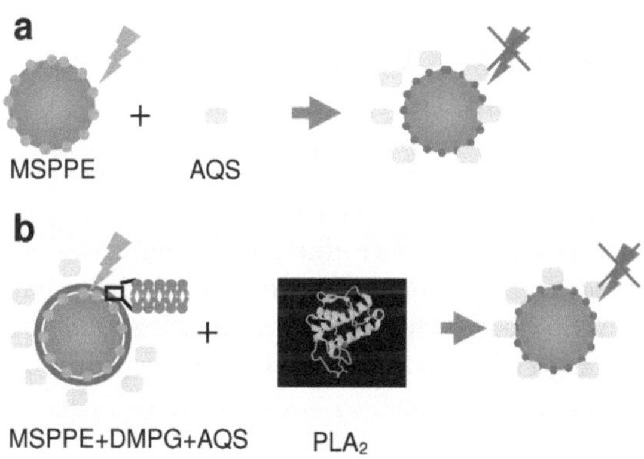

Fig. 2.6. (**a**) Microsphere supported PPE (MSPPE) with desired concentration of AQS. This results in the turning off of the fluorescence of PPE. (**b**) MSPPE is coated with a layer of DMPG lipid bilayer and incubated with AQS. The DMPG layer protects the fluorescence of PPE from being quenched by AQS. When PLA_2 is added, it disrupts the lipid bilayer by cleaving the *sn*-2 ester bond of the lipid bilayer, resulting in the exposure of the fluorescence of PPE, which is then quenched by AQS

incubated with a mixture of PLA_2 and AQS, the cleaving of the *sn*-2 acyl ester bond of the DMPG is catalyzed by PLA_2, exposing the fluorescent polymer to AQS to be quenched (Fig. 2.6b). This reaction occurring on the surface of the bead is detected by a flow cytometer. Compared with previously developed radiometric, [30] calorimetric, [31] fluorogenic, [32, 33] or spectrophotometric [34] assays reported in the literature, this assay demonstrates the use of a simple homogeneous platform to detect PLA_2 enzymatic activity directly by using a natural substrate of PLA_2 without any modification.

The assay for PLA_2 activity described above, using frustrated super-quenching, is a fluorescence turn off assay. This assay can also be adapted as a fluorescence turn on assay for screening of potential inhibitors of PLA_2. A fluorescence turn on assay using the principle of FRET has also been developed using solid supported polymer and rhodamine (Rh)-labeled DMPG. As described previously, the solid supported polymer is coated with DMPG. In this particular assay, however, the head group of the DMPG beads is labeled with rhodamine. Polymer and rhodamine are bought in close contact with each other when the solid supported conjugated polyelectrolyte is coated with Rh-DMPG lipid bilayer, enabling FRET to occur. When these lipobeads are incubated with PLA_2, the Rh-DMPG bilayer is disrupted, resulting in an increase in the intermolecular distance between the polymer and Rh. An increase in the fluorescence of conjugated polyelectrolyte can thus be observed.

The principle of frustrated super-quenching as a detection technique for quantifying the catalytic activity of hydrolytic enzymes can also be extended by using solid supported polymer and enveloping it with anionic biopolymers like carboxymethyl amylose (CMA) or carboxymethyl cellulose (CMC) to shield its fluorescence from electron transfer or energy transfer quenchers.

2.2.5 Supramolecular Self-Assembly and Scaffold Disruption/Destruction Assays

As described in Sect. 2.1, cyanine dyes can form J-aggregates that are characterized by sharp red-shifted absorbance and fluorescence compared with the monomer. As discussed in the introduction, J-aggregates can form on inorganic nanoparticles or microspheres as well as within polymeric ensembles. Other studies have indicated that cyanines can self-assemble on biopolymeric scaffolds to form aggregates with properties that depend on the structure of the host.

2.2.6 Cyanines and Supra-Molecular Self-Assembly

At the core of the scaffold formation and disruption/destruction assay is the electronic exciton concept that is relevant to molecular aggregation [35–41]. Cyanine dyes and other aromatic or heteroaromatic organic compounds can form molecular aggregates with varying structural and physical properties, which are often very sensitive to medium, potential host structure and chromophore substituents. Two common forms of dye aggregates are the so-called J-type and H-type. H-type dimers or aggregates are arrays of molecules where the principal transition moments are largely parallel and individual chromophores are arranged either face-to-face or edge-to-face. H-type aggregates are usually characterized by absorption spectra that show a strong maximum at higher energy than the principal transition of the monomer and a much weaker, red-shifted absorption [35]. H-aggregates are nonfluorescent or weakly fluorescent. In contrast, J-type dimers or aggregates have transition moments of monomers either end-on-end or in an otherwise extended structure, wherein the aggregate principal absorption is sharp and red-shifted compared with the monomer transition. J-type aggregate fluorescence is typically sharp and only modestly red-shifted compared with the J-type aggregate absorption [39–41]. Self-assembling cyanine dyes aggregate on chem-bio-helices (helicophilic cyanines) or assemble onto linear polymers, transforming the latter to adopt helical or supra-helical structures (helicogenic cyanines). During this process, nonluminescent molecules become highly fluorescent [35–39]. Helices of proteins, peptides, carbohydrates, lipids, and nucleic acids and nonhelices such as phospholipids, membranes, or liposomes may mediate J-type aggregate fluorescence [42–44]. Finally, J-type aggregate transformations are observed with several types of solid supports and nanoparticles [4, 6, 7, 45]. Thus, the spectral properties of J-type aggregates make them attractive candidates for developing a variety of chem-bio-sensing applications.

2.2.7 Cyanine Chemistry

The structures of representative examples of cyanines with the potential to form helicogenic and/or helicophilic J-type aggregates are shown in Fig. 2.7. The syntheses of some of these cyanines were described previously [45]. These

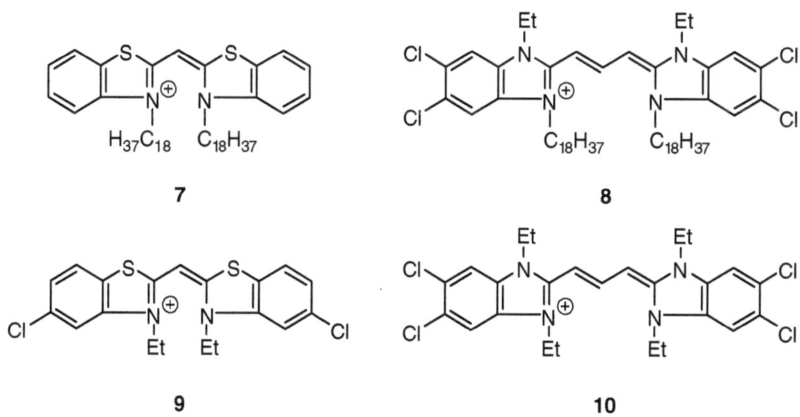

Fig. 2.7. Chemical structures of cyanine dyes that may form J-type aggregates

water soluble and amphiphilic cyanines are anionic or cationic and span a range of absorption from 375–600 nm. The cyanines, depending on their structures, can form either J- or H-type aggregates. However, since normally only J-type aggregates exhibit strong fluorescence, the detection is dependent on the use of a cyanine that forms a fluorescent J-type aggregate. Cyanines **9** and **10** are cationic and they both form J-type aggregates when associated with anionic supports such as the biopolymer CMA or CMC, as well as nanoparticles of clay, anionic microspheres, or silica microparticles [4, 6–8, 18, 45]. Anionic structures having the same chromophores as **9** and **10** have also been prepared and may be anticipated to form aggregates when exposed to chemical and biological helices that are cationic. We have already observed that clay nanoparticles (anionic) "overcoated" with cyanines **9** and **10** associate with some of the anionic cyanines [4, 6, 7]. Both **9** and **10** are slightly water soluble and we have shown that they "exchange" with anionic sites on different nanoparticles or polymeric supports. We have also studied amphiphilic cyanines **7** and **8** with the same chromophores as **9** and **10**. These are nearly water insoluble and our investigations indicate that once bound to an anionic binding site (clay, silica, or biopolymer) they do not dissociate. These amphiphilic structures offer the possibility of preparing a precoated sensor that will be useful in continuous (one-step, real time, kinetic) assays. Sensors prepared thus may be highly sensitive and are subject to very few nonspecific effects, which might be triggered by other components in the complex milieu that offer binding sites for the cyanine monomer and/or the J-type aggregate. Several physico-chemical properties of cyanines might be involved in forming fluorescent J-type aggregates including side-chain length, side-chain substituents, and dye chirality [46]. The wide range of absorption for the cyanines (375–600 nm) offers the possibility of multiplexing the assays. Some cyanines form J-type aggregates with low fluorescence efficiency; yet when there is no background from the fluorescence of the monomer or the H-type aggregate, the fluorescence may still be sharp and strong enough to provide a useful assay. Furthermore, even where the cyanine fluorescence efficiency is low, if it has virtually zero background emission from monomers, then it will yield high signal/background; a valuable assay metric.

2.2.8 Glycosidases and Scaffold Disruption/Destruction Assay

We described a new optical sensing system for the high-throughput screening (HTS) of a broad range of chemical and biological molecules, based on the principle of scaffold formation or disruption/destruction [8, 19]. The technology offers the flexibility of configuring the assays in *label free* or *labeled* formats and operating in continuous or discontinuous (multistep, endpoint) modes with complex sample formulations, under operationally relevant testing conditions [8]. The new assay platform was first demonstrated using carbohydrates and carbohydrate metabolizing glycosidase enzymes. Cellulose is the most abundant plant carbohydrate. It is a linear polysaccharide composed of β-D-glucose linkage – (1 → 4). Cellulose is of interest to the bioenergy/biofuel industries. Amylose is a major component of starch and the second major plant carbohydrate. It is a linear polymer of α-D-glucose linkage – (1 → 4). Amylase converts glycogen and starch into sugars. We used the cyanine **9** that undergoes a cooperative self-assembly with CMA and CMC. The water soluble **9** exists as a nonfluorescent monomer in the absence of CMA or CMC, but is converted to a highly fluorescent, red-shifted J-aggregate when it self-assembles on the anionic polymers; the induced circular dichroism (CD) spectrum for the achiral cyanine, as well as other evidence, indicates that the linear, chiral CMA polymers convert to supramolecular helical assemblies as the cyanine and polymer self-assemble [8, 18]. We utilized this property to develop an assay for amylase, based on the concept of "scaffold destruction," wherein the destruction of CMA by amylase enzyme is accompanied by attenuation of light emission from the J-aggregate [8]. The extent of light attenuation was an index of amylase activity (Fig. 2.8). A similar assay format can be applied to monitor the enzymatic digestion of cellulose by cellulases.

Hyaluronidase is involved in bacterial and fungal infections because of virulence factors evoked by tissue degradation and mediates host–pathogen interactions [47]. Since hyaluronic acid (HA) is a major component of the extracellular matrix involved in joint lubrication, a sensitive hyaluronidase assay is important. Current hyaluronidase assays rely on turbidimetric techniques that require high levels of the enzyme and are relatively inaccurate [47]. HA was previously shown to bind cyanines [48, 49]. The detection scheme designed for CMA, CMC, and amylase enzyme described earlier was also applicable to HA and hyaluronidase activity [19]. "Scaffold Destruction"

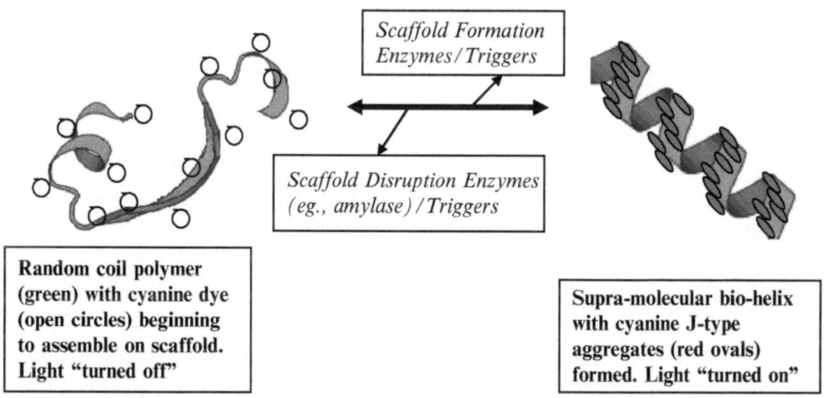

Fig. 2.8. Scaffold formation/disruption assays based on supramolecular self-assemblies. In this scheme, a cyanine is helicogenic. Molecular self-assembly of cyanine upon linear chiral polymer such as CMA, CMC, or HA results in a conformation transition of the polymer to adopt a super-helix structure. Scaffold disrupting glycosidases (hyaluronidase, amylase, cellulase) trigger fluorescence attenuation by disruption of the helical scaffold

hyaluronidase assay has applications in pathogen/clinical diagnostics [19]. Other polymers of carbohydrates, proteins, nucleic acids, and chemical polymers might provide similar scaffolding for helicogenic cyanines upon which molecular aggregation occurs. This is in addition to – and different from – the previously demonstrated helicophilic property of cyanines [8,18,19,50,51]. By controlling and regulating the interactions of cyanines with various helices and nonhelices in a solvent-directed fashion, a variety of chem-bio-sensing may be accomplished, in either optical "turn on" or "turn off" modes (Fig. 2.8).

We described earlier fundamental investigations for developing optical sensors with broad, interdisciplinary applications to the fields of biology, chemistry, and physics. Novel, *label-free* or *labeled*, rapid, HTS assays were described that are suitable for screening drugs in optical "turn on" or "turn off" modes for a variety of chem-bio helices. Since J-aggregation is accompanied by spectral shift toward longer (red) wavelengths compared with monomeric cyanine, these assays could avoid the effects of interfering substances that absorb in the blue region [52, 53]. The above assay formats offer the additional advantages of orthogonality, redundancy, and degrees of freedom to maximize the specificity of detection for biosensor applications, besides exploiting nature's own selectivity [54]. These assay metrics are achieved through ratiometric analysis of light absorption and fluorescence emission intensities as well as spectral shifts of the monomer dye relative to cooperatively self-assembled aggregates, as a consequence of scaffold formation or scaffold destruction. By appropriately tuning the interactions between the scaffold and the cyanine, a variety of chem-bio-sensing applications are possible. Thus, a broad range of solid-phase, solution-phase, and interfacial sensing applications are feasible using this novel fluorescence platform technology.

2.3 Conjugated Polyelectrolyte Surface-Grafted Colloids

Conjugated polyelectrolytes can also be synthesized as covalently attached coatings on nano- or micro-spheres such as silica [55] or latex [56] particles. In a general approach, grafting points are introduced to the particles as an aryl halide group. Polymers can be surface grafted via metal-catalyzed step growth polymerization. These conjugated polymer grafted colloids show promise as a new fluorescence sensor platform, with several advantages over other forms of coatings.

In our recently developed approach, introduction of aryl iodide groups on the surface of silica particles was accomplished by reacting the surface with a trialkoxysilane bearing an aryl iodide group (Fig. 2.9). One advantage of this method is that the surface density of the grafting points can be controlled by varying the amount of silane, thus allowing the number of polymer brushes on the surface to be adjusted accordingly. A PPE-type polymer was grown from these aryl iodide groups via Pd(0) catalyzed A-B type polymerization under Sonogashira conditions. Although a large amount of free polymer

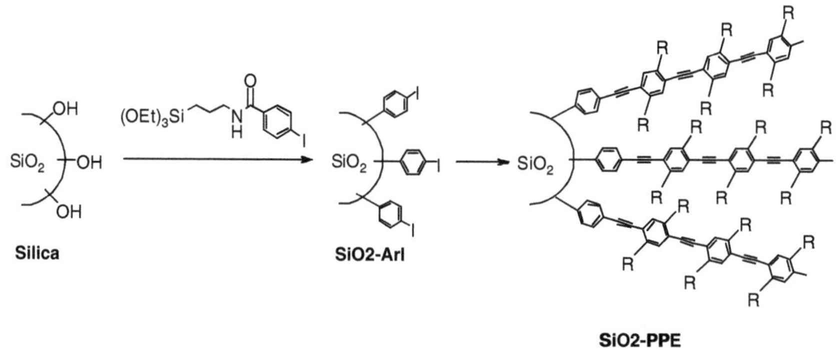

Fig. 2.9. Synthesis of polymer and surface-grafted silica particles. Reproduced from ref. [55] with permission. Copyright 2007, American Chemical Society

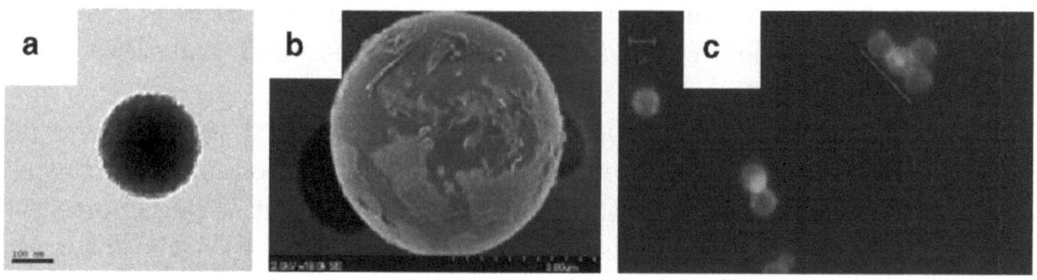

Fig. 2.10. Electron microscope images of silica particles: (**a**) 300 nm CP-grafted particle (TEM); (**b**) 5 μm CP-grafted particle (SEM); (**c**) Confocal fluorescence microscope image of 5 μm CP-grafted particles. Reproduced from ref. [55] with permission. Copyright 2007, American Chemical Society

was obtained as a byproduct of the reaction, isolated silica particles (SiO_2-PPE) exhibit yellow coloring with bright green fluorescence upon irradiation with a long wavelength UV lamp, which are characteristics of PPE. In a control experiment, the identical polymerization reaction was performed with unmodified silica particles in place of aryl iodide modified silica particles. Free polymer and unmodified silica particles without coloring were obtained after a standard work-up procedure. This suggests the importance of aryl iodide functionality on the surface for covalent attachment of the polymer. The surface modification process can be monitored by use of infrared (IR) spectroscopy to confirm the presence of functional groups on the surface [55]. Although the signals are suppressed by a strong peak from Si-O-Si stretch, the presence of the polymer can be confirmed by comparing IR spectra of polymer-grafted particles and free polymer.

The presence of polymer brushes on the surface of silica particles can also be confirmed by electron microscopy and confocal fluorescence microscopy. As seen in Fig. 2.10a, transmission electron microscope (TEM) images of SiO_2-PPE particles show rough surface texture. It should be noted that no significant change in particle size was observed after grafting the polymer. The thickness of the polymer layer was estimated using thermal gravimetric analysis data to be approximately 12 nm [55]. Although the particle size was not affected, the TEM data clearly show a uniformly covered surface with a thin layer-grafted polymer. Scanning electron microscope (SEM) images also show the presence of polymer on the surface of the particles. As shown in Fig. 2.10b, the polymer-grafted particles exhibit an "orange peel"-like appearance, which is associated with the polymer. In addition to a thin layer of polymer, some large aggregates are also observed on the surface. The origin of the material is unclear; however, one possible explanation is that the material was initially produced in solution during the polymerization and then became either chemically or physically adsorbed onto the surface. The confocal images (Fig. 2.10c) of polymer-grafted particles clearly show the green fluorescence from the polymer on the surface of the particles. Although the emission seems to be from the entire surface of the particles, some clustering of the fluorescent material can also be observed. Such observation correlates with the SEM images of the surface of the particles, which suggests that the polymer graft layer is not completely uniform.

Photophysical characterization of the polymer-grafted particles was performed to investigate their potential for fluorescent sensor material. Suspensions of SiO_2-PPE particles in water and methanol exhibit strong fluorescence with $\lambda_{max} = 535$ nm. Attempts were made to measure UV–vis absorption property of the particles; however, the colloidal nature of the suspension caused strong interference in the near UV region because of light scattering. This also precluded fluorescence quantum efficiency measurement of the particles. Although direct measurement of the absorption spectrum was not possible, the absorption profile can be approximated by the fluorescence excitation spectrum of colloidal suspensions of particles. It was found that the excitation spectrum closely resembles the absorption spectrum of the free polymer [55].

A series of fluorescence quenching studies of SiO_2-PPE particles were conducted using a variety of quenchers including methyl viologen (**5**), Cu^{2+}, diethyldicarbocyanine (DEDCC), diethylcyanine (DEC), and diethylthiadicarbocyanine (DETDCC). Among these quenchers, **5** and Cu^{2+} quench via a charge-transfer mechanism, while cyanine dyes quench via a Förster energy-transfer mechanism. As illustrated in Fig. 2.11a, charge-transfer type quenchers only quench ~70% of the fluorescence emission from the particle. Energy-transfer type quenchers, on the other hand, exhibit complete quenching, with Stern–Volmer constants (K_{SV}) ranging from 10^5 to $10^7\,M^{-1}$ as seen in Fig. 2.11b. For charge-transfer type quenchers, K_{SV} values were calculated using a modified Stern–Volmer expression [57–59] to give values in the $10^6\,M^{-1}$ range. Such large K_{SV} constants indicate that the grafted polymers retain the amplified quenching property, which is desirable for sensor materials. Although the K_{SV} values are in a similar range, there exists a distinct difference in quenching behavior between these different types of quenchers. The exciton and quencher must be in proximity (<1 nm) for charge-transfer

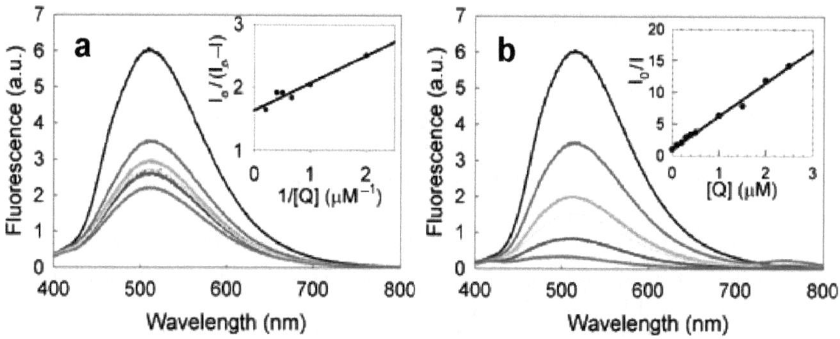

Fig. 2.11. Stern–Volmer quenching of CP-grafted particles suspended in water: (**a**) Electron transfer quenching by MV^{2+} (0–5 μM) and (**b**) energy transfer quenching by DEDCC (0–2.5 μM). Reproduced from ref. [55] with permission. Copyright 2007, American Chemical Society

type quenching to be efficient, whereas energy-transfer type quenching can be efficient even when the exciton and quencher are separated by distances greater than 5 nm. Considering the difference in quenching mechanism, the fact that charge-transfer type quenchers only quench a portion of the fluorescence from the particle suggests that the grafted polymers on the colloid surface exist in a strongly aggregated state and that some fraction of excitons are trapped within the aggregates. Such aggregates prevent quenchers from penetrating into the polymer matrix. Therefore, energy-transfer quenchers with longer effective distances can quench excitons that are trapped deeply inside the aggregates, while charge-transfer quenchers only quench excitons on the exterior of the grafted layer. In addition to such quenching effects, a very efficient sensitized fluorescence emission from DETDCC was observed at 700 nm [55].

As illustrated above, conjugated polymers can be covalently attached to silica particles to yield colloidal particles with bright fluorescence emission. The surface-grafted polymer retains its amplified quenching property, in which the fluorescence emission can be quenched efficiently with K_{SV} constants in 10^5–$10^7\,M^{-1}$ range. Such characteristics suggest that conjugated polymer-grafted colloids may be useful for applications in fluorescence sensors for biological targets.

2.4 Summary and Conclusions

This chapter describes a broad range of optical biosensor technologies, based on interchromophore interactions and the super-quenching phenomenon. Several specific biosensor schemes are described that are able to sense the activity of kinase, protease, lipase, and cellulase enzymes. Although the systems are disparate, the underlying sensor mechanisms are related, in that they all rely on the unique optical and photophysical properties of excitons that are present in these polymer-based or self-assembled chromophore aggregates. The excitons are strongly delocalized and highly mobile within the polychromophore assemblies; in this way these solution chromophore-based assemblies are analogous to solid-state semiconductors, which display optical and electronic properties that are quite sensitive to the presence of traps at low concentration.

We believe that there is great promise for the use of polyelectrolyte-based fluorescence sensors in optical biosensor applications. The systems are flexible in their design and format, they are highly sensitive and can be quite specific, and they are easily adapted to high-throughput screening. Although we have described a number of possible applications in the present chapter, many other applications have been described elsewhere, and there are clearly many more potential target analytes and biosensor systems that undoubtedly will be developed as a result of ongoing and future investigations.

Acknowledgments

The work at UNM was partially funded by the National Science Foundation (NSF) under Award CTS-0332315 awarded to Dr. Gabriel P. Lopez. The work at UF was partially funded by an award (W911NF-07-1-0079) from the Defense Threat Reduction Agency. Sandia is a multiprogram laboratory operated by Sandia Corporation, a Lockheed Martin Company for the United States Department of Energy's National Nuclear Security Administration under contract DE-AC04-94AL85000. This work was partially funded by the Defense Threat Reduction Agency (DTRA) under Contract MIPRG089XR076 awarded to Komandoor Achyuthan. Thanks are due to Dr. Stephen Casalnuovo for facilities and support.

References

1. L.H. Chen, D.W. McBranch, H.L. Wang, R. Helgeson, F. Wudl, D.G. Whitten, Proc. Natl. Acad. Sci. USA **96**(22), 12287 (1999)
2. J.C. Russell, D.G. Whitten, A.M. Braun, J. Am. Chem. Soc. **103**(11), 3219 (1981)
3. R.M. Jones, T.S. Bergstedt, D.W. McBranch, D.G. Whitten, J. Am. Chem. Soc. **123**(27), 6726 (2001)
4. R.M. Jones, T.S. Bergstedt, C.T. Buscher, D. McBranch, D. Whitten, Langmuir **17**(9), 2568 (2001)
5. L.D. Lu, R. Helgeson, R.M. Jones, D. McBranch, D. Whitten, J. Am. Chem. Soc. **124**(3), 483 (2002)
6. R.M. Jones, L.D. Lu, R. Helgeson, T.S. Bergstedt, D.W. McBranch, D.G. Whitten, Proc. Natl. Acad. Sci. USA **98**(26), 14769 (2001)
7. L.D. Lu, R.M. Jones, D. McBranch, D. Whitten, Langmuir **18**(20), 7706 (2002)

8. D.G. Whitten, K.E. Achyuthan, G.P. Lopez, O.K. Kim, Pure Appl. Chem. **78**(12), 2313 (2006)
9. M.R. Pinto, C. Tan, M.B. Ramey, J.R. Reynolds, T.S. Bergstedt, D.G. Whitten, K.S. Schanze, Res. Chem. Intermed. **33**(1, 2), 79 (2007)
10. B.S. Harrison, M.B. Ramey, J.R. Reynolds, K.S. Schanze, J. Am. Chem. Soc. **122**(35), 8561 (2000)
11. M.R. Pinto, K.S. Schanze, Proc. Natl. Acad. Sci. USA. **101**(20), 7505 (2004)
12. K.E. Achyuthan, T.S. Bergstedt, L. Chen, R.M. Jones, S. Kumaraswamy, S.A. Kushon, K.D. Ley, L. Lu, D. McBranch, H. Mukundan, F. Rininsland, X. Shi, W. Xia, D.G. Whitten, J. Mater. Chem. **15**(27, 28), 2648 (2005)
13. S.A. Kushon, K.D. Ley, K. Bradford, R.M. Jones, D. McBranch, D. Whitten, Langmuir **18**(20), 7245 (2002)
14. S.A. Kushon, K. Bradford, V. Marin, C. Suhrada, B.A. Armitage, D. McBranch, D. Whitten, Langmuir **19**(16), 6456 (2003)
15. S. Kumaraswamy, T. Bergstedt, X.B. Shi, F. Rininsland, S. Kushon, W.S. Xia, K. Ley, K. Achyuthan, D. McBranch, D. Whitten, Proc. Natl. Acad. Sci. USA **101**(20), 7511 (2004)
16. F. Rininsland, W.S. Xia, S. Wittenburg, X.B. Shi, C. Stankewicz, K. Achyuthan, D. McBranch, D. Whitten, Proc. Natl. Acad. Sci. USA **101**(43), 15295 (2004)
17. R. Zeineldin, M.E. Piyasena, T.S. Bergstedt, L.A. Sklar, D. Whitten, G.P. Lopez, Cytom Part A, **69A**(5), 335 (2006)
18. O.K. Kim, J. Je, G. Jernigan, L. Buckley, D. Whitten, J. Am. Chem. Soc. **128**(2), 510 (2006)
19. K.E. Achyuthan, L.D. Lu, G.P. Lopez, D.G. Whitten, Photochem. Photobiol. Sci. **5**(10), 931 (2006)
20. J.A. Nicholson, T.J. Peters, Anal. Biochem. **87**(2), 418 (1978)
21. D.H. Porter, H.E. Swaisgood, G.L. Catignani, Anal. Biochem. **123**(1), 41 (1982)
22. J. Dalvi-Malhotra, L.H. Chen, J. Phys. Chem. B **109**(9), 3873 (2005)
23. L.H. Chen, D. McBranch, R. Wang, D. Whitten, Chem. Phys. Lett. **330**(1, 2), 27 (2000)
24. L.H. Chen, S. Xu, D. McBranch, D. Whitten, J. Am. Chem. Soc. **122**(38), 9302 (2000)
25. J.H. Exton, J. Biol. Chem. **265**(1), 1 (1990)
26. J.J. Schrijen, A. Omachi, W.A.H.M. Vangroningenluyben, J.J.H.H.M. Depont, S.L. Bonting, Biochim. Biophys. Acta **649**(1), 1 (1981)
27. T. Takahashi, T. Sugahara, A. Ohsaka, Methods Enzymol. **71** (Lipids, Pt. C), 710 (1981)
28. S. Zhang, Y. Xu, B. Wang, W. Qiao, D. Liu, Z. Li, J. Contr. Release **100**(2), 165 (2004)
29. R.I. Zhdanov, O.V. Podobed, V.V. Vlassov, Bioelectrochemistry **58**(1), 53 (2002)
30. K. Kugiyama, Y. Ota, K. Takazoe, Y. Moriyama, H. Kawano, Y. Miyao, T. Sakamoto, H. Soejima, H. Ogawa, H. Doi, S. Sugiyama, H. Yasue, Circulation **100**(12), 1280 (1999)
31. M. Rigoni, G. Schiavo, A.E. Weston, P. Caccin, F. Allegrini, M. Pennuto, F. Valtorta, C. Montecucco, O. Rossetto, J. Cell Sci. **117**(16), 3561 (2004)
32. T.M. Rose, G.D. Prestwich, Acs Chem. Biol. **1**(2), 83 (2006)
33. J.U. Eskola, T.J. Nevalainen, T.N.E. Lovgren, Clin. Chem. **29**(10), 1777 (1983)
34. M. Jimenez, J. Cabanes, F. Gandia-Herrero, J. Escribano, F. Garcia-Carmona, M. Perez-Gilabert, Anal. Biochem. **319**(1), 131 (2003)

35. M. Bednarz, J. Knoester, J. Phys. Chem. B **105**(51), 12913 (2001)
36. C. Spitz, J. Knoester, A. Ouart, S. Daehne, Chem. Phys. **275**(1–3), 271 (2002)
37. J. Moll, W.J. Harrison, D.V. Brumbaugh, A.A. Muenter, J. Phys. Chem. A, **104**(39), 8847 (2000)
38. G.D. Scholes, Annu. Rev. Phys. Chem. **54**(57), (2003)
39. T. Kobayashi (ed.), *J-Aggregates* (World Scientific, Singapore, 1996)
40. X.D. Song, C. Geiger, M. Farahat, J. Perlstein, D.G. Whitten, J. Am. Chem. Soc. **119**(51), 12481 (1997)
41. D.G. Whitten, L.H. Chen, H.C. Geiger, J. Perlstein, X.D. Song, J. Phys. Chem. B **102**(50), 10098 (1998)
42. M. Reers, T.W. Smith, L.B. Chen, Biochemistry **30**(18), 4480 (1991)
43. C. McCullough, M. Heywood, H. Samha, Am. J. Undergrad. Res. **4**(3), 1 (2005)
44. N. Kato, J. Prime, K. Katagiri, F. Caruso, Langmuir **20**(14), 5718 (2004)
45. I. Place, J. Perlstein, T.L. Penner, D.G. Whitten, Langmuir **16**(23), 9042 (2000)
46. A. Mishra, R.K. Behera, P.K. Behera, B.K. Mishra, G.B. Behera, Chem. Rev. **100**(6), 1973 (2000)
47. R. Stern, M.J. Jedrzejas, Chem. Rev. **106**(3), 818 (2006)
48. K. Sakurai, S. Shinkai, J. Inclusion Phenom. Macrocyclic Chem. **41**(1–4), 173 (2001)
49. T. Sagawa, H. Tobata, H. Ihara, Chem. Commun. **39**(18), 2090 (2004)
50. M.M. Wang, G.L. Silva, B.A. Armitage, J. Am. Chem. Soc. **122**(41), 9977 (2000)
51. K.M. Sovenyhazy, J.A. Bordelon, J.T. Petty, Nucleic Acids Res. **31**(10), 2561 (2003)
52. K.L. Vedvik, H.C. Eliason, R.L. Hoffman, J.R. Gibson, K.R. Kupcho, R.L. Somberg, K.W. Vogel, Assay Drug Dev. Technol. **2**(2), 193 (2004)
53. D.M. Olive, Expert Rev. Proteomics **1**(3), 327 (2004)
54. J.J. Davis, Chem. Commun. **40**(28), 3509 (2005)
55. K. Ogawa, S. Chemburu, G.P. Lopez, D.G. Whitten, K.S. Schanze, Langmuir **23**(8), 4541 (2007)
56. M. Beinhoff, A.T. Appapillai, L.D. Underwood, J.E. Frommer, K.R. Carter, Langmuir **22**(6), 2411 (2006)
57. S.S. Lehrer, Biochemistry **10**(17), 3254 (1971)
58. Y.K. Gong, T. Miyamoto, K. Nakashima, S. Hashimoto, J. Phys. Chem. B, **104**(24), 5772 (2000)
59. J.R. Lakowicz, *Principles of Fluorescent Spectroscopy*, 2nd edn. (Plenum Publishers, New York, 1999)
60. Y. Liu, K. Ogawa, K.S. Schanze, Anal. Chem. Web ASAP, (2007), DOI: 10.1021/ac 701672a.

3

Structurally Integrated Photoluminescent Chemical and Biological Sensors: An Organic Light-Emitting Diode-Based Platform

R. Shinar and J. Shinar

3.1 Introduction

3.1.1 Photoluminescence-Based Sensors

The field of photoluminescence (PL)-based chemical and biological sensors continues to grow rapidly [1–8]. Such sensors, which are usually utilized for monitoring a single analyte, are sensitive, reliable, and suitable for a wide range of applications, such as medical, biological (including biodefense), environmental, and industrial. The sensors are typically composed of a luminescent sensing component, whose PL is monitored before and during exposure to the analyte, a light source that excites the PL, a photodetector (PD), a power supply, and the electronics for signal processing. Typical light sources include lasers, lamps, and inorganic light-emitting diodes (LEDs). Such excitation sources are either bulky and/or costly, cannot be integrated with the other components due to size, geometrical, or operational constraints, or require intricate integration procedures for their incorporation in a structurally integrated, compact device [9]. Recently, organic LEDs (OLEDs) have been introduced as promising light sources for PL-based sensing applications. In contrast to the typical excitation sources, the structural integration of an OLED with a sensing component is *uniquely simple*, resulting in small-size and potentially very low-cost devices [10–20] that are promising for developing miniaturized sensor arrays for the above-mentioned applications, including for multianalyte analysis. The latter is motivated by the need for high throughput and low-cost analyses of complex samples. Hence, it is not surprising that numerous studies on multianalyte detection in a single sample have been reported. The multianalyte detection methods include electrochemical [21, 22], piezoelectric [23], electrical resistance [24,25], and optical [7,8,26–32]. The sensor arrays are fabricated using photolithography and soft lithography [22, 25, 33], inkjet-, screen-, and pin-printing [31, 32, 34], or photodeposition [26, 27, 33]. These techniques are typically complex multistep procedures, and/or require sophisticated image analysis and pattern-recognition codes. The OLED-based platform would drastically simplify the fabrication and application of multianalyte PL-based sensor arrays.

3.1.2 Structurally Integrated OLED/Sensing Component Modules

OLEDs

The basic structure of a typical OLED is shown in Fig. 3.1 [35]. It consists of a transparent conducting anode, typically indium tin oxide (ITO) coated on a glass or plastic mechanical support, the organic layers, and a metal cathode. The thickness of OLEDs (excluding the mechanical support) is typically $<0.5\,\mu m$. Under forward bias electrons are injected from the low-workfunction cathode into the electron-transport layer (ETL). Similarly, holes are injected from the high-workfunction ITO into the hole-transport layer (HTL). Due to the applied bias, the electrons and holes drift toward each other, and typically recombine in a recombination zone near, or at, the ETL/HTL interface. A fraction of the recombination events forms radiative excited states. The radiative decay of these states provides the electroluminescence (EL) of the device.

OLEDs are unique in their simple and versatile design, low cost and ease of fabrication, whether by spin-coating or inkjet printing of the polymer-containing solutions, or by thermal vacuum evaporation of the small molecules [35, 36]. They have dramatically improved over the past decade [37–41], and commercial products incorporating them are proliferating rapidly [41]. They are inherently advantageous as low-voltage [37–41], miniaturizable [42], and

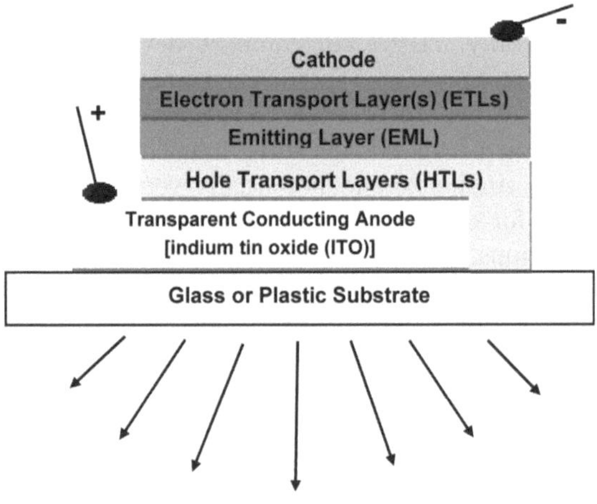

Fig. 3.1. Basic structure of an OLED

flexible light sources [43]; electrophosphorescent red-to-green OLEDs with external quantum efficiencies $17 \leq \eta_{ext} \leq 20\%$ [44] and blue OLEDs with $\eta_{ext} \approx 11\%$ [40] have been reported. Additionally, OLEDs can be operated at an extremely high brightness, and they are relatively easily scalable from micron-sized pixels to large emissive areas; 60 nm diameter OLEDs were demonstrated recently [45]. The rapid development and inherent advantages of OLEDs present an opportunity to develop a new platform of integrated (micro)sensor arrays, including for multianalytes.

3.1.3 Structural Integration of the OLED Array/Sensing Film

The structural integration of an OLED array with a sensing film is unique in its simplicity. It is achieved by fabricating these components on opposite sides of a common glass or plastic mechanical support, or, alternatively, on two separate glass slides that are attached back-to-back. This geometry eliminates the need for components such as optical fibers, lens, and mirrors, resulting in a compact OLED/sensor film module, whose ~2 mm thickness is determined by that of the mechanical supports [10–18]. Additionally, the nearly ideal coupling between the excitation source and the sensor film enables operation at relatively low power, which minimizes heating that is a critical issue for sensor materials and analytes involving heat-sensitive bio(chemical) compounds.

The PD, e.g., a photomultiplier tube or a Si photodiode, can be placed in front of the sensing film ("front detection") or behind the OLED array ("back detection"). The basic structure of the integrated OLED/sensor film in the back-detection geometry is shown in Fig. 3.2. In this configuration, the PD collects the PL that passes through the gaps between the OLED pixels. The

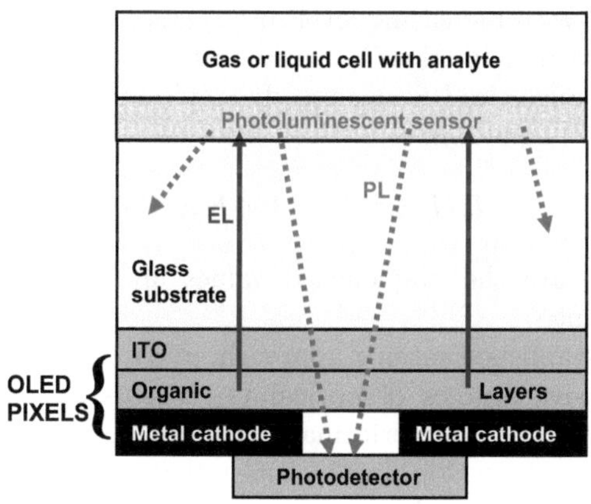

Fig. 3.2. Basic structure of an integrated OLED pixel array/luminescent sensor film module in the "back detection" geometry

resulting sensor is consequently very compact when using a Si photodiode, which is <2 mm thick.

The OLED structure, as an array of small-size pixels, presents an opportunity to fabricate multiple sensors in this compact design. Any number of pixels is individually addressable and can be associated with a different sensor film. Single- or multi-color OLEDs can be used for generating such sensors.

As the volume of manufactured OLEDs increases, their eventual cost will drop to the point of disposability. Thus, the integrated OLED platform will address the need for field deployable, compact, low cost, user-friendly, and autonomous sensors.

3.2 Single Analyte Monitoring

In the examples provided below, typically 2–4 OLED pixels, $2 \times 2\,\mathrm{mm}^2$ each, were utilized as the excitation source; $0.3 \times 0.3\,\mathrm{mm}^2$ pixels were also successfully employed. Importantly, the OLEDs were often operated in a pulsed mode, enabling monitoring the effect of the analytes not only on the PL intensity I, but also on the PL decay time τ. The τ-based detection mode is advantageous since moderate changes in the intensity of the excitation source, dye leaching, or stray light have essentially no effect on the measured value of τ. Thus, the need for a reference sensor or frequent sensor calibration is avoided.

3.2.1 Gas-Phase and Dissolved Oxygen

The viability of the OLED-based sensor platform was demonstrated first for the extensively studied O_2 sensors. A well-known approach for gas-phase and dissolved O_2 (DO) sensing is based on the dynamic quenching of the PL of oxygen-sensitive dyes such as Ru complexes and Pt or Pd porphyrins [1–6, 46–55]. Collisions with increasing levels of O_2 result in a decrease in I and τ. In a homogeneous matrix, the O_2 concentration can be determined ideally by monitoring changes in I under steady-state conditions or in τ using the Stern–Volmer (SV) equation:

$$I_0/I = \tau_0/\tau = 1 + K_{\mathrm{SV}}[\mathrm{O}_2], \tag{3.1}$$

where I_0 and τ_0 are the unquenched values and K_{SV} is a temperature-dependent constant.

Despite the established sensing approach, in particular for gas phase measurements, extensive studies of optical O_2 sensors are still continuing in an effort to enhance sensor performance, reduce sensor cost and size, simplify fabrication, and develop an O_2 sensor that is compatible with *in vivo* biomedical monitoring [56]. Development of field deployable, compact sensors such as those envisioned for the structurally integrated OLED-based platform is therefore expected to be beneficial for the varied needs of gas phase and DO monitoring in biological, medical, environmental, and industrial applications [46–58].

OLED arrays were fabricated by thermal vacuum evaporation, as detailed elsewhere [13, 17, 35, 59–61]. The organic layers consisted of a 5-nm thick copper phthalocyanine (CuPc) hole injecting layer that is also believed to reduce the surface roughness of the ~100-nm thick treated ITO [62] and a 50-nm thick N,N'-diphenyl-N,N'-bis(1-naphthyl phenyl)-1,1′-biphenyl-4,4′-diamine (NPD) HTL. For blue OLEDs, with peak emission at ~460–470 nm, suitable for tris(4,7-diphenyl-1,10-phenanthroline) ruthenium(II) (Ru(dpp)) excitation, the 40-nm thick emitting layer was 4,4′-bis(2,2′-diphenylvinyl)-1,1′-biphenyl (DPVBi), or perylene (Pe)-doped 4,4′-bis(9-carbazolyl) biphenyl (Pe:CBP) [63], typically followed by a 4–10-nm thick tris(quinolinolate) aluminum (Alq_3) ETL. For green OLEDs with peak emissions at ~535 and ~545 nm, suitable for Pt octaethylporphyrin (PtOEP) and the Pd analog (PdOEP) excitation, respectively, the ~40-nm thick emitting layer and ETL were Alq_3 and rubrene-doped Alq_3, respectively. In all cases, an 8–10 Å CsF buffer layer was deposited on the organic layers [65], followed by the ~150-nm thick Al cathode. The total thickness of the OLEDs, excluding the glass mechanical support, was thus <0.5 μm.

Structurally integrated OLED-based sensors with Ru(dpp) embedded in a sol–gel film and PtOEP and PdOEP embedded in polystyrene (PS) were used to monitor gas-phase O_2 and DO in water, ethanol, and toluene [17]. The results indeed demonstrated the viability of the OLED-based sensor platform for high-sensitivity O_2 monitoring. The performance of the sensors in the gas phase and in solution was evaluated in terms of the dynamic range and the detection sensitivity at different temperatures, the effect of the temperature on τ_0 and $\tau(100\%\ O_2)$, and the effect of the preparation procedure of the sensing elements on these metrics. In the gas phase, the detection sensitivity is defined as $S_g \equiv \tau_0/\tau(100\%\ O_2)$; in solution, S_{DO} is defined as the ratio of τ measured in a deoxygenated solution to that of an oxygen-saturated solution.

For an O_2-sensitive dye in solution, the predicted linear SV dependence was confirmed [66]. It was also observed for the solid matrix of the [DPVBi OLED]/[Ru(dpp) sensor film] [17]. But as shown in Figs. 3.3 and 3.4, which show SV plots of PtOEP and PdOEP-based sensors prepared from solutions containing a dye:PS ratio of 1:50, that is not always the case. However, the τ_0 of Ru(dpp), which is only ~8 μs, is relatively short and therefore results in a lower detection sensitivity (~2.3–4 in the gas phase) than that of dyes with longer PL lifetimes, such as PtOEP and PdOEP. Indeed, the SV plots shown in Figs. 3.3 and 3.4 demonstrate that Alq_3-based OLEDs, together with a sensing element based on a PS film doped with PtOEP or PdOEP, result in much higher sensitivities.

Figure 3.5 shows a linear gas-phase SV plot obtained for a film prepared from a solution with a 1:10 PtOEP:PS ratio. The detection sensitivity of ~30 is somewhat lower than that obtained for the 1:50 PtOEP:PS films, which exhibited nonlinear SV plots. Figure 3.5 also demonstrates the performance

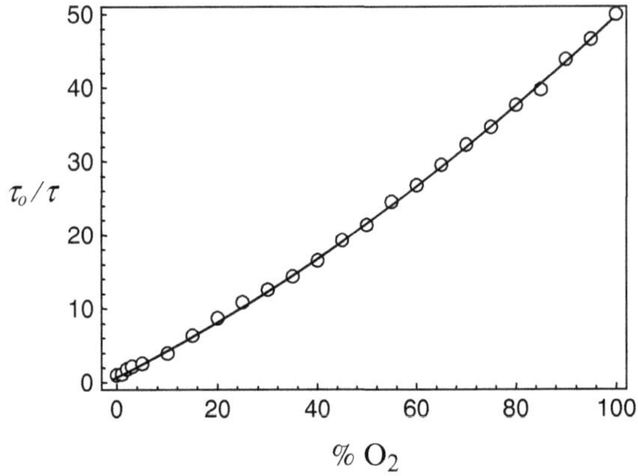

Fig. 3.3. Stern–Volmer plot obtained at 23°C for a film prepared by drop casting 50 µL of a solution containing PtOEP:PS at a ratio of 1:50. The line represents a best quadratic fit

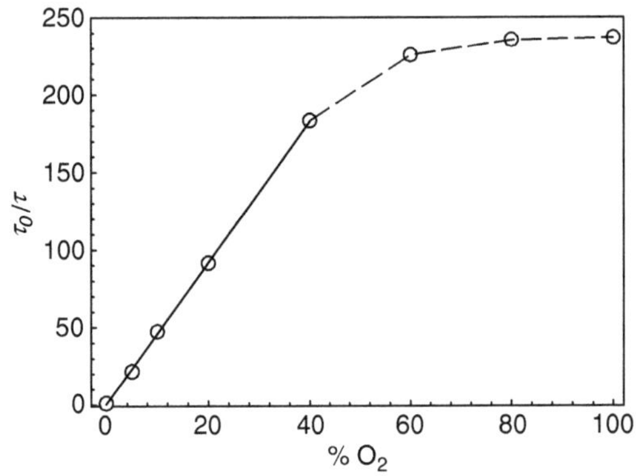

Fig. 3.4. The SV plot at 23°C of a PdOEP-based gas-phase sensor excited by a rubrene-doped Alq_3 OLED [17]. The sensing film was prepared by drop casting. The SV plot is linear up to ~40% O_2. Copyright 2006, with permission from Elsevier

of the DO sensors in water, ethanol, and toluene. The sensitivities are ~9.5 and 10.8 for water and ethanol, respectively. In water, the values of τ increased from ~8.9 µs in O_2-saturated solution to ~84 µs in Ar-saturated solution; in ethanol, they increased from ~7.9 to ~86 µs. For DO in toluene, the values of τ increased from ~2.2 µs for 5% gas-phase oxygen in Ar to ~60 µs in an Ar-saturated solution. This relatively short τ_0 probably reflects either incomplete deoxygenation of the solvent or solvent-related PL quenching [55]. The observed high sensitivity in toluene clearly results from the lowest value of τ in the O_2-saturated solvent. The larger values of τ in the O_2-saturated water and ethanol in comparison to that in toluene may be due to the more limited

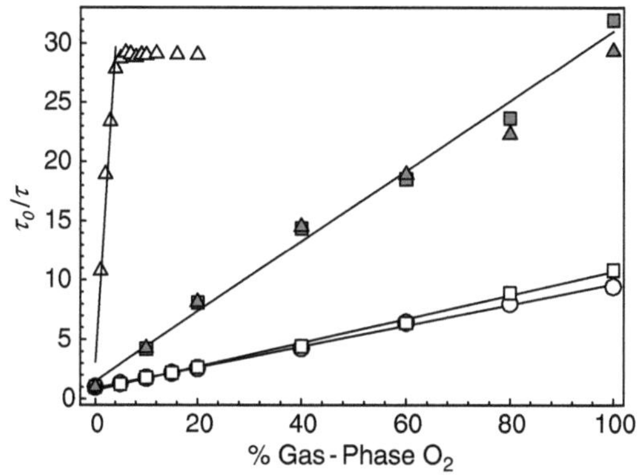

Fig. 3.5. Linear Stern–Volmer plots obtained for gas phase O_2 at 23°C (*filled square*) and 55°C (*filled triangle*) for a PtOEP:PS film prepared by drop casting 50 µL of a solution containing 1 mg mL^{-1} PtOEP and 10 mg mL^{-1} PS, and SV plots for DO in water (*open circle*), ethanol (*open square*), and toluene (*open triangle*). For water and ethanol, the sensing element was a thin film of PtOEP embedded in polystyrene; for toluene, it was solution-based with 0.01 mg mL^{-1} PtOEP in toluene

solubility of oxygen in the former. Overall, however, the difference in the sensor responses between toluene and the other solutions is probably largely a result of the enhanced accessibility of the dye molecules in solution relative to their accessibility in the moderately porous solid films.

The similarity of the responses and detection sensitivities for water and ethanol, despite the larger solubility of oxygen in ethanol (~8 times higher at 25°C) [67], may indicate that the oxygen concentration in the PS host, which is in equilibrium with the DO, is comparable for both liquids. We note that the detection sensitivity of DO in water measured with the OLED-based sensors is among the highest reported for PtOEP:PS.

The OLED-based sensors were tested in the 23–60°C temperature range. The values of τ are expected to generally decrease with increasing temperature, as the PL quenching is enhanced at elevated temperatures [47]. However, in the 23–60°C studied range, the phosphorescence of porphyrins is only slightly dependent on the temperature [68]. Indeed, the temperature effect on the SV plots was minimal; small reductions in τ_0 and τ (100% O_2) were observed as the temperature increased, e.g., for one film, τ_0 decreased from ~91 to 84 µs with S_g varying from ~36.5 to ~37.5.

Figure 3.6 shows the evaluation of an OLED-based sensor performance over a 30-day period [17]. In that test, the [Alq_3 OLED]/[PtOEP-doped PS film] module was biased continuously at 8 V dc; the PtOEP-doped PS film, prepared by drop casting a toluene solution containing 1 mg mL^{-1} PtOEP and 15 mg mL^{-1} PS, was exposed to air and excited continuously by the OLED. This brightness level was comparable to the average brightness of the OLED in the pulsed mode during the measurement. Once every 24 h, the bias to the

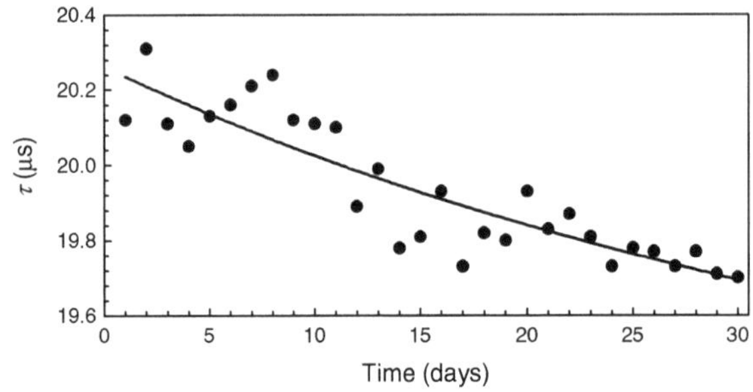

Fig. 3.6. PL decay time τ of a PtOEP-doped PS film in ambient air; the PL was excited by an Alq_3 OLED. The OLED was biased continuously at 8 V for 30 days, and τ was measured once every 24 h by switching the bias to a pulsed mode. The line is a guide to the eye

OLED was switched to a pulsed mode, and τ (corresponding to the 21% O_2 in air) was determined. In this pulsed mode, the pulse amplitude, width, and repetition rate were 20 V, 100 μs, and 50 Hz, respectively. The displayed value of τ was determined by averaging the decay curves over 1,000 sweeps.

As Fig. 3.6 shows, τ slowly decreased from 20.2 ± 0.1 to 19.7 ± 0.05 μs during this 30 day test. In other words, the relative error actually decreased with time, from ~0.5% to ~0.25%. These and later results demonstrated that the lifetime of this sensor module is well beyond the 30 day test.

To monitor O_2 levels accurately over the entire 0–100% O_2 range, an array of sensors with various PtOEP and PdOEP films can be used simultaneously. By preparing the films under different conditions (e.g., PS:dye ratio, film thickness), different SV plots and detection sensitivities are expected. For example, one such simple array could comprise two sensing films: a 1:10 PtOEP:PS film that exhibits a near-linear SV plot over the whole 0–100% range (Fig. 3.5), which would be excited by Alq_3 OLED pixels, and a PdOEP:PS film that is very sensitive to low levels of O_2 and exhibits a linear behavior up to ~40% O_2 (Fig. 3.4), which would be excited by rubrene-doped Alq_3 OLED pixels. Alternatively, both sensor films could be excited by CBP or Spiro-CBPOLEDs, whose EL peaks at ~380 nm. Hence, by using the OLED-based sensing platform, it is possible to easily fabricate a small-size array of OLED pixels, where 2–4 pixels correspond to a given sensing film in the sensor array. Thus, through consecutive or simultaneous excitation of such small groups of OLED pixels, O_2 can be detected by different sensing films that exhibit linear calibration plots and sensitivities suitable for different regions of O_2 levels. This approach will also result in redundancy in determining the O_2 level, thus providing a more accurate and reliable result. Moreover, it will provide the basis for sensor (micro)arrays for multianalyte detection, using an array of OLEDs emitting at various wavelengths. Such arrays were recently fabricated using combinatorial methods [69].

In summary, the example of the O_2 sensor demonstrates that the use of OLEDs as excitation sources in PL-based chemical and biological sensors is promising. The ease of OLED fabrication and OLED/sensing component structural integration result in compact modules, which are expected to be inexpensive and suitable for real-world applications. The example of oxygen sensing demonstrates the viability of the approach for using the PL intensity and decay time detection modes. The latter is advantageous over the former, as moderate changes in the intensity of the excitation source, dye leaching, or stray light have a minimal effect on the sensor response. The results also demonstrate the promise of the OLED pixel platform for developing sensor arrays for multiple analyses of a single analyte or for detection of multiple analytes. In evaluating the OLED-based platform for oxygen sensing, a new assessment of PtOEP-based sensors in terms of PtOEP aggregation, and the effects of film composition and measurement temperature on the PL lifetime was obtained [17].

3.2.2 Enhanced Photoluminescence of Oxygen-Sensing Films Through Doping with Titania Particles [70]

In spite of the potential widespread applications of PL-based sensors, simple, low-cost approaches that further improve sensor performance are highly desirable. Enhancement of the detection sensitivity is often achieved with new matrices that are more permeable to the analyte, with dyes with a high PL quantum yield and/or a long τ, and by optimizing the dye concentration when embedded in a thin host film, immobilized on a surface, or dissolved in solution. These approaches improve the signal-to-noise ratio (S/N) in the measurement, enabling more accurate determination of analyte levels. However, new methods are desired for signal enhancement beyond the incremental increases obtained by the approaches mentioned above.

Embedding nanoparticles in sensor films has been employed to enhance the performance of oxygen sensors. Typically, modified silica nanospheres or mesoporous silica were used [71, 72]. Silica fillers embedded in a sensor film improve the mechanical strength of films made of soft (low-glass transition temperature) polymers and/or serve as carriers for the dye molecules. The particles can affect the oxygen diffusion rate and path in the film, as well as aggregation of dye molecules. Moreover, O_2 can adsorb onto the particles' surface, resulting in complex oxygen transport and PL quenching mechanisms. A large filler volume fraction and mesoporous silica can also generate voids in the films. Such voids scatter light, affect the optical properties of the film, increase the surface area exposed to the gas phase, and can result in faster sensor response [72].

We have developed a uniquely simple approach to increase the intensity of the PL of PtOEP- and PdOEP-doped PS sensor films [70]. In this approach, sensor films were additionally doped with small-size particles that have a high dielectric constant, such as 360-nm diameter titania (TiO_2) particles. When

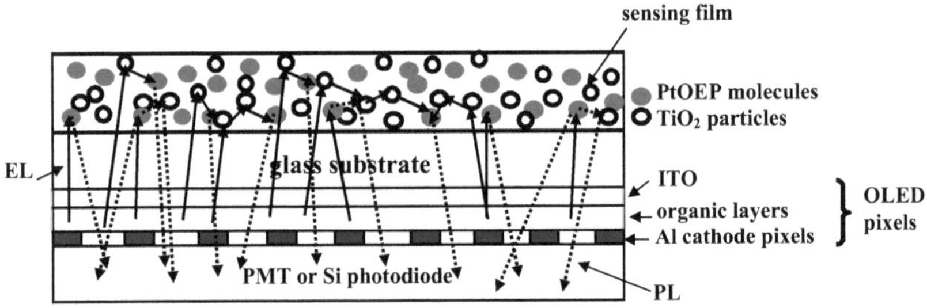

Fig. 3.7. Schematic of the structurally integrated OLED-based oxygen sensor (not to scale). The photodetector, a PMT or Si photodiode, is behind the OLED pixel array. The TiO_2 nanoparticles, which are embedded in the dye:PS-sensing film, act as a scattering medium, increasing the absorption of the EL by the dye

excited by an OLED, the dye PL intensity increases up to ~10 fold, depending on the TiO_2 concentration and the excitation source. The enhanced PL is attributed to light scattering by the embedded particles and possibly by voids in the film. The particles scatter the light that excites the PL, increasing the optical path of the exciting light and consequently the absorption of that light and the PL. The particles can also result in an increase in the PL outcoupling, reducing waveguiding to the film edges. A schematic of the structurally integrated OLED-based oxygen sensor, where the embedded TiO_2 nanoparticles act as a scattering medium, increasing the absorption of the EL by the dye, is shown in Fig. 3.7.

Figure 3.8 shows the effect of titania doping on the PL decay curve and on the EL of the Alq_3 OLED in a 100% (gas phase) Ar environment at room temperature. It shows the intensity measured by the PD during and following the 50 µs Alq_3 OLED pulse at different concentrations of titania particles. A 610 nm long-pass filter was used, so that only a small fraction of the EL was detected by the PD. As seen, for a given film, there is an initial fast increase in the intensity, which is due to the EL. This sharp increase is followed by a gradual increase due to the PtOEP PL. As $\tau_0 \sim 90\,\mu s$, the PL intensity did not reach a steady-state value after the 50 µs of the pulse duration. As the OLED pulse is turned off, a fast decrease in the intensity, due to the fast decay of the EL, is first seen, followed by the PtOEP PL decay. The significant enhancement of the PL intensity (by up to a factor of ~8) due to the addition of TiO_2 is clearly demonstrated in the inset of this figure.

The increased PL results in an improved S/N ratio in oxygen monitoring, without any deterioration or change in the response time or the long-term stability of the sensor films (the particles affect τ, independently of their concentration) [70]. The improved S/N can improve the analyte limit of detection (LOD), allow shortened data acquisition times, and enable the use of low-intensity excitation sources to minimize potential dye photobleaching.

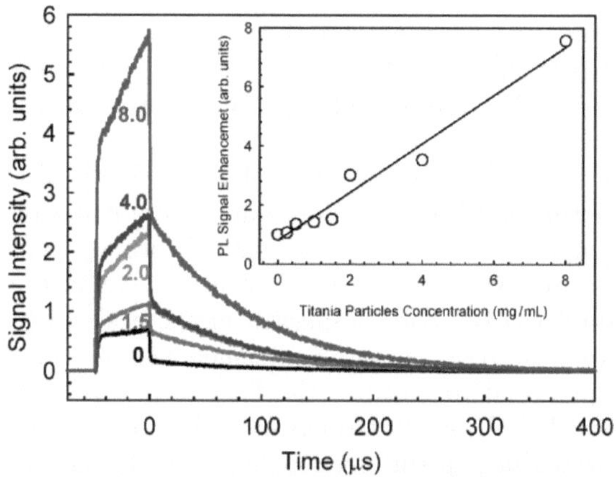

Fig. 3.8. The effect of titania particles on the PtOEP:PS PL decay curves and on the EL of the Alq_3 OLED in a 100% gas-phase Ar environment at room temperature. Shown is the intensity measured by the PD during and following the 50 μs Alq_3 OLED pulse at different concentrations of titania particles (the numbers are the concentrations in mg/mL). A 610 nm long-pass filter was used, so that only a small fraction of the OLED emission reached the PD

3.2.3 Glucose

Glucose-sensing methods often rely on the enzymatic oxidation of glucose in the presence of glucose oxidase (GOx):

$$\text{glucose} + O_2 \xrightarrow{\text{GOx}} H_2O_2 + \text{gluconic acid}. \tag{3.2}$$

The glucose concentration c_{Gl} can be determined by analyzing the reaction products [26], or by measuring the PL intensity I or PL decay time τ of an oxygen-sensitive dye, co-embedded with GOx in a thin film or dissolved in solution. In the presence of glucose, the PL quenching of the dye molecules is reduced (i.e., I and τ increase) due to the enzymatic oxidation-induced reduction in the local O_2 level [4,68,73,74].

Issues associated with this PL-based detection of glucose, utilizing an oxygen-sensitive dye, include fluctuations in the light intensity of the excitation source, instability of the oxygen-sensitive dye, variations in the local oxygen level in the samples, and leaching of the dye and/or GOx from the films. Several approaches have been used to address these issues. For example, to address problems due to changes in the excitation source intensity, two fluorescent dyes, an indicator oxygen-sensitive dye, and a reference oxygen-insensitive dye, were incorporated together with GOx in a ratiometric nanosensor [74]. To account for possible variations in the oxygen concentration, $[O_2]$, which can compromise accurate glucose monitoring, two sensors, i.e., an oxygen sensor and a glucose biosensor, were used simultaneously [4,26].

The excitation sources for investigating PL-based glucose biosensors have included Ar^+ lasers operated at 20–40 mW [4, 26], Hg lamps [4], or a fluorometer equipped with a 150 W Xe lamp [73]. c_{Gl} was usually monitored via changes in I.

The glucose biosensor is an example of the viability of the structurally integrated platform for PL-based biosensing, where c_{Gl} is determined from either I or τ. The individually addressable OLED pixels, operable as the light source, are integrated with the sensing element. The OLEDs and the sensing element are fabricated on glass or plastic mechanical supports, which are attached back-to-back, resulting in an extremely simple and compact device, which has the potential to advance the field of glucose sensing in various applications as well as PL-based biosensing technology for other analytes.

The results described in this section build on the successful integration strategy for the oxygen sensors, where the sensing elements consist of Ru(dpp) immobilized in a sol–gel matrix, which is spin-coated on a glass slide, or PtOEP embedded in PS and drop-cast on the mechanical support [10–14, 17]. As for the oxygen sensors, a blue DPVBi OLED [35, 60, 61, 65] is structurally integrated with the Ru(dpp)-based sensing element, and a green Alq_3 OLED with the PtOEP-based sensor. Importantly, the use of the OLED-based platform results in negligible, otherwise often damaging, sensor or analyte heating.

As mentioned, most reports on PL-based glucose biosensors are based on monitoring the effect of c_{Gl} on I [4, 9, 26, 74, 75]. Though not often used, τ measurements are expected to provide similar information [76] with the advantage of being independent of changes in the intensity of the light source or moderate dye degradation (e.g., leaching or photodegradation [66]). This advantage eliminates the need for frequent sensor calibration, which remains an issue in real-world applications. In the case of Ru(dpp), the reported τ values range from 0.3 to 8 µs, depending on the matrix and $[O_2]$ [4, 77–80]. The far longer phosphorescence τ_0 of PtOEP, ~100 µs [81], renders this dye ideal for monitoring glucose using the decay time mode, in particular in conjunction with the stable green Alq_3 OLED. The fast decay of the EL pulse, which has a decay constant <0.1 µs [82], enables monitoring the significantly longer τs of the O_2-sensitive dyes used. The effect of glucose concentration on I and τ is shown in Figs. 3.9–3.11 for various sensing elements. For comparison, results for a solution-based sensing element are also shown.

The glucose sensor performance, including the dynamic range and LOD, is affected by the nature and configuration of the sensing element, the amount of the indicator dye and GOx, and the oxygen level in the analyte sample [4, 68, 73, 74]. A dynamic range $0.1 \leq c_{Gl} \leq 15$ mM ($1.8 \leq c_{Gl} \leq 270$ mg dL^{-1}) was reported for a PL-based sensor in which GOx was sandwiched between two sol–gel layers, one of which contained Ru(dpp) [73]. It was possible to extend the dynamic range to 20 mM (360 mg dL^{-1}) by increasing the oxygen partial pressure or decreasing the level of the immobilized GOx, though the latter situation resulted in decreased sensitivity [26]. The data in

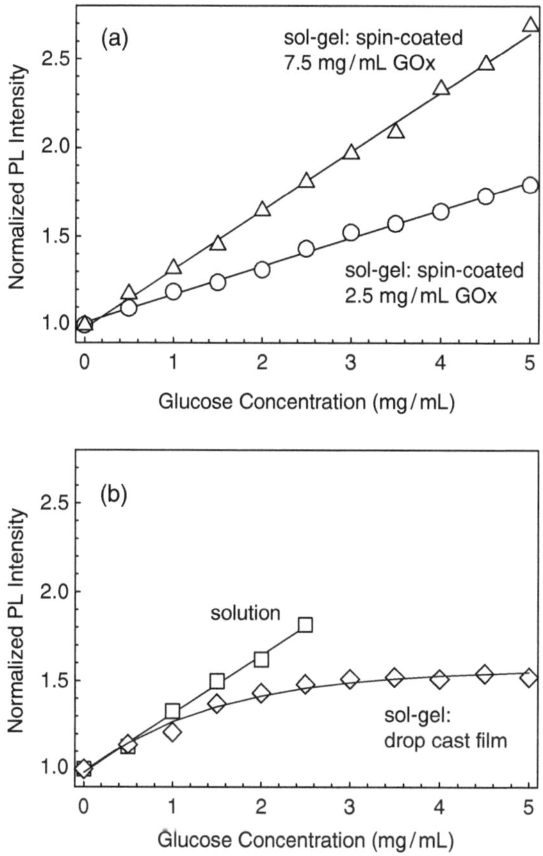

Fig. 3.9. Normalized PL intensity I of a DPVBi/Ru(dpp) glucose sensor as a function of glucose concentration c_{Gl}: (**a**) Sol–gel-based sensing elements. The films were fabricated by spin coating at 2,000 rpm with a sol containing 7.5 mg mL^{-1} Ru(dpp) and 2.5 mg mL^{-1} GOx (*open circle*) or 7.5 mg mL^{-1} GOx (*open triangle*). (**b**) 20 µl solution-based sensing element in PDMS wells, containing 7.5 mg mL^{-1} each of Ru(dpp) and GOx (*open square*), and a drop cast, sol–gel film (*open diamond*) prepared from 30 µL solution containing 7.5 mg mL^{-1} Ru(dpp) and 7.5 mg mL^{-1} GOx. Excitation was obtained using a blue DPVBi OLED. In the solution-based sensing element, c_{Gl} is the concentration in the 40 µL sample; combined with the 20 µL sensing element, the total solvent amount is 60 µL

Figs. 3.9–3.11 include results for sensing films that show additional extension of the dynamic range for the OLED-based sensors.

Blue DPVBi OLED/Ru(dpp)-Based Sensor

Figure 3.9a shows I vs. c_{Gl} of typical glucose biosensors with sol–gel films differing in the level of embedded GOx. The films were prepared by spin

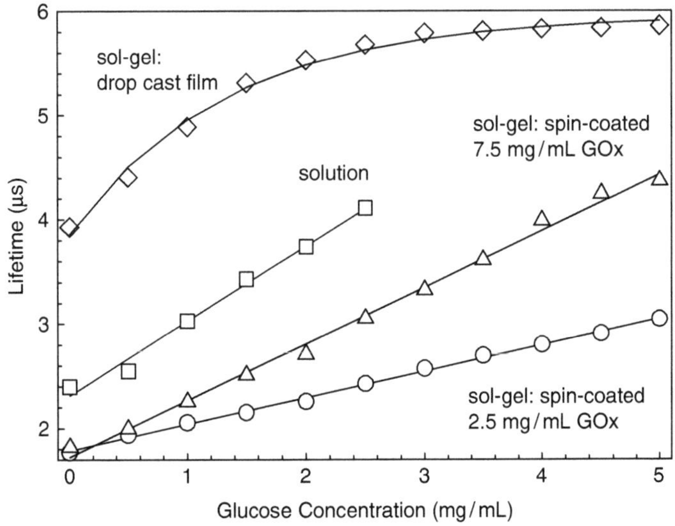

Fig. 3.10. PL lifetime of DPVBi/Ru(dpp) sensors with different sensing elements. (*open circle*) spin-coated sol–gel prepared from a sol containing 2.5 mg mL^{-1} GOx, (*open triangle*) spin-coated sol–gel prepared from a sol containing 7.5 mg mL^{-1} GOx, (*open square*) solution-based sensing element, and (*open diamond*) drop cast film prepared from 30 µl containing 7.5 mg mL^{-1} Ru(dpp) and 7.5 mg mL^{-1} GOx. Excitation was obtained using a blue DPVBi OLED

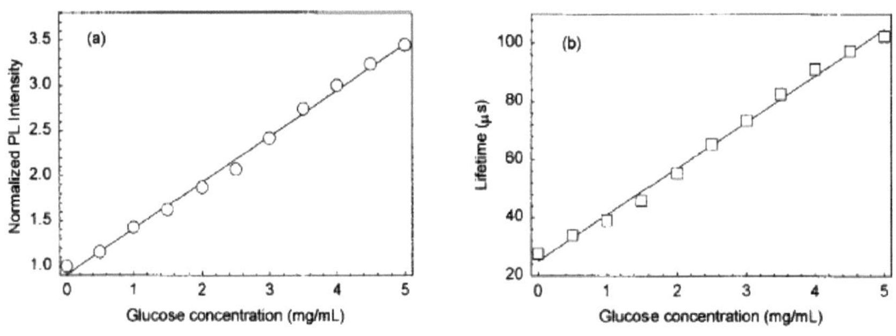

Fig. 3.11. Effect of glucose concentration on the PL normalized intensity (**a**) and decay time (**b**) for an Alq3/PtOEP-based sensor

coating at 2,000 rpm. The Ru(dpp) concentration in the solution used for preparation of the films was 7.5 mg mL^{-1}; that of GOx was 2.5 mg mL^{-1} in one and 7.5 mg mL^{-1} in the other. As seen in Fig. 3.9a, by increasing c_{Gl} from 0 to 5 mg mL^{-1} (500 mg mL^{-1}), I increased linearly by a factor of ~1.8 and ~2.7, respectively. However, further increase in the level of the GOx to 25 mg mL^{-1}, while maintaining the Ru(dpp) level, deteriorated the sensor performance, significantly decreasing the dynamic range. Figure 3.9b shows similar plots for (1) a drop-cast sol–gel film prepared from 30 µl containing

7.5 mg mL^{-1} Ru(dpp) and 7.5 mg mL^{-1} GOx and (2) a 20 µl solution-based sensing element containing 5 mg mL Ru(dpp) and 2.5 mg mL^{-1} GOx, to which a 40 µl buffered solution of glucose was added. The drop-cast film was significantly thicker than the spin-coated films with larger amounts of embedded Ru(dpp) and GOx. As seen, the dynamic range for the drop cast film is limited to c_{Gl} ~2 mg mL^{-1}, with I (and τ) values saturating at higher c_{Gl}. The response is highest for the spin-coated sol–gel film prepared from 7.5 mg mL^{-1} GOx and the solution-based sensing element; their LOD is estimated to be ~0.2 mg mL^{-1}.

Figure 3.10 shows the effect of c_{Gl} on τ for the different sensing elements. In the films prepared from solutions containing 2.5 mg mL^{-1} or 7.5 mg mL^{-1} GOx, increasing c_{Gl} from 0 to 5 mg mL^{-1} resulted in a linear increase in τ from 1.8 µs to 3 and 4.4 µs, respectively.

The increased τ (and I) for a given c_{Gl} in the presence of a larger amount of GOx is consistent with a stronger reduction in the local oxygen level. A similar increase in I was previously observed for a fiber-optic glucose sensor [26]. However, increasing the GOx level further, to 25 mg mL^{-1} in the solution used for film preparation, resulted in an increased $\tau \sim$ 6 µs in the absence of glucose, and a limited dynamic range. This situation is suspected to result from a reduced local oxygen level, possibly due to reduced film porosity, a consequence of incorporation of higher GOx levels.

Figure 3.10 also shows that the PL lifetimes for a given c_{Gl} for the solution-based sensing element are longer, ranging from 2.4 to 3.7 µs for a c_{Gl} range of 0–2 mg mL^{-1}, in comparison to those of the spin-coated films, where they vary from 1.8 to 2.25 µs in the same concentration range. The τs for the drop cast film are the longest, ranging from 4 µs for a buffer solution without glucose to 5.5 µs for a solution containing 2 mg mL^{-1} glucose. We speculate that the shorter PL lifetimes in the spin-coated films are due to small air bubbles trapped within the porous matrix, resulting in a higher local level of oxygen in comparison to the oxygen dissolved in solution or trapped in the drop cast film. SEM images were consistent with this assumption [13]. They showed that the drop cast film is less porous in comparison to the spin-coated film. The denser drop cast film is believed to have a lower level of trapped oxygen. Additionally, its denser texture inhibits glucose and DO from penetrating the film. This situation, together with the larger amount of GOx that can further result in depletion in the oxygen level, is believed to contribute to the measured longer lifetimes and the reduced dynamic range.

In summary, the dynamic range of up to ~1.5–2.0 mg mL^{-1} glucose obtained using the OLED-based glucose sensor with the drop cast film is comparable to or larger than those reported using other excitation sources and device configurations [4, 26, 74]. It was possible, however, to extend the dynamic range of the PL-based sensor to 5 mg mL^{-1} (500 mg dL^{-1}) glucose, which is comparable to the dynamic range in commercial electrochemical-based sensors. The extension of the dynamic range was achieved, without reducing the response magnitude of the OLED-based sensors, by spin coating the sensing films. An additional increase in the detection sensitivity, without limiting the dynamic range, was obtained by using an OLED/PtOEP pair, as shown below.

Green Alq_3 OLED/PtOEP-Based Sensor

Figure 3.11 shows the effect of c_{Gl} on the PL intensity (a) and decay time (b) of a typical Alq_3/PtOEP sensor. The best results were obtained from a mixture containing 3.5 mg mL^{-1} PtOEP and 10 mg mL^{-1} polystyrene in toluene. The sensing film was prepared by drop-casting the solution on the mechanical support and allowing it to dry for 1 h. Next, a sol–gel containing 7.5 mg mL^{-1} GOx was drop-cast over the PS film and allowed to dry in the dark at 4°C for 3–4 days. As seen in the figure, the PL decay time increased linearly from 28 to 100 μs when c_{Gl} increased from 0 to 5 mg mL^{-1}; the normalized intensity increased by a factor of 3.5. The sensitivity of the Alq_3/PtOEP-based sensor is higher (LOD ∼ 0.1 mg/mL^{-1}) than that of the DPVBi/Ru(dpp)-based sensors over a similar dynamic range. The large dynamic range and long PL decay times, together with the stability of the Alq_3 OLED render this type of sensor very attractive. The detection sensitivity of PL-based sensors excited by OLEDs may be further increased by increasing the porosity of the sensing element and using brighter OLEDs. Recently developed phosphorescent OLEDs with a higher external quantum efficiency of ∼20% are attractive for sensor applications [38, 44, 83]. However, when using the PL decay time mode of operation, usable dyes are only those with significantly longer τs in comparison to the OLED EL decay time.

The response of the aforementioned glucose sensors depends on the level of the oxygen in the analyte sample in contact with the sensing element. To determine this level, a second sensor, similarly prepared without GOx, and in close proximity to the glucose sensor, can be used simultaneously with the glucose detection. We note that the Alq_3/PtOEP oxygen sensor, like other PtOEP-based sensors, and, in particular, the rubrene-doped Alq_3/PdOEP oxygen sensor, is suitable for detection of trace O_2 levels [17]. An additional reference dye that is insensitive to oxygen can also be used to compensate for variations in light intensity when glucose or oxygen are monitored by measuring changes in I. However, the need for the reference dye is eliminated when operating the sensor in the PL decay time mode. We note that due to the individually addressable pixel design of the OLED, the two sensors, i.e., for oxygen and glucose, can be fabricated next to each other on the same mechanical support, enabling simultaneous monitoring of different analytes in the same sample.

Figure 3.12 shows the glucose sensor in back-detection operation. Six $\sim 2 \times 2\,\text{mm}^2$ green Alq_3 OLEDs pixels are lit simultaneously (the use of individually addressable array pixels, any number of which can be lit simultaneously, is unique to OLEDs in its simplicity). As seen in the image, the EL of the OLED pixels is very bright. Consequently, the pixels appear white;

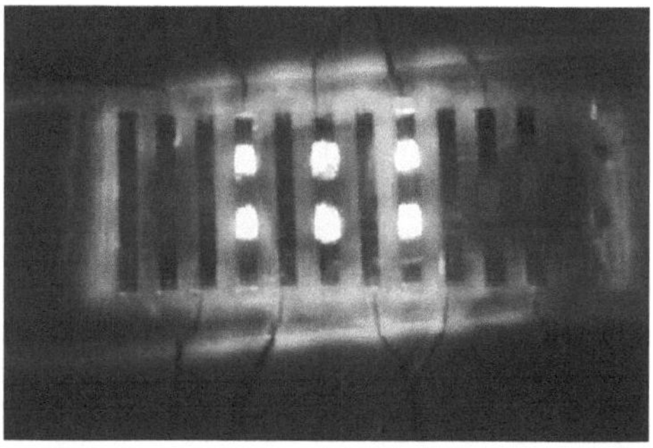

Fig. 3.12. Demonstration of the glucose sensor "back detection" design and operation. The green Alq_3 OLED array is behind the PtOEP-based sensor film, which is confined to a region in front of the middle two OLED pixels. The green emission from these pixels combines with the red PL of the PtOEP dye to produce yellowish spots when viewed in color. The PD is located behind the OLED array

however, the green emission is evident from the green shades of the background, when viewed in color. The analyte sample is placed in a region above the two middle pixels; the combined green emission of the OLED pixels and the red PL from the dye result in the yellowish appearance of the middle pixels when viewed in color.

3.2.4 Hydrazine (N_2H_4)

This section describes recent results toward the realization of the new OLED-based sensor platform for detection of hydrazine [15]. The results demonstrate the high-detection sensitivity that can be achieved using this new platform. The development of OLED-based compact, field-deployable hydrazine sensors is motivated by the use of hydrazine, a highly toxic and volatile compound, as a powerful monopropellant in NASA space shuttles and a common precursor in the synthesis of some polymers, plasticizers and pesticides. Specifically, its melting and boiling points are 2.0°C and 113.5°C, respectively; its vapor pressure is 14.4 Torr at 25°C. The American Conference of Governmental Industrial Hygienist has recommended that the threshold limit value (TLV) for hydrazine exposure (i.e., the time-weighted average concentration of permissible exposure within a normal 8-h workday) be lowered from 100 to 10 ppb in air [84]. The OSHA recommended skin exposure limit is 0.1 ppm (0.1 mg m^{-3}), and the Immediately Dangerous to Life or Health concentration is 50.77 ppm [85].

The hydrazine sensor is based on the reaction between N_2H_4 and anthracene 2,3-dicarboxaldehyde (ADA) (Fig. 3.13) [15, 86]. The reaction product is excited at 476 nm by a blue OLED (see Fig. 3.14) and emits at 549 nm; the signal is proportional to the N_2H_4 level. The sensor can be operated in air or solution.

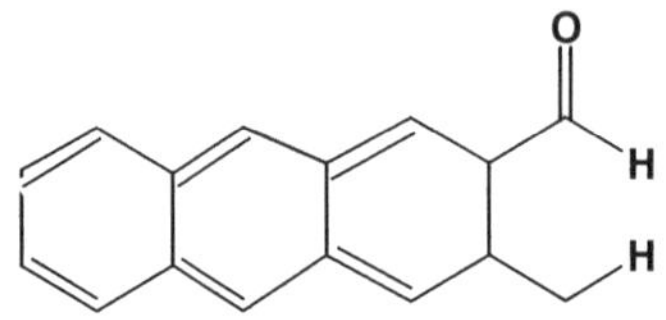

Fig. 3.13. Structure of anthracene 2,3-dicarboxaldehyde (ADA)

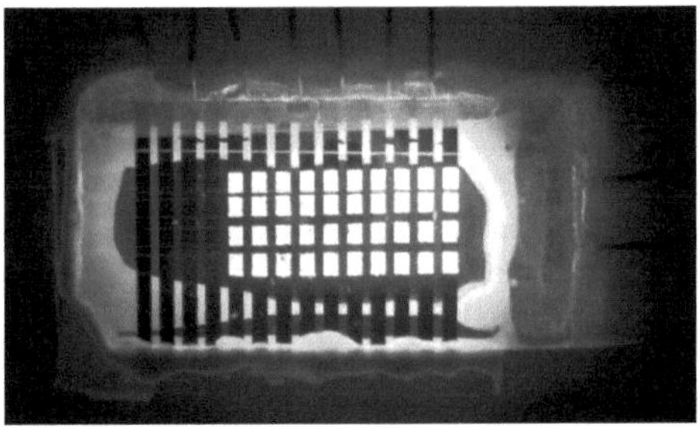

Fig. 3.14. A 4 × 14 array of blue DPVBi OLEDs used to excite the PL of the reaction product of ADA and hydrazine. 4×10 pixels are lit; each pixel is ~2×2 mm^2

The blue OLEDs were based on DPVBi [13, 17, 35, 60, 61, 65, 69]. The OLEDs were prepared as a small encapsulated matrix array of ~2 × 2 mm^2 square pixels resulting from perpendicular stripes of etched ITO and evaporated Al for back-detection, as described above [15,17]. A typical encapsulated array, with 4 × 10 pixels lit simultaneously, is shown in Fig. 3.14. The OLEDs were operated in a dc mode with a forward bias of 9–20 V, or in a pulsed mode with a forward bias of up to 35 V. The photodetector (PD) was a photomultiplier tube (PMT).

For monitoring the hydrazine gas, disposable hydrazine permeation tubes were used. Figure 3.15 shows the measurement system assembled to determine the LOD of the sensor for hydrazine in air. The disposable-permeation tube was heated to 80°C by a Kin Tek CO395 Certification Oven. At that temperature, the emission rate of the hydrazine is 2,750 ng min^{-1}. That emission rate corresponds to a concentration of 3.84 ppm in a carrier gas flowing at 500 sccm. Since the emission rate decreases by about 50% for every 10°C decrease of the oven temperature, a hydrazine concentration of ~60 ppb is obtained when the oven is at room temperature.

To optimize the OLED-based hydrazine sensor performance, the OLED pulse width and voltage were varied. The optimal values were found to be 20 μs and 30 V, respectively. The corresponding results of the change in the PL of the hydrazine/ADA solution are shown in Fig. 3.16. They clearly show that the LOD of hydrazine by this system is ~60 ppb in 1 min, i.e., roughly equivalent to 1 ppb in 1 h. That is, the sensitivity of this system exceeds the OSHA requirements by a factor of ~80. Methods to develop ADA-based solid state sensors are currently being explored.

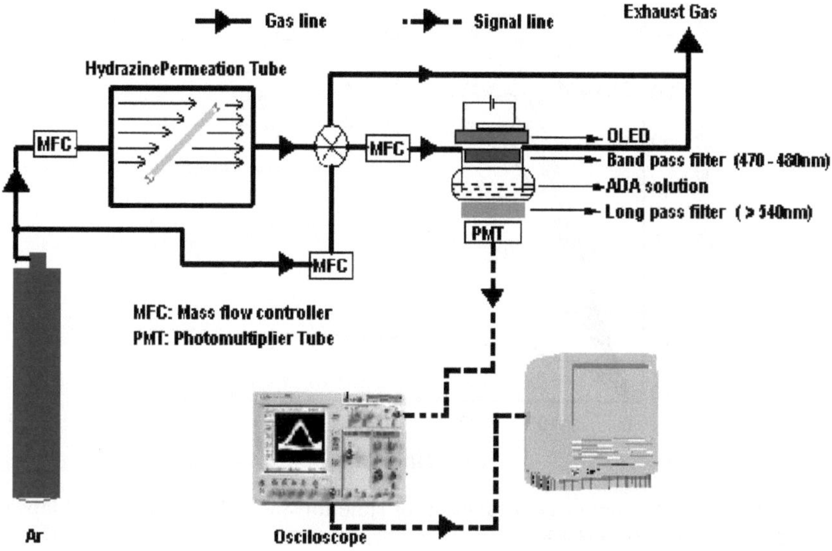

Fig. 3.15. Schematic of the trace hydrazine generation and detection system

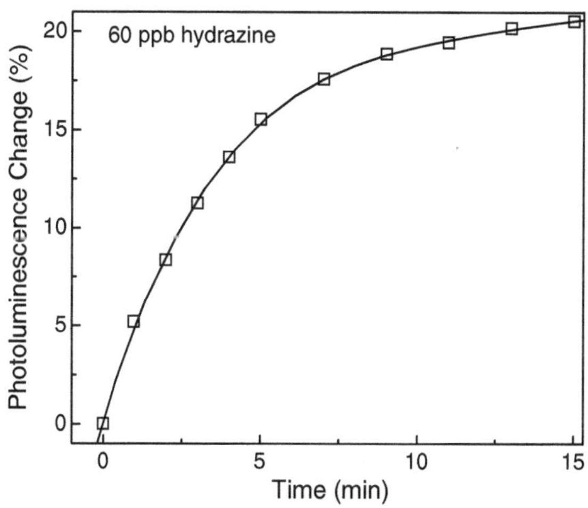

Fig. 3.16. PL change of the hydrazine/ADA solution upon exposure to 60 ppb hydrazine in Ar, bubbled at 500 sccm, vs time. The PL was excited by a 20 μs pulse of a DPVBi OLED biased at 30 V, at a repetition rate of 50 Hz

3.2.5 Anthrax Lethal Factor (LF)

The development of compact, field-deployable sensors is highly desirable for prompt, on-site detection of active *Bacillus anthracis* bacteria, which would eliminate the need to send samples for diagnosis to a remote laboratory. As a first step toward this goal, we describe the development of an OLED-based

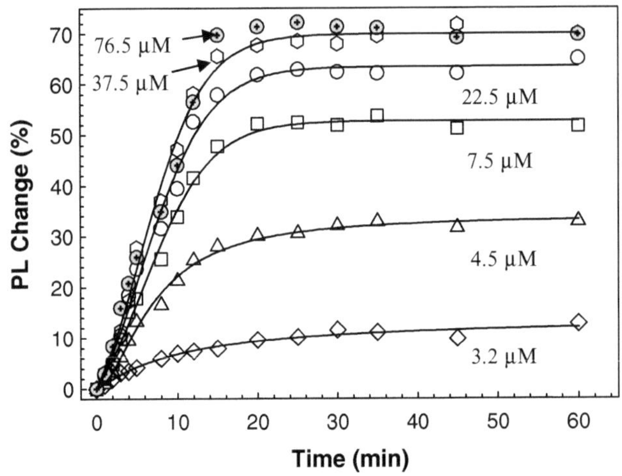

Fig. 3.17. The PL change as a function of time for 3.2 to 76.5 mM FRET-labeled peptide exposed to 25 nM LF

sensor for anthrax LF, which is one of the three enzymes secreted by the live anthrax bacterium. The anthrax LF sensor is based on the cleavage of certain peptides by LF [87–89]. As the LF cleaves a peptide labeled with a fluorescence resonance energy transfer (FRET) donor–acceptor pair, and the two cleaved segments are separated, the PL of the donor, previously absorbed by the acceptor, becomes detectable by the photodetector.

The labeled peptide synthesized for this work was:

(**donor**)–**Nle–K–K–K–K–V–L–P–** | **–I–Q–L–N–A–A–T–D–K–**(**acceptor**) **G–G–NH$_2$**.

As shown, the cleavage site should appear under the vertical line between the proline (**P**) and isoleucine (**I**) residues [87]. The donor was a rhodamine-based dye and the acceptor a Molecular Probes QSY7 dark quencher. Preliminary results at room temperature, using green ITO/CuPc/NPD/Alq_3/CsF/Al OLEDs, are shown in Fig. 3.17 for 25 nM LF and varying levels of the peptide. As seen in the figure, the PL change reached its maximal value after an incubation time of ~15 min; a peptide concentration of ~40 μM was needed to achieve the maximal PL change. However, the overall PL change was relatively low (a maximal increase of ~100% was observed at 37°C with an optimal OLED bias of 28 V), pointing to an issue associated with OLED excitation in conjunction with the specific dye employed. Indeed, using a green inorganic LED with a bandpass filter centered around 530 nm, and a longpass filter of 550 nm in front of the PMT, resulted in an increase by a factor of 6 in the PL following peptide cleaving by LF.

In developing OLED-based oxygen and glucose sensors, oxygen-sensitive Ru, Pt, and Pd-based dyes, with large Stokes shifts (>100 nm), were used. However, the rhodamine-based dye has a Stokes shift of only ~20 nm. As a result, the Alq_3 OLED, which has a broad EL spectrum with a peak emission at ~535 nm, overlaps the PL of the donor. This overlap strongly reduces the fractional change in the light monitored by the photodetector following exposure of the peptide to LF. To overcome this issue, we are exploring the synthesis of the peptide with a Ru dye that has a large Stokes shift as donor, and the dark quencher QSY21 (Molecular Probes) as acceptor, so that the OLED EL tail would be easily filtered out. To that end, unlabeled N-α-9-fluorenylmethyloxy-carbonyl (Fmoc)-peptide-resin with appropriate protecting side groups was first synthesized by a 432A Peptide Synthesizer using Fmoc-chemistry on a 25 μmol scale. Next, the QSY21 quencher was introduced by reacting its succinimidyl ester with a lysine residue, following selective removal of a 4-methyltrityl protecting group. After removal of the Fmoc groups, the QSY21-labeled peptide was cleaved from the resin. The crude QSY21-labeled peptide was lyophilized after its precipitation from cold diethyl ether. Finally, the N-terminus was labeled by bis(2, 2′-bipyridine)-4′-methyl-4-carboxybipyridine-Ru N succinimidyl ester-bis(hexafluorophosphate) (Sigma-Aldrich) [90]. The initial purification step was performed using a HPLC Beckman Gold System to fractionize the crude peptide. The PL amplitude and total integrated intensity, measured using a SpectraMax Gemini EM dual scanning microplate spectrafluorometer, increased by a factor of ~2.5 following a 10 min incubation of the labeled peptide with LF at 37°C. To enhance the PL change, the peptide needs to be further purified to reduce the background PL from the fluorphores that are not bonded to the peptide. Developing OLEDs with narrower EL bands, thus reducing interfering background light is also expected to enhance the anthrax LF sensor performance. To that end, microcavity OLEDs are being developed and tested for such sensing applications.

3.3 Advanced Sensor Arrays

3.3.1 OLED-Based Multiple Analyte Sensing Platform [91]

As mentioned in the introduction, multianalyte detection in a single sample using sensor arrays of various designs and sizes has been studied extensively using various techniques. These techniques frequently involve labor-intensive multistep fabrication, and/or require sophisticated image analysis and pattern-recognition codes [7,8,21–33]. The OLED-based sensing platform presents an opportunity to develop compact PL-based sensor arrays for simultaneous or sequential monitoring of multiple analytes in a single sample. These arrays are unique in their simplicity. The individually addressable OLED pixels comprising the excitation source can be of identical color or a combinatorial array of blue-to-red pixels [69]. Beyond such simple arrays, typically composed of $\sim 2 \times 2\,\text{mm}^2$ pixels, nm-scale pixels may also become plausible sources for nm-scale OLED-based single- or multianalyte sensors [45].

This section reviews some recent developments on a multiple analyte sensor that is based on identical Alq_3-based OLED pixels and monitors the serum analytes glucose, lactate, ethanol, and oxygen, which are of clinical, health, industrial, and environmental importance. These analytes were monitored sequentially or simultaneously; comparable results were obtained using both approaches.

The sensing elements were based on PtOEP-doped PS films; the films served as the base of wells that contained the solution of the analytes and GOx, lactate oxidase (LOx), or alcohol oxidase (AOx). The analyses were performed typically in sealed cells; more complex responses were obtained when the analyses were performed in cells open to air. As shown below, the LOD for glucose, lactate, and ethanol was ~0.02 mM. The sensors were operated in the back-detection geometry and τ mode.

In O_2 sensors, the relation between the oxygen concentration $[O_2]$, I, and τ is ideally given by the SV equation (Eq. (3.1)). This SV equation was modified, to generate calibration curves for glucose, lactate, and ethanol monitored in sealed containers, where there is no supply of DO beyond the initial concentration. The modified calibration curves were obtained for the oxidation reactions of these analytes in the presence of their respective oxidase enzymes. For each analyte and its enzyme, the reaction is similar to that given in Eq. (3.2) for glucose. The calibration curves are based on the following considerations: If the initial analyte concentration $[\text{analyte}]_{initial}$ is smaller than the initial DO concentration $[\text{DO}]_{initial}$, and assuming that the analyte conversion into products is complete, the final DO concentration $[\text{DO}]_{final}$ is given by

$$[\text{DO}]_{final} = [\text{DO}]_{initial} - [\text{analyte}]_{initial}. \tag{3.3}$$

The SV equation becomes, accordingly:

$$I_0/I = \tau_0/\tau = 1 + K_{\text{SV}} \times \{[\text{DO}]_{initial} - [\text{analyte}]_{initial}\}. \tag{3.4}$$

Thus, a plot of $1/\tau$ vs. the initial analyte concentration in the test solution will ideally be linear. Note that this equation is valid also for measurements performed with containers open to air, where the oxidation reaction is significantly faster than the in-diffusion of gas-phase oxygen.

The foregoing sealed-cell sensor limits the dynamic range of the analyte concentration in the test solution to ~0.25 mM in water, as that value approaches the ~8.5 ppm DO concentration in equilibrium with air at ~23°C. However, this apparent limited dynamic range corresponds only to the analyte concentration in the final test solution, where it is diluted. Thus, the actual dynamic range is wide, covering the concentration range of medical/industrial interest.

Sensor Array Design

The OLED pixel array is similar to that shown in Fig. 3.12, i.e., the OLED pixels are defined by the overlap between the mutually perpendicular ITO and

Al stripes. The pixel size is typically $2 \times 2\,\text{mm}^2$; $0.3 \times 0.3\,\text{mm}^2$ pixels were also tested and found similarly adequate for use as the excitation source for sensing applications. We note that there is no cross talk between the OLED pixels. Two pixels were used for each of the four analytes: O_2, glucose, lactate, and alcohol.

As mentioned, the multianalyte sensing was performed using two approaches. In one, the analytes were detected sequentially using a single PD. That is, mixtures of the analytes were added sequentially above the different OLED pixels to the enzyme-containing wells; the OLED pixels were energized in succession to monitor each analyte separately. Such a measurement requires only a single PD, and takes a few minutes to determine the level of all of the four analytes. In a second approach, four $5 \times 5\,\text{mm}^2$ Si photodiodes were assembled in an array compatible with the OLED pixel array. The analyte mixtures were analyzed simultaneously on the OLED array, with the output of each PD corresponding to a different analyte. We note that various Si-based PD arrays with rectangular 1–4 mm^2 elements are available commercially (from, e.g., Hamamatsu) and can be used in conjunction with OLED arrays for monitoring a larger number of analytes. Additionally, as detailed in the next section, work is in progress to develop μm-thick thin film PDs, such as those based on amorphous or nanocrystalline Si, that are fabricated together with OLED pixels on a common mechanical support, to generate a compact, fully integrated PL-based sensor array for multiple analytes [18].

As described in Sect. 3.2.3, the GOx could be successfully doped in a sol–gel film that is spin-coated above the PtOEP:PS film, and its activity was stable, i.e., it did not degrade appreciably over time. In contrast, while LOx and AOx could also be embedded in a sol–gel film, reduced enzyme activity within a few measurements limited their usefulness. The sensing elements for lactate and alcohol therefore consisted of the respective enzymes dissolved in a buffer solution to which the analytes were added. In this geometry, the container with the reaction mixture was typically sealed from air. This approach was adopted due to the more complex responses observed in a cell open to air, where the DO was constantly replenished.

Next, the OLED-based single-analyte sensors are described briefly, followed by the multianalyte sensor arrays for detection in analyte mixtures.

Lactate Sensor

The oxidation reaction of lactate in the presence of LOx at 23°C and 37°C was completed in ~10 s to ~4 min, depending, as expected, on the LOx and lactate concentrations. Despite the relatively long time needed for completion of the oxidation reaction at low enzyme concentrations, a calibration curve for a given film can be obtained at shorter times by monitoring τ after a constant reaction time for different analyte concentrations. A calibration curve can be obtained also by plotting the initial rate of change of τ, $(\mathrm{d}\tau/\mathrm{d}t)_0$, which is related to the initial reaction rate, vs. lactate concentration. Such a calibration curve is shown in Fig. 3.18. In this example, $(d\tau/dt)_0$ was obtained from the first 4–5 data points (up to 83 s; linear part) of τ vs. the reaction time.

Figure 3.19 shows a calibration curve of $1/\tau$ at the completion of the reaction (<30 s) vs. lactate concentration (see Eq. (3.4)) using 7.5 units mL^{-1} stabilized enzyme. As shown below, this type of linear calibration curve was adopted for monitoring analyte mixtures.

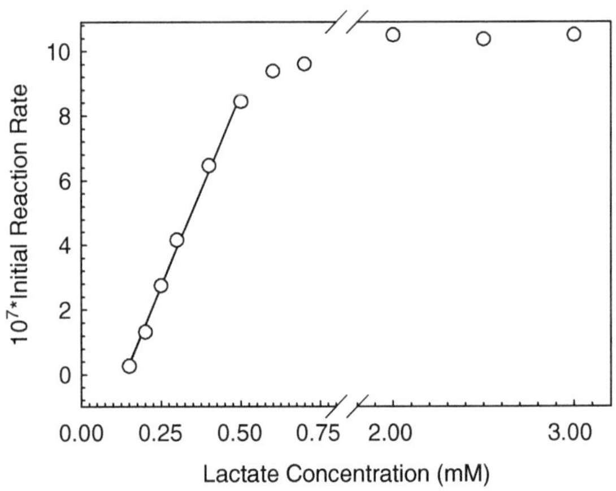

Fig. 3.18. The initial change in τ as a function of lactate concentration. The enzyme concentration was 1 unit mL^{-1}. The measurement was performed at 37°C in a sealed cell

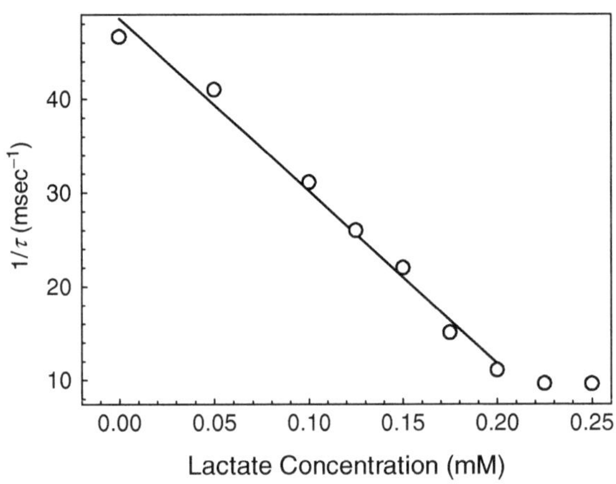

Fig. 3.19. $1/\tau$ as a function of lactate concentration at the completion of the reaction. The enzyme concentration was 7.5 unit mL^{-1}. The measurement was performed at 23°C in a sealed cell

Ethanol and Glucose Sensors

Figure 3.20 shows the initial reaction rate vs. ethanol concentration when using a reaction cell that was open to air [16]. As seen, the dynamic range extends to ~40 mM, though the calibration curve is not linear.

Plotting $1/\tau$ as a function of ethanol or glucose concentration (Fig. 3.21) at the completion of the reaction in buffered solutions with AOx and GOx, respectively, in sealed cells, yielded calibrations for these analytes that were comparable to that shown for lactate in Fig. 3.19. As clearly seen from Fig. 3.21, the LOD is ~0.02 mM.

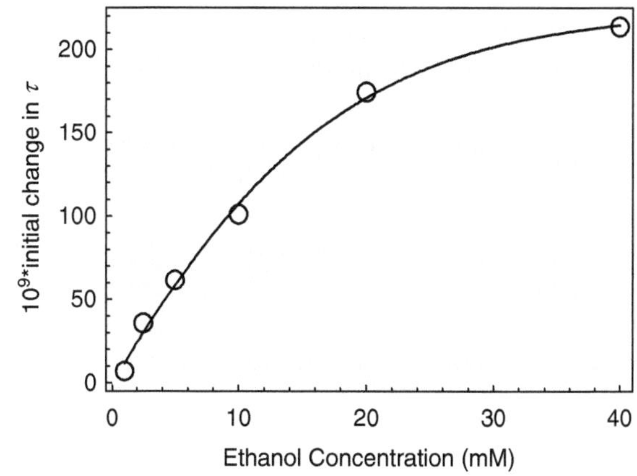

Fig. 3.20. The initial change in τ as a function of ethanol concentration. The enzyme concentration was 3 units mL^{-1}. The measurement was performed at 23°C in an open cell

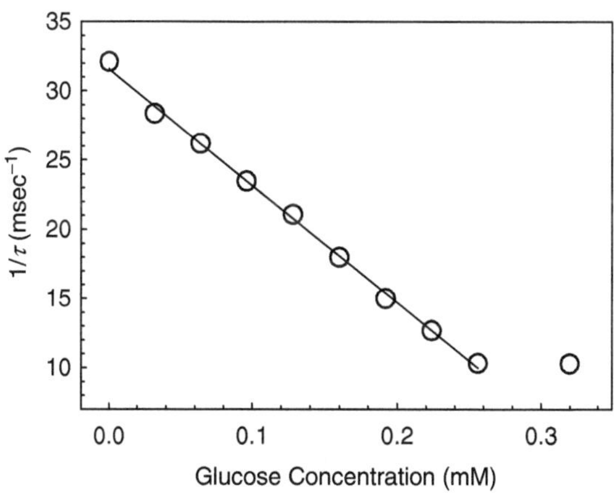

Fig. 3.21. $1/\tau$ vs. glucose concentration. The measurement was performed at ~22°C in a sealed container

Sequential and Simultaneous Monitoring of Multiple Analytes

As mentioned, in monitoring multianalytes sequentially, the OLED pixels associated with the different analytes were lit sequentially, monitoring separately each of the analytes in mixtures [16].

Figure 3.22 shows such a multianalyte sensor in operation, where six pairs of Alq_3 OLED pixels are lit. The leftmost and rightmost pairs are lit for illustration purposes only.

A PS:PtOEP film was drop cast on the second, third, fourth, and fifth pairs; when viewed in color, these pairs are orange due to the superposition of the green Alq_3 OLED EL and the red PtOEP PL. A GOx-doped sol–gel film was deposited on the third pair, while AOx and LOx were contained each in a separate glass well, whose base was the PtOEP-doped PS film, above pairs # 4 and 5, respectively. The measurement was performed using a PMT.

In sealed containers, calibration curves for the three analytes were obtained using their mixtures at comparable concentrations. Figure 3.23 shows $1/\tau$ vs. the analyte concentration for lactate, ethanol, and glucose, using a single PS:PtOEP film.

As seen, the data for all three analytes can be presented by a single line, with τ being independent of the analyte, whether glucose, lactate, or ethanol. This is expected in view of the similar oxidation reactions. No interference between the analytes was observed and the reaction for a given analyte occurred only in the presence of the specific corresponding enzyme. For example, τ for, e.g., 0.15 or 0.35 mM glucose was not affected by the presence of

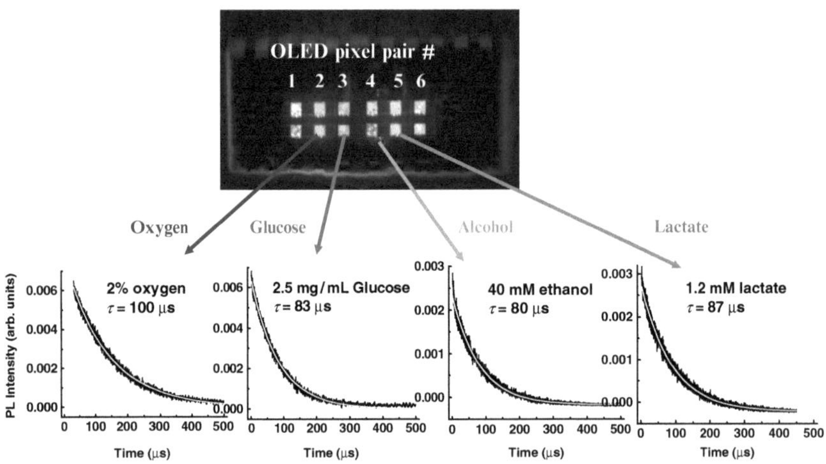

Fig. 3.22. Structurally integrated OLED-based multianalyte sensor for oxygen, glucose, alcohol, and lactate. All of the sensing elements were based on a PS:PtOEP film, and the analytes were monitored via the PL decay time τ of the PtOEP. The figure shows the intensity as a function of time (*black lines*) and the exponential fit (*white lines*). Measurements were conducted in air at ~23°C

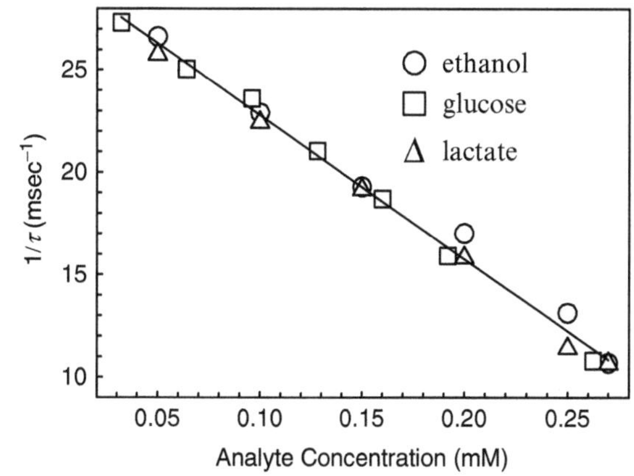

Fig. 3.23. $1/\tau$ as a function of analyte concentrations. The measurement was performed at ~22°C in a sealed container

0.1 or 0.2 mM lactate, and 0.15 or 0.35 mM ethanol. The deviations of the individual values from the average are <2%.

In the second approach, in which an array of PDs compatible with the OLED pixels was used to simultaneously monitor the analytes in a mixture, the results obtained were, as expected, comparable to those obtained using the sequential monitoring. Importantly, the results show that use of the compact OLED-based sensor array is a viable approach for simultaneous monitoring of these multiple analytes.

3.3.2 Extended Structural Integration: OLED/Sensing Component/Photodetector Integration

In attempts to develop field-deployable sensors, efforts focus on, e.g., enhancing sensor performance, reducing sensor cost and size, and simplifying fabrication. We are therefore developing a compact PL-based O_2 sensor to evaluate a fully integrated platform, where the PL excitation source is an OLED array and the PD is a p–i–n structure based on thin films of hydrogenated amorphous Si (a-Si:H) and related materials, or nanocrystalline Si [18].

Like OLEDs, a-(Si,Ge):H-and nanocrystalline-based PDs can be easily fabricated on glass or plastic mechanical supports. Modifying the layer thickness and composition of the p–i–n device enables tuning the PD to detect efficiently the emission at the desired wavelength.

For this purpose, a-Si:H and a-(Si,Ge):H PDs were evaluated by measuring their quantum efficiency (QE) vs. wavelength [18]. Figure 3.24 shows the results for selected PDs. In fabricating these PDs, the goal was to construct a PD with a maximal response at the ~635 nm peak emission of PtOEP and a minimal response at the shorter wavelengths of the OLED EL. As seen in the figure, a-(Si,Ge):H PDs with 10–13% gas-phase germane used for fabrication

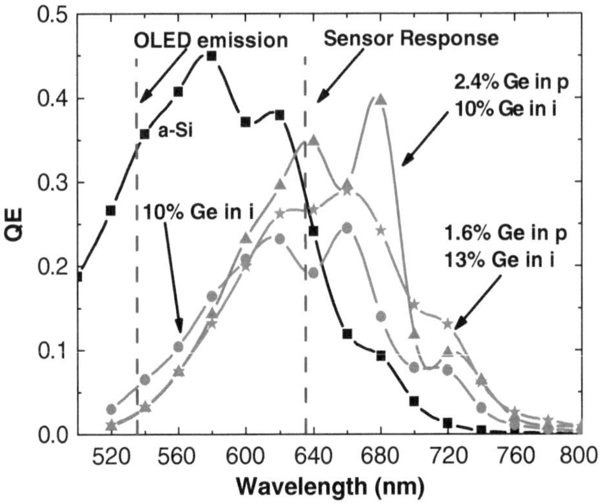

Fig. 3.24. Effect of Ge incorporation in electron cyclotron resonance (ECR)-grown devices on the QE; the percentages are of gas-phase germane

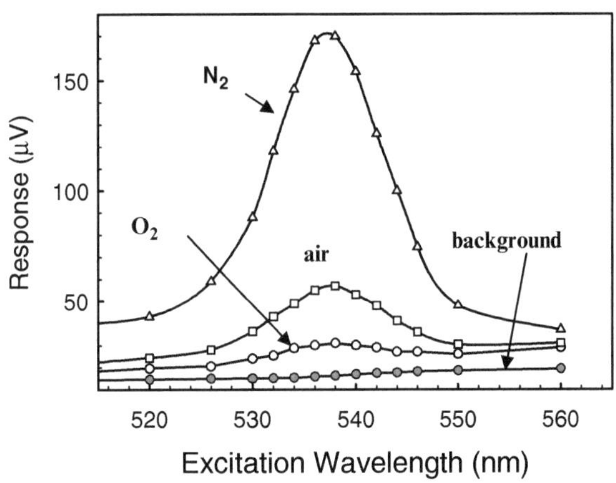

Fig. 3.25. The effect of O_2 concentration on the PD response using a tungsten–halogen lamp/monochromator excitation source and a 600 nm long-pass filter with the integrated PS:PtOEP/a-(Si.Ge):H PD

of the *i*-layer are suitable; however, their dark current is higher than that of PDs with no Ge.

The thin-film PDs were tested in conjunction with the O_2 sensing film, using a tungsten–halogen lamp/monochromator (set at 535 nm) as the excitation source. Figure 3.25 shows the PD response to the PL of the sensor film exposed to 100% O_2, air, and 100% N_2 environments, as well as to the background light; a 600 nm long-pass filter was placed in front of the PD to reduce

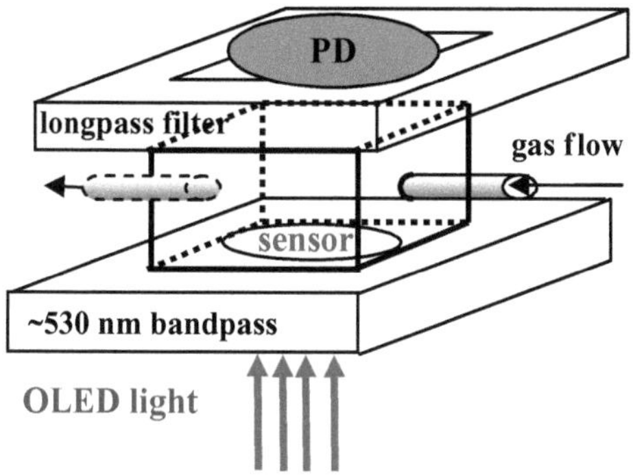

Fig. 3.26. Schematic of the OLED/sensing film/thin film PD set-up in a front detection geometry. The drawing is not to scale

scattered light. Without the sensing film, but with a 600 nm long-pass filter, the background response at an excitation wavelength of 535 nm was ~15 μV. However, with the addition of the sensing film, the excitation light at 535 nm is partially absorbed, resulting in PtOEP emission at ~635 nm. This emission varies with the O_2 level; as expected, it is quenched by oxygen and increases in an N_2 atmosphere. Reducing the dark current of the PD should improve its response.

As a first step toward integration of the three sensor components, an OLED/sensing film/PD structure was assembled in a front detection mode, where the PD is in front of the sensor film (see Fig. 3.26). Initial results showed a small but reproducible response to O_2 with a poly(*N*-vinyl carbazole) (PVK) UV-violet OLED, whose EL overlaps the strong PtOEP absorption band at ~380 nm; the PD response changed by ~20% when switched between O_2 and N_2 atmospheres. The small response is suspected to be due to the broad EL band, the relatively large PD dark current (1.5×10^{-8} A cm^{-2}), and unoptimized device design. Recent improvements in the OLEDs and PDs designs enabled an increase in S to ~7 using an Alq_3 excitation source, a PtOEP:PS film, and an n–i–p a-Si,Ge-based PD. A sensitivity of ~26 was obtained for the PdOEP:PS film.

As the measurement of τ is beneficial over I, the development of the thin-film PDs focuses currently on shortening their response time, and understanding the factors affecting their cutoff frequency f_c. It was observed that f_c is affected by the device structure, with boron diffusion during growth from the $p+$ to the i-layer in p–i–n PDs resulting in a reduction in f_c. Incorporating a SiC buffer layer and a superstrate structure, where the $p+$ layer was grown last, resulted in an increase in f_c. The effects of Ge, nanocrystallinity, and grain passivation, bandgap, defect states, as well as the illumination wavelength and reverse bias on f_c are also considered in these current studies. Additionally, efforts are centered on improving the design of the sensor and the PD, as well as devising structurally integrated a-Si:H-based filters to block

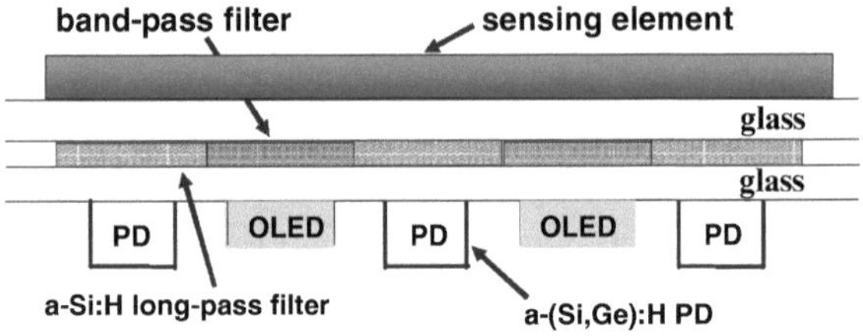

Fig. 3.27. Envisioned fully integrated OLED/sensing film/thin film PD array in a back detection configuration. Thin isolated Al layers between the OLEDs and PDs will block the edge EL

the EL at $\lambda > 600$ nm. The goal is to reduce the dark current of the PD by a factor >100 and the EL at the PL emission band by a factor >4.

Figure 3.27 shows the envisioned fully integrated OLED/sensing elements/ PD array in a back detection geometry, where the PD is coplanar with the OLED pixels; thin isolated Al layers (not shown) between the OLEDs and PDs will block the edge EL of the OLED. Such an array, which will also include thin film filters, would be miniaturizable and should enable realization of multianalyte (micro)sensor arrays, potentially badge-size.

3.4 Future Directions

3.4.1 Improved OLEDs

To further enhance the performance of the OLED-based sensing platform, approaches to narrow the EL emission bands and enhance the brightness, efficiency, and lifetime of the OLEDs are desired. Such approaches will include the following:

1. Tandem stacked phosphorescent OLEDs (SPhOLEDs) (with EL decay times $\leq 1\,\mu$s) [92]. The SPhOLEDs' structure is

   [anode]/[organic layers (OLs)]/[charge generation layer (CGL)]/[OLs]/[CGL]/.../[OLs]/[cathode].

   Each CGL injects electrons into the preceding layer and holes into the next layer. Thus, a SPhOLED is effectively a stack of PhOLEDs connected in series with each other. Its primary advantage is operation at lower current (but higher overall voltage), where each of the PhOLEDs in the stack emits at its highest efficiency. Thus, high brightness SPhOLEDs operate at much higher efficiency than regular PhOLEDs.
2. OLEDs with a mixed hole, electron, and emitting material layer [93]. Several studies have shown that such mixed layers enhance the stability of the devices.

3. Microcavity OLEDs fabricated on distributed Bragg reflectors (quarter wave stacks) [94]. Such OLEDs are fabricated on dielectric layers with significant dielectric contrast, so they narrow the emission spectrum by constructive interference. The narrow emission spectra also result in more efficient and more stable devices than regular OLEDs. The emission spectrum can be tailored to the specific sensor requirements (i.e., the absorption peak of the sensing element).

3.4.2 Sensor Microarrays

One of the goals in sensor development is to generate sensor microarrays for multianalyte detection. OLED pixel arrays present an opportunity to develop such sensors in a uniquely simple approach. Recently, nanosize OLEDs have been fabricated [45], which are promising for future micro/nanosize sensor arrays. Importantly, using combinatorial fabrication, OLED pixels of different colors can be fabricated next to each other. Such OLEDs can be usable for PL-based sensors with analyte-sensitive markers that emit at different wavelengths.

3.4.3 Autonomous Field-Deployable Sensors for Biological Agents

Beyond the specific sensors described above, initial efforts to develop the OLED-based platform for various biological sensors are underway. Notable among these are sensors for food-borne pathogens, e.g., *Salmonella spp.*, *E. coli* O157:H7, and *Listeria monocytogenes*. Indeed, in 1999 the Center for Disease Control estimated that there were 76 million illnesses, 325,000 hospitalizations, and 5,000 deaths due to food-borne pathogens in the US each year [95], and in 2000 the US Department of Agriculture projected the resulting economic burden at ∼\$7 billion per year [96]. In response to this issue, initial steps toward OLED-based sensors that would monitor such food-borne pathogens and food spoilage have recently been taken. These steps focused on Cu-methanobactin (Cu-mb), a novel luminescent chromopeptide recently isolated from the food grade methanotroph *Methylosinus trichosporium* OB3b, a source of single cell protein from methane [97]. DiSpirito and coworkers have recently confirmed the Cu-mb's bacteriocidal activity toward *L. monocytogenes*, and likely other aerobic bacteria [98]. Cu-mb fluoresces at 473 nm [99, 100], when excited at 394 nm, but that PL is quenched by the bacteriocidal activity. Since that activity also terminates the consumption of DO by these bacteria, OLED-based DO sensors with Cu-mb could be used to monitor some food-borne pathogens and bacteria-induced food spoilage continuously and autonomously. Indeed, preliminary results on such an OLED-based platform are very promising [101].

Other biological agents will require additional components for a field-deployable autonomous OLED-based sensors. For example, the OLED-based anthrax LF sensor will require a concentrator, germination, and amplification station, which would convert the *B. anthracis* spore to a live bacterium, and amplify that bacterium to a level that could be detected by the OLED-based sensor. As a consequence, efforts to design and build such stations are highly desirable.

3.5 Summary and Concluding Remarks

The new OLED-based platform for luminescent chemical and biological sensors was reviewed. The advantages of the structurally integrated OLED-excited sensors include their ease of fabrication, unique simplicity of integration with the sensing elements, consequent low cost of the sensor module, and promise as efficient light sources in sensor (micro)arrays for multianalyte detection. The viability of this platform was demonstrated for gas-phase and dissolved oxygen, glucose, lactate, ethanol, hydrazine, and anthrax lethal factor, and for monitoring of multiple analytes on a compact structure. Additionally, initial results indicate that structural integration of the three major sensor components, i.e., the OLED excitation source, the thin-film sensing element, and a thin-film photodetector, would result in badge-size monitors. Thin film filters using a-Si:H technology in conjunction with the OLED-based sensors are currently being tested. The OLEDs could be operated in a pulsed mode, which enables their use not only in the PL intensity mode, but also in a PL decay-time τ mode, where τ, which is often significantly longer than the EL lifetime, is related to the analyte concentration. This mode of operation is advantageous because it eliminates the need for frequent sensor calibration and optical filters, as τ is independent of moderate changes in the sensing film, light source intensity, and background light. Finally, some future directions for the development of this platform were described (a) Fabrication of microcavity stacked mixed layer phosphorescent OLEDs. These OLEDs would be brighter, more efficient, and longer-lived than the conventional fluorescent OLEDs, and their narrow EL spectra could be tailored to the absorption spectrum of the sensing element, resulting in drastic improvement of the sensor platform. (b) Development of the platform for various biological sensors, notably sensors for food-borne pathogens, which are a major ongoing issue in the global food supply chain. (c) Development of concentrator, germination, and amplification stations for biological agents, e.g., *B. anthracis*, which would enable the realization of field-deployable autonomous sensors for such agents.

References

1. O.S. Wolfbeis, L. Weis, M.J.P. Leiner, W.E. Ziegler, Anal. Chem. **60**, 2028 (1988)
2. B.H. Weigl, A. Holobar, W. Trettnak, I. Klimant, H. Kraus, P. O'Leary, O. Wolfbeis, J. Biotech. **32**, 127 (1994)

3. P. Hartmann, W. Ziegler, Anal. Chem. **68**, 4512 (1996)
4. Z. Rosenzweig, R. Kopelman, Sens. Actuators B **35–36**, 475 (1996)
5. B.D. MacCraith, C. McDonagh, A.K. McEvoy, T. Butler, G. O'Keeffe, V. Murphy, J. Sol-Gel Sci. Tech. **8**, 1053 (1997)
6. A.K. McEvoy, C.M. McDonagh, B.D. MacCraith, Analyst **121**, 785 (1996)
7. V.K. Yadavalli, W.-G. Koh, G.J. Lazur, M.V. Pishko, Sens. Actuators B **97**, 290 (2004)
8. M. Vollprecht, F. Dieterle, S. Busche, G. Gauglitz, K.-J. Eichhorn, B. Voit, Anal. Chem. **77**, 5542 (2005)
9. E.J. Cho, F. V. Bright, Anal. Chem. **73**, 3289 (2001)
10. W. Aylott, Z. Chen-Esterlit, J.H. Friedl, R. Kopelman, V. Savvateev, J. Shinar, US Patent No. 6,331,438 (December 2001)
11. V. Savvate'ev, Z. Chen-Esterlit, C.-H. Kim, L. Zou, J.H. Friedl, R. Shinar, J. Shinar, R. Kopelman, Appl. Phys. Lett. **81**, 4652 (2002)
12. B. Choudhury, R. Shinar, J. Shinar, in *Organic Light Emitting Materials and Devices VII*, ed. by Z.H. Kafafi and P.A. Lane, SPIE Conference Proceedings, vol. 5214, 2004, p. 64
13. B. Choudhury, R. Shinar, J. Shinar, J. Appl. Phys. **96**, 2949 (2004)
14. R. Shinar, B. Choudhury, Z. Zhou, H.-S. Wu, L. Tabatabai, J. Shinar, in *Smart Medical and Biomedical Sensor Technology II*, ed. by B.M. Cullum, SPIE Conference Proceedings, vol. 5588, 2004, p. 59
15. Z. Zhou, R. Shinar, B. Choudhury, L.B. Tabatabai, C. Liao, J. Shinar, in *Chemical and Biological Sensors for Industrial and Environmental Security*, ed. by A.J. Sedlacek III, S.D. Christensen, R.J. Combs, and T. Vo-Dinh, SPIE Conference Proceedings, vol. 5994 (SPIE, Bellingham, WA, 2005) p. 59940E-1
16. R. Shinar, C. Qian, Y. Cai, Z. Zhou, B. Choudhury, J. Shinar, in *Smart Medical and Biomedical Sensor Technology III*, ed. by B.M. Cullum, J.C. Carter, SPIE Conference Proceedings, vol. 6007 (SPIE, Bellingham, WA, 2005) p. 600710–1
17. R. Shinar, Z. Zhou, B. Choudhury, J. Shinar, Anal. Chim. Acta **568**, 190 (2006)
18. R. Shinar, D. Ghosh, B. Choudhury, M. Noack, V.L. Dalal, J. Shinar, J. Non Crystalline Solids **352**, 1995 (2006)
19. S. Camou, M. Kitamura, J.-P. Gouy, H. Fujita, Y. Arakawa, T. Fujii, Proc. SPIE **4833**, 1 (2002)
20. O. Hofmann, X. Wang, J.C. deMello, D.D.C. Bradley, A.J. deMello, Lab on a Chip **5**, 863–868 (2005)
21. H. Frebel, G.-C. Chemnitius, K. Cammann, R. Kakerow, M. Rospert, W. Mokwa, Sens. Actuators B **43**, 87–93 (1997)
22. M.S. Wilson, W. Nie, Anal. Chem. **78**, 2507–2513 (2006)
23. E.T. Zellers, M. Han, Anal. Chem. **68**, 2409 (1996)
24. M.S. Freund, N.S. Lewis, Proc. Nat. Acad. Sci. USA **92**, 2652–2656 (1995)
25. A. Carbonaro, L. Sohn, Lab on a Chip **5**, 1155 (2005)
26. L. Li, D.R. Walt, Anal. Chem. **67**, 3746 (1995)
27. T.A. Dickinson, J. White, J.S. Kauer, D.R. Walt, Nature **382**, 697 (1996)
28. D.R. Walt, T. Dickinson, J. White, J. Kauer, S. Johnson, H. Engelhardt, J. Sutter, P. Jurs, Biosens. Bioelectron. **13**, 697–699 (1998)
29. K.L. Michael, L.C. Taylor, S.L. Schultz, D.R. Walt, Anal. Chem. **70**, 1242–1248 (1998)
30. M.D. Marazuela, M.C. Moreno-Bondi, Anal. Bioanal. Chem. **372**, 664–682 (2002)

31. E.J. Cho, F.V. Bright, Anal. Chem. **74**, 1462 (2002)
32. E.J. Cho, Z. Tao, E.C. Tehan, F.V. Bright, Anal. Chem. **74**, 6177 (2002)
33. G. McGall, J. Labadie, P. Brock, G. Wallraff, T. Nguyen, W. Hinsberg, Proc. Natl. Acad. Sci. (USA) **93**, 13555–13560 (1996)
34. V. Lemmo, J.T. Fisher, H.M. Geysen, D.J. Rose, Anal. Chem. **69**, 543–551 (1997)
35. J. Shinar and V. Savvate'ev, in *Organic Light-Emitting Devices: A Survey*, ed. by J. Shinar, Chap. 1 (Springer, Berlin Heidelberg New York, 2003)
36. J. Shinar, (ed.), *Organic Light-Emitting Devices: A Survey* (Springer, Berlin Heidelberg New York, 2003)
37. C. Adachi, M.A. Baldo, S.R. Forrest, S. Lamansky, M.E. Thompson, R.C. Kwong, Appl. Phys. Lett. **78**, 1622 (2001)
38. C. Adachi, R.C. Kwong, P. Djurovich, V. Adamovich, M.A. Baldo, M.E. Thompson, S.R. Forrest, Appl. Phys. Lett. **79**, 2082 (2001)
39. R.H. Friend, R.W. Gymer, A.B. Holmes, J.H. Burroughes, R.N. Marks, C. Taliani, D.D.C. Bradley, D.A. Dos Santos, J.L. Brédas, M. Lögdlund, W.R. Salaneck, Nature **397**, 121 (1999)
40. http://www.universaldisplay.com
41. http://www.cdt.co.uk
42. http://www.emagincorp.com
43. G. Gustafsson, Y. Cao, G.M. Treacy, F. Klavetter, N. Colaneri, A.J. Heeger, Nature **357**, 447 (1992); C.C. Wu, S.D. Theiss, G. Gu, M.H. Lu, J.C. Sturm, S. Wagner, S.R. Forrest, IEEE Electron. Device Lett. **18**, 609 (1997); H. Kim, J.S. Horwitz, G.P. Kushto, Z.H. Kafafi, D.B. Chrisey, Appl. Phys. Lett. **79**, 284 (2001); E. Guenther, R.S. Kumar, F. Zhu, H.Y. Low, K.S. Ong, M.D.J. Auch, K. Zhang, S.J. Chua, in *Organic Light-Emitting Materials and Devices V*, ed. by Z.H. Kafafi, SPIE Conference Proceedings, vol. 4464 (SPIE, Bellingham, WA, 2002), p. 23; T.-F. Guo, S.-C. Chang, S. Pyo, Y. Yang, *ibid.* 34
44. C. Adachi, M.A. Baldo, M.E. Thompson, S.R. Forrest, J. Appl. Phys. **90**, 5048 (2001)
45. J.G.C. Veinot, H. Yan, S.M. Smith, J. Cui, Q. Huang, T.J. Marks, Nano Lett. **2**, 333–335 (2002); F.A. Boroumand, P.W. Fry, D.G. Lidzey, Nano Lett. **5**, 67 (2005); H. Yamamoto, J. Wilkinson, J.P. Long, K. Bussman, J.A. Christodoulides, Z.H. Kafafi, Nano Lett. **5**, 2485 (2005)
46. W. Trettnak, W. Gruber, F. Reininger, I. Klimant, Sens. Actuators B **29**, 219 (1995)
47. D.B. Papkovsky, G.V. Ponomarev, W. Trettnak, P. O'Leary, Anal. Chem. **67**, 4112 (1995)
48. Z. Rosenzweig, R. Kopelman, Anal. Chem. **67**, 2650 (1995)
49. S.-K. Lee, I. Okura, Anal. Commun. **34**, 185 (1997)
50. C. McDonagh, B.D. MacCraith, A.K. McEvoy, Anal. Chem. **70**, 45 (1998)
51. D. Garcia-Fresnadillo, M.D. Marazuela, M.C. Moreno-Bondi, G. Orellana, Langmuir **15**, 6451 (1999)
52. Y. Amao, K. Asai, T. Miyashita, I. Okura, Polym. Adv. Technol. **11**, 705 (2000)
53. Y. Amao, Michrochim. Acta **143**, 1 (2003)
54. Y.-E.L. Koo, Y. Cao, R. Kopelman, S.M. Koo, M. Brasuel, M.A. Philbert, Anal. Chem. **76**, 2498 (2004)
55. R.N. Gillanders, M.C. Tedford, P.J. Crilly, R.T. Bailey, Anal. Chim. Acta **502**, 1 (2004)

56. P.A.S. Jorge, P. Caldas, C.C. Rosa, A.G. Oliva, J.L. Santos, Sens. Actuators B **103**, 290 (2004)
57. C. Preininger, I. Klimant, O.S. Wolfbeis, Anal. Chem. **66**, 1841 (1994)
58. Y. Amao, T. Miyashita, I. Okura, Analyst **125**, 871 (2000)
59. V. Savvate'ev, J.H. Friedl, L. Zou, J. Shinar, K. Christensen, W. Oldham, L.J. Rothberg, Z. Chen-Esterlit, R. Kopelman, Appl. Phys. Lett. **76**, 1501 (2000)
60. K.O. Cheon, J. Shinar, Appl. Phys. Lett. **81**, 1738 (2002)
61. G. Li, J. Shinar, Appl. Phys. Lett. **83**, 5359 (2003)
62. L. Zou, V. Savvate'ev, J. Booher, C.-H. Kim, J. Shinar, Appl. Phys. Lett. **79**, 2282 (2001)
63. V.G. Kozlov, G. Parthasarathy, P.E. Burrows, S.R. Forrest, Y. You, M.E. Thompson, Appl. Phys. Lett. **72**, 144 (1998)
64. L.S. Hung, C.W. Tang, M.G. Mason, Appl. Phys. Lett. **70**, 152 (1997)
65. S.E. Shaheen, G.E. Jabbour, M.M. Morell, Y. Kawabe, B. Kippelen, N. Peyghambarian, M.-F. Nabor, R. Schlaf, E.A. Mash, N.R. Armstrong, J. Appl. Phys. **84**, 2324 (1998)
66. Z.J. Fuller, D.W. Bare, K.A. Kneas, W.-Y. Xu, J.N. Demas, B.A. DeGraff, Anal. Chem. **75**, 2670 (2003)
67. W.F. Linke, *Solubilities of Inorganic and Metal-Organic Compounds* (American Chemical Society, Washington, DC, 1958); J.M. Hale, *Technical Notes* (Orbisphere Laboratories, NJ, 1980); M. Vadekar, PTQ Winter 2002/03, p. 87 (`http://www.eptq.com/Pages/Articles/Winter0203/Winter0203_pdf/PTQ03107.pdf`)
68. S. de Marcos, J. Galindo, J.F. Siera, J. Galban, J.R. Castillo, Sens. Actuators B **57**, 227 (1999)
69. K.O. Cheon, J. Shinar, Appl. Phys. Lett. **83**, 2073 (2003)
70. Z. Zhou, R. Shinar, A.J. Allison, J. Shinar, Adv. Func. Mater. (to be published)
71. X. Lu, I. Manners, M.A. Winnik, Macromol. **34**, 1917 (2001)
72. X. Lu, M.A. Winnik, Chem. Mater. **13**, 3449 (2001)
73. O.S. Wolfbeis, I. Oehme, N. Papkovskaya, I. Klimant, Biosens. Bioelec. **15**, 69 (2000)
74. H. Xu, J.W. Aylott, R. Kopelman, Analyst **127**, 1471 (2002)
75. A.N. Ovchinnikov, V.I. Ogurtsov, W. Trettnak, D.B. Papkovsky, Anal. Lett. **32**, 701 (1999)
76. D.B. Papkovsky, T. O'Riordan, A. Soini, Biochem. Soc. Trans. **28**, 74 (2000)
77. G.A. Crosby, R.J. Watts, J. Amer. Chem. Soc. **93**, 3184 (1971)
78. E.R. Carraway, J.N. Demas, B.A. DeGraff, J.R. Bacon, Anal. Chem. **63**, 337 (1991)
79. K. Mongey, J.G. Vos, B.D. MacCraith, C.M. McDonagh, J. Sol-Gel Sci. Tech. **8**, 979 (1997)
80. J.M. Kurner, I. Klimant, C. Krause, H. Preu, W. Kunz, O.S. Wolfbeis, Bioconjugate Chem. **12**, 883 (2001)
81. K. Kalyanasundaram, *Photochemistry of Polypyridine and Porphyrin Complexes* (Academic Press, New York, 1992)
82. Z. Zhou, J. Shinar, unpublished results
83. M.A. Baldo, S. Lamansky, P.E. Burrows, M.E. Thompson, S.R. Forrest, Appl. Phys. Lett. **75**, 4 (1999)

84. American Conference of Governmental Industrial Hygienists (ACGIH). 1999 TLVs and BEIs. Threshold Limit Values for Chemical Substances and Physical Agents, Biological Exposure Indices (Cincinnati, OH 1999). See also http://www.epa.gov/ttn/atw/hlthef/hydrazin.html#ref12
85. National Institute for Occupational Safety and Health (NIOSH). Pocket Guide to Chemical Hazards. U.S. Department of Health and Human Services, Public Health Service, Centers for Disease Control and Prevention (Cincinnati, OH. 1997, <http://www.cdc.gov/niosh/npg/npg.html>).
86. S. Rose-Pehrsson, G.E. Collins, US Patent 5,719,061, 17 Feb 1998
87. R.T. Cummings, S.P. Salowe, B.R. Cunningham, J. Wiltsie, Y.W. Park, L.M. Sonatore, D. Wisniewski, C.M. Douglas, J.D. Hermes, E.M. Scolnick, Proc. Nat. Acad. Sci. **99**, 6603 (2002)
88. F. Tonello, P. Ascenzi, C. Montecucco, J. Biol. Chem. **278**, 40075 (2003)
89. B.E. Turk, T.Y. Wong, R. Schwarzenbacher, E.T. Jarrell, S.H. Leppla, R.J. Collier, R.C. Liddington, L.C. Cantley, Nat. Struc. Mol. Biol. **11**, 60 (2004)
90. B.M. Peek et al., Int. J. Pep. Prot. Res. **38**, 114 (1991); B. Geisser, A. Ponce, R. Alsfasser, Inorg. Chem. **38**, 2030 (1999)
91. Y. Cai, R. Shinar, J. Shinar, unpublished results
92. L.S. Liao, K.P. Klubek, C.W. Tang, Appl. Phys. Lett. **84**, 167 (2004); T.-Y. Cho, C.-L. Lin, C.-C. Wu, Appl. Phys. Lett. **88**, 111106 (2006)
93. V.-E. Choong, S. Shi, J. Curless, C.-L. Shieh, H.-C. Lee, F. So, J. Shen, J. Yang, Appl. Phys. Lett. **75**, 172 (1999); V.-E. Choong, S. Shi, J. Curless, F. So, Appl. Phys. Lett. **76**, 958 (2000); H. Aziz, Z.D. Popovic, N.-X. Hu, Appl. Phys. Lett. **81**, 370 (2002)
94. A. Dodabalapur, L.J. Rothberg, T.M. Miller, E.W. Kwock, Appl. Phys. Lett. **64**, 2486 (1994); A. Dodabalapur, L.J. Rothberg, T.M. Miller, Appl. Phys. Lett. **65**, 2308–2310 (1994); A. Dodabalapur, L.J. Rothberg, R.H. Jordan, T.M. Miller, R.E. Slusher, J.M. Phillips, J. Appl. Phys. **80**, 6954 (1996)
95. P.S. Mead, L. Slutsker, et al., Emerg. Infect. Dis. **5**, 607 (1999)
96. www.ers.usda.gov/briefing/FoodborneDisease/features.htm, *Economics of Foodborne Disease: Estimating the Benefits of Reducing Foodborne Disease.* 2000
97. A. Skrede, G.M. Berge, et al., Ann. Food Sci. Tech. **76**, 103 (1998)
98. A.A. DiSpirito, J.A. Zahn, et al., USPTO, Editor. 2007: USA
99. D.W. Choi, C.J. Zea, Y.S. Do, J.D. Semrau, W.E. Antholine, M.S. Hargrove, N.L. Pohl, E.S. Boyd, G.G. Geesey, S.C. Hartsel, P.H. Shafe, M.T. McEllistrem, C.J. Kisting, D. Campbell, V. Rao, A.M. de la Mora, A.A. DiSpirito, Biochemistry **45**, 1442 (2006)
100. D.W. Choi, Y.S. Do, et al., J. Inorgan. Biochem. **100**, 2150 (2006)
101. Y. Cai, R. Shinar, et al., in *Second Symposium on Food Safety and Security*, Iowa State University, Ames, IA, 2007

4

Lab-on-a-Chip Devices with Organic Semiconductor-Based Optical Detection

O. Hofmann, D.D.C. Bradley, J.C. deMello, and A.J. deMello

4.1 Introduction

4.1.1 Microfluidics and Lab-on-a-Chip

On 29 December 1959, the Nobel Laureate Richard Feynman delivered his prophetic talk before the American Physical Society in which he contemplated the potentials of miniaturization in the physical sciences. His vision, based on known technology, scrutinized the limits set by physical principles and suggested a range of new "nano-tools," including the concept of "atom-by-atom" fabrication. As we now know, over the intervening decades, many of these predictions have become reality. For example, microelectronic circuits have shrunk to sizes that approach the molecular level, scanning probe microscopes can image and manipulate individual atoms, and the molecular machinery of living systems is now being more fully understood and harnessed.

It is therefore surprising that it is only within the last 20 years that the concepts of miniaturization have been seriously applied to the fields of chemical and biological analysis. Of particular interest has been the development and application of microfluidic or *lab-on-a-chip* technology. In simple terms, microfluidics describes the study and development of systems that manipulate and process small instantaneous amounts of fluid (typically on the picoliter to nanoliter scale) using features whose characteristic dimensions are most conveniently measured in microns. Interest in such technology has been driven by a range of fundamental features that accompany system miniaturization. Such features include the ability to process and handle small volumes of fluid, improved analytical performance when compared to their macroscale analogues, reduced instrumental footprints, low unit cost, facile integration of functional components, and the exploitation of atypical fluid dynamics to control molecules in both time and space [1]. Based on these advantageous characteristics, microfluidic chip devices (such as those shown in Figs. 4.1–4.4) have been used to good effect in a wide variety of applications, including nucleic acid separations, protein analysis, process control, small-molecule

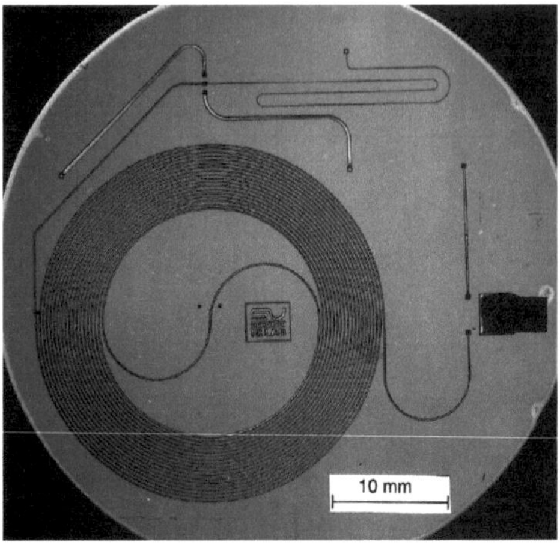

Fig. 4.1. Lab-on-a-chip gas chromatograph on silicon substrate (From [4] – Copyright 1979 IEEE)

organic synthesis, DNA amplification, immunoassays, DNA sequencing, cell manipulations, nanomaterial synthesis, and medical diagnostics [2,3].

Much of the early pioneering work centered on the transfer of established analytical methods from conventional (macroscale) to microfluidic (chip-based) formats. Specifically, a primary focus of these early studies was the creation of chip-based devices for performing analytical separations (based on chromatographic and electrophoretic phenomena). For example, it is generally recognized that the first lab-on-a-chip system, a gas chromatograph, was presented by Terry and coworkers in 1979 (Fig. 4.1) [4]. Structured on a 2-in. silicon wafer, the device integrated a spiral separation column, carrier-gas and sample injectors, and a thermal conductivity detector. In 1992, Manz and Harrison reported the first chip-based electrophoretic separations [5]. These and the subsequent studies marked an explosion in both academic and industrial interest in the field, due to the fact that by miniaturizing column dimensions and creating monolithic fluidic networks on planar substrates, huge enhancements in analytical performance in terms of analytical speed, component resolution, reproducibility, and analytical throughput could be achieved. Indeed, a survey of the current literature demonstrates that almost all separation methods (based on electrophoretic or chromatographic partitioning) have been successfully transferred to micromachined formats with enhanced performance characteristics. Such techniques include liquid chromatography, micellar electrokinetic capillary chromatography, synchronized cyclic capillary electrophoresis, free flow electrophoresis, and open channel electrochromatography.

A subsequent theme of much effort in the microfluidics community has been the development of integrated microfluidic systems. In conceptual form, the success of a microfluidic system is defined by its ability to rapidly and efficiently extract required information from a chemical or biological system. This almost always involves performing a series of systematic operations on an analytical sample, which includes sample and reagent introduction, fluid motivation, reagent mixing and reaction, sample separation and purification, analyte detection, and product isolation. Indeed, one of the principal advantages of using microfabrication methods to create analytical instruments is the ease with which large-scale integration of functional components can be achieved. In recent years some elegant examples of functional integration of components have begun to revolutionize the way in which microfluidic tools are developed and used to solve fundamental problems. A good example has been the introduction of soft lithography in poly(dimethylsiloxane) as a fast and efficient route to microfluidic device manufacture. The elastomeric nature of this material has facilitated the development of sophisticated valve technologies (based on the restriction of fluidic channels via pneumatic actuation), which in turn have enabled the creation of highly integrated microfluidic circuits that can perform complex biological and chemical assays. An example of such an integrated system can be seen in Fig. 4.2. However, it should be realized that more traditional micromachining methods and materials (such as glass and silicon) can still be used to create microfluidic systems with high levels of integration and operational excellence. A good example, shown in Fig. 4.3, is a nanoliter-scale microfabricated bioprocessor that integrates thermal cycling, sample purification, and capillary electrophoresis for Sanger DNA sequencing [6].

Since microfluidic systems can rapidly manipulate, process, and analyze small volumes of complex fluids with high efficiency, their use as basic tools in medical and clinical diagnostics is an attractive prospect [7]. At the current time, a wide diversity of diseases such as infectious or cardiac diseases are

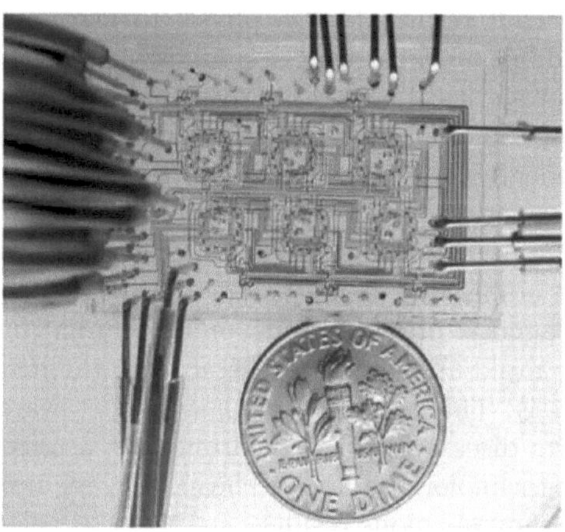

Fig. 4.2. Complex microfluidic circuit on polydimethylsiloxane (PDMS) substrate

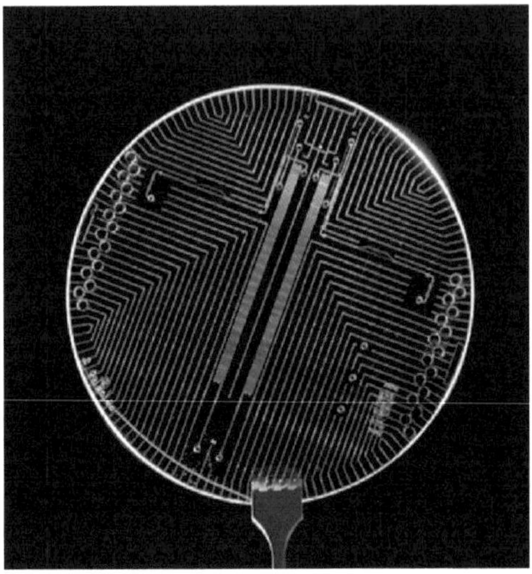

Fig. 4.3. Microfabricated bioprocessor for integrated DNA sequencing (From [6] – Copyright 2006 National Academy of Sciences, U.S.A.)

diagnosed using immunoassay methods. In simple terms, an immunoassay is a biochemical test that quantitatively measures the presence of a specific analyte in a biological fluid (typically serum, urine, or saliva) using the specific reaction of an antibody to its antigen. In the developed world, basic blood chemistry panels that integrate 10–20 separate tests of this kind are typically analyzed using automated, benchtop instruments located in centralized laboratories. Patient samples (e.g., 5 mL of whole blood) are normally collected at a hospital or physician's office and then transferred via a logistics chain to the centralized laboratory for testing. After a period of between 24 and 96 h, results are electronically returned to the physician who can then make a therapeutic decision and initiate treatment at a follow-up visit. For many tests, the time-to-result is not a critical factor; however, in cases where the patient may be experiencing, for example, acute cardiovascular disease, myocardial infarction, or a thromboembolic event, such large time delays are unacceptable. According to the American Heart Association, 64 million individuals in the US currently have acute cardiovascular disease, with over eight million emergency department admissions being associated with chest pains (and thus requiring rapid diagnosis of arrhythmia or myocardial infarction). Accordingly, the need for rapid and accurate diagnosis at the initial *point-of-care* is compelling (Fig. 4.4). This combined with a trend towards monitoring and control of long-term disease in home environments necessitates the development of diagnostic technologies that are cheap, fast, require minimal operator input, and provide unambiguous results.

Although significant advancements continue in public healthcare within the developed world, more than one billion people still lack the most basic

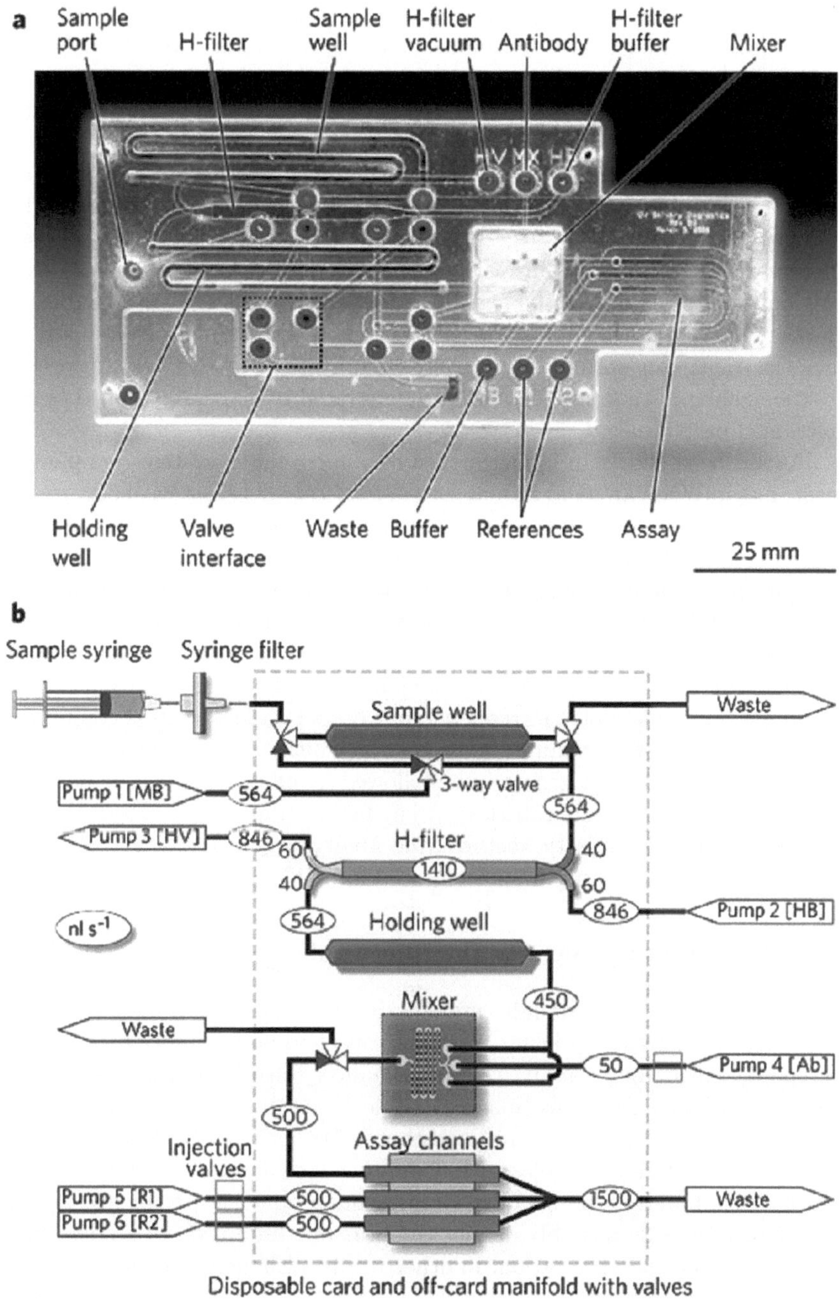

Fig. 4.4. Diagnostic test based on microfluidic lab-on-a-chip technology (From [7] – Copyright 2006 Nature Publishing Group)

healthcare facilities within developing and third world countries. As noted by Yager and coworkers, accomplishments such as the eradication of smallpox in 1979 have unfortunately been accompanied by the reemergence of infectious diseases such as tuberculosis and the appearance of new diseases such as HIV/AIDS. Since more than 50% of all deaths in the world's poorest regions are a result of infectious disease, the availability of appropriate diagnostic technologies could have a profound impact on healthcare in the twenty-first century. Nevertheless, the successful and widespread implementation of *point-of-care* diagnostic devices in a developing country is a nontrivial challenge. Not only must the devices be accurate, inexpensive, robust, and reliable, but they are likely to be used in settings that can be hot, humid, dirty, dusty, and lack electricity or running water. These complications are further exacerbated by the likelihood that the tests will be performed by partially trained or even untrained individuals.

As has been shown in numerous studies, microfluidic systems, on paper at least, may provide an ideal format for *point-of-care* diagnostics. They are able to process and manipulate small amounts of sample, can be mass-fabricated at low unit cost, can be operated by untrained personnel, and are able to operate in resource-poor environments. In recent years, a number of prototype microfluidic diagnostic devices have been reported. These developments have been reviewed in detail by Yager and coworkers, who conclude that the primary challenge associated with the successful deployment of microfluidic diagnostic systems relates to reducing unit costs to a level that is competitive with current lateral flow immunoassays (which although suitable for crude on-the-spot diagnosis must generally be followed by more quantitative testing at remote or centralized laboratories). A key factor in achieving this goal within the *point-of-care* diagnostic sector is the ability to perform sensitive analyte detection at low cost and in an integrated format.

4.1.2 Detection Problem at the Microscale

Clearly the use of microfluidic systems in chemical and biological analysis engenders considerable advantages with respect to performance markers such as speed, throughput, analytical performance, selectivity, automation, and control. All such gains are made possible by system downscaling and the associated improvements in mass and thermal transfer. Nonetheless, handling and processing fluidic samples with instantaneous volumes ranging from a few picoliters to hundreds of nanoliters represents a nontrivial challenge for analyte detection, and arguably defines the principle limitation of a microfluidic system in a specific application. In other words, although a reduction in scale may afford a significant improvement in analytical performance, this benefit is offset by a progressive decrease in the number of molecules available for detection. This issue is highlighted by a simple calculation: when performing capillary electrophoresis within a chip-based microfluidic system, analyte injection volumes are commonly no larger than 50 pL. This means that for an analyte concentration of 10^{-9} mol L^{-1}, only about 30,000 molecules are present in the system for separation and detection!

Small volume detection within microfluidic environments has typically been based around optical measurements. Optical detection is well suited for most microfluidic systems due to the favorable optical characteristics of most substrate materials (including glasses, quartz, and plastics) in the visible region of the electromagnetic spectrum. Emission based optical techniques are usually the most sensitive, in particular, fluorescence- and chemiluminescence-based detection, in which analytes emit light in response to optical or chemical excitation, respectively. The most common chip-based detection methods include laser-induced fluorescence (LIF), UV–vis absorbance, chemiluminescence, refractive index variation, and thermal lens microscopy. Although attractive for the detection of small molecules, absorbance measurements are compromised due to the difficulty of probing small sample volumes while maintaining a sufficiently long optical pathlength to achieve appreciable absorption. By far, the most popular optical detection technique for chip-based analysis is LIF, which affords exceptional sensitivity and low mass detection limits at the expense of relatively sophisticated equipment.

As few as 10^5 molecules are routinely detected using LIF, and recent developments in ultra-high sensitivity fluorescence detection have allowed single molecule detection to be performed within microchannel environments [8]. A detailed evaluation of detection methods used for small-volume environments is provided elsewhere [9, 10]; however, it is important to appreciate that effective detection within microfluidic environments is clearly defined by a close interrelationship of factors such as detector sensitivity, optical coupling efficiencies, response times, detection limits, and information content.

The specific needs of *point-of-care* diagnostic devices in relation to analyte detection are unmistakable. Small sample volumes and low analyte concentrations typical in microfluidic systems make high sensitivity detection a prerequisite. While in the laboratory these demands may be met by sophisticated detection schemes using benchtop optical components such as lasers, photomultiplier tubes (PMTs), or fluorescence microscopes, the realization of portable diagnostic devices for *point-of-care* testing necessitates the development of integrated, inexpensive, versatile, and miniaturized detection modules. To date, few if any diagnostic systems with fully integrated, low-cost, and versatile optical detectors have been reported.

4.2 Fabrication

4.2.1 Microfluidic Systems

A cursory glance through microfluidic literature in the early 1990s reveals that almost all early microfluidic systems were constructed from glass, quartz, or silicon. This is unsurprising since standard photolithography and wet-etching

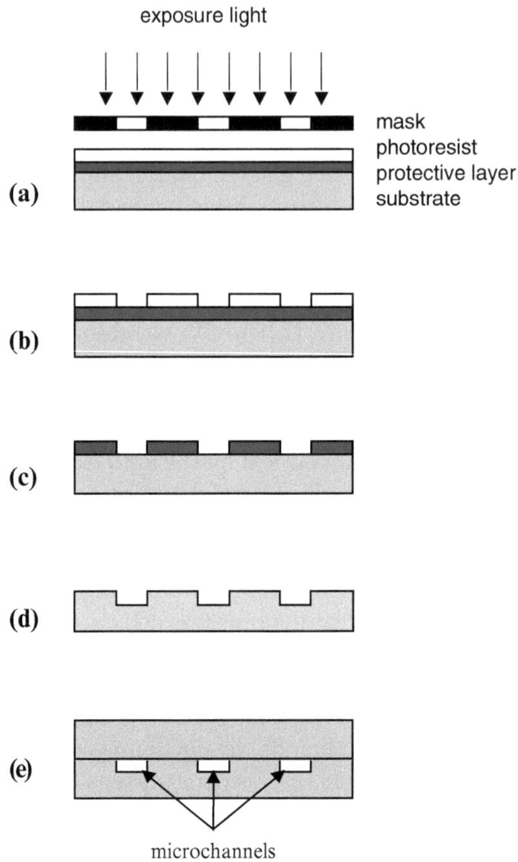

Fig. 4.5. Microfabrication process. (**a**) Photolithographic photoresist patterning. (**b**) Photoresist development. (**c**) Protective layer etching. (**d**) Substrate etching. (**e**) Bonding to cover plate

techniques perfected in the microelectronics and semiconductor industries could be used to efficiently structure all these materials to produce microchannel networks [11].

The generic microfabrication process used can be separated into three major stages; patterning, etching, and closure (Fig. 4.5). For silicon substrates a protective layer of silicon oxide is thermally grown at 200°C prior to processing. For photolithographic patterning, a positive photoresist is then spun onto the protective layer (Fig. 4.5a). UV exposure of the photoresist through a mask (defining the desired channel pattern) results in locally increased photoresist solubility. After dissolving such exposed regions of the photoresist (Fig. 4.5b), the uncovered silicon oxide can be etched (Fig. 4.5c). Depending on the required channel geometry, wet or dry etching techniques are applied to structure the underlying substrate (Fig. 4.5d). Etching silicon with a mixture of hydrofluoric and nitric acid results in semicircular channels due to a nondirectional (or isotropic) etching action. For basic etchants such as potassium hydroxide the shape of the resulting channels depends on the crystallographic orientation of the silicon substrate. If etching occurs along the crystallographic planes, pyramidal pits or V-shaped grooves are obtained depending on the size of the etched area and the etching time. Dry etching techniques are based on the formation of reactive ion species in the gas phase. With guided ion beams, near-vertical side walls and high aspect ratios can be achieved at the expense of high instrumental costs and low etching rates.

After etching, the remaining protective layer and photoresist is removed and a second unstructured silicon wafer is bonded on top to enclose fluidic channels (Fig. 4.5e). Thermal fusion bonding above 600°C is based on the temperature-induced softening of the silicon–silicon interface and rehardening upon cooling. Anodic bonding can also be employed to bond a glass cover plate to the structured silicon substrate (Fig. 4.6). This process requires the application of a voltage of 700–1,200 V between the substrates (with silicon serving as the anode) and temperatures between 300 and 400°C. While micromachining techniques for silicon are well-established, there remain drawbacks with its use as a substrate material in microfluidic applications. For example, the low breakdown-voltage of silicon necessitates the growth of insulating layers (e.g., thermally grown silicon oxide) for electrophoretic applications. Furthermore, silicon is opaque in the visible wavelength range employed for most optical detection schemes.

In basic terms, glass microfabrication is identical to silicon microfabrication. Glass substrates are commonly coated with chromium to serve as the protective layer for etching. Wet etching techniques are almost always used for substrate structuring, with typical etching solutions comprising hydrofluoric acid and ammonium fluoride. Since glass has a noncrystalline structure, etching is isotropic, which limits the aspect ratios obtainable. The most common

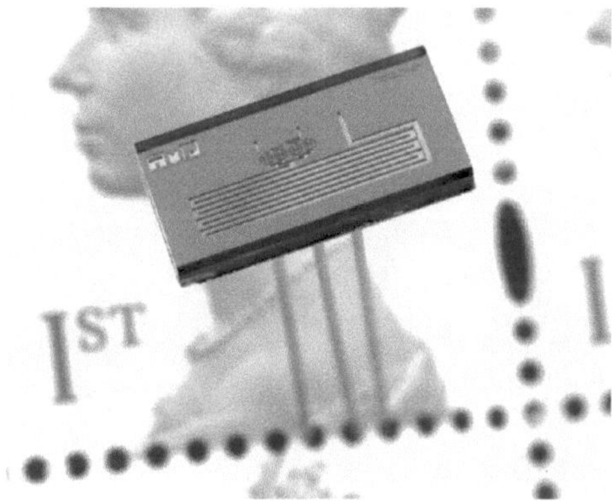

Fig. 4.6. Typical glass–silicon microfluidic device for diffusive mixing of two components

bonding technique for enclosing glass structures is thermal bonding, which requires application of temperatures of 600°C over several hours. However, recently alternative low temperature methods have been developed. These are based on the formation of an additional bonding layer, such as epoxy, sodium silicate, or a thin layer of HF between the glass surfaces resulting in temporary dissolution of the glass boundaries and subsequent rehardening. Glass microfluidic devices offer adequate electrical insulation and are optically transparent in the visible range, rendering them amenable for a much wider range of *lab-on-a-chip* applications compared to silicon. However, a limitation that silicon and glass microfabrication techniques have in common is that they are based on batch processing, i.e., only one substrate is processed at a time.

As previously discussed, a recent trend in microfluidic device fabrication is the use of polymers and elastomers that are amenable to replication technologies and allow for high-throughput and low-cost manufacture [12]. Photoablation is a method used to produce channels in a variety of polymeric substrates by means of "microexplosions" induced by pulsed UV-laser light. Hot Embossing is a common technique for imprinting micron-sized features on polymer substrates using a master mold (Fig. 4.7). Microfabricated Ni-shims are typically used for printing a layout in a thin polymer film, which is heated up to its softening point. This simple technique enables the fabrication of microstructures with high structural accuracy at low cost. Currently, the most advanced replication technology is injection molding. It enables the fabrication of very deep, high-quality microstructures in thick thermoplastic substrates. Because of the need for a high-pressure injection chamber into which the microfabricated Ni-shim is fixed, costs are slightly higher than for hot embossing. The described polymer micromachining methods typically offer straightforward bonding techniques. Taking advantage of the strongly temperature-dependent properties of most polymers, thermal

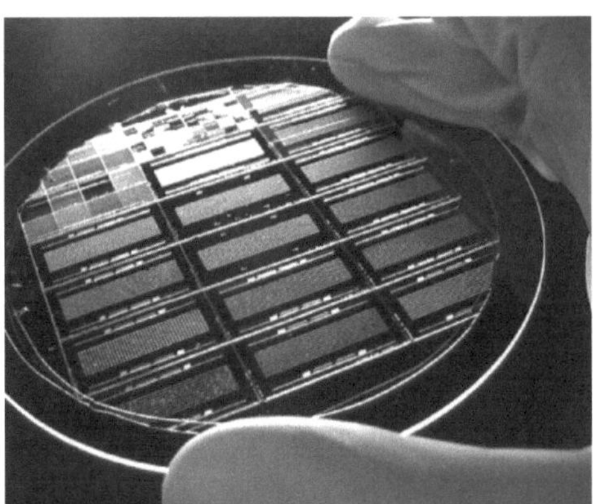

Fig. 4.7. Plastic microchip fabricated by hot embossing (Photograph courtesy of CSEM)

bonding or lamination can be used to seal microchannels. These processes have in common that they are fast, with typical processing times of less than 10 min. They are typically performed at temperatures around 100°C, depending on the softening point of the polymers employed. Substrate materials for hot embossing or injection molding include polycarbonate (PC), polypropylene (PP), polymethylmethacrylate (PMMA), and cyclic olefin copolymers (COCs) such as Topas®.

In contrast to machine-based molding techniques, conventional molding can be performed with a simple master comprising a surface relief, placed in a molding dish (Fig. 4.8). This approach, termed *soft lithography*, has become highly popular over the past few years, offering a rapid, flexible, and inexpensive route to the creation of microfluidic components on planar substrates [13]. Soft lithographic methods describe the molding of elastomeric polymers using master templates. Elastomeric siloxane polymers such as PDMS are

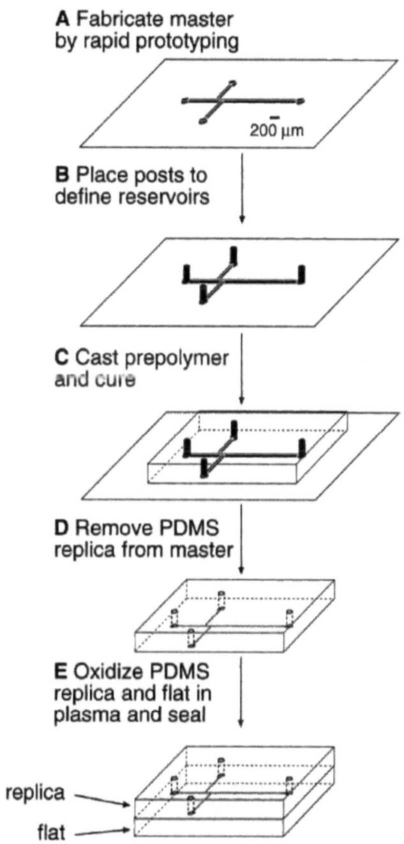

Fig. 4.8. Rapid prototyping approach for polydimethylsiloxane (PDMS) elastomer devices (Reprinted with permission from [14]. Copyright 1998 American Chemical Society)

easily molded, optically transparent (in the visible and UV regions of the electromagnetic spectrum), robust, cheap, nontoxic, and stable over wide temperature ranges. In the basic process, high aspect ratio microfluidic structures are formed by simply pouring a mixture of the elastomer precursor and a curing agent over a template. After curing, the structured polymer layer is peeled away from the template and an enclosed fluidic structure is created by contacting the elastomer with a planar surface. The seal between surfaces need not be permanent (facilitating fluidic cleaning and removal of blockages), although treatment of surfaces with an oxygen plasma allows siloxanes to be irreversibly bound to a variety of substrate materials [14].

4.2.2 Organic Semiconductor-Based Light Sources and Detectors

Lab-on-a-chip devices have shown themselves to be highly effective for laboratory-based research, where their superior analytical performance has established them as efficient tools for complex tasks in genetic sequencing, proteomics, and drug-discovery applications. However, to date they have not been well suited to *point-of-care* or *in-the-field* applications, where cost and portability are of primary concern. Although the chips themselves are cheap and small, they must generally be used in conjunction with bulky optical detectors that are needed to identify or quantify the analytes or reagents present. Furthermore, most existing detectors are limited to the analysis of a single analyte at a predetermined location on the chip. The lack of an integrated, versatile detection scheme (one which is miniaturized, integrated, wavelength-selective, and able to monitor multiple locations on the chip) is a major obstacle to the deployment of diagnostic devices in the field, and has prevented the development of more complex tests where rapid, kinetic, or multi-point analysis is required. Although there have been a few attempts to integrate optics within the chip structure itself, few have demonstrated the levels of integration demanded for *point-of-care* diagnostics.

One promising option for creating integrated light sources and photodetectors is the use of organic semiconductors. There are two main classes of organic semiconductor – the first based on small molecules [15] and the second based on polymers [16]. Broadly speaking, small molecule and polymer semiconductors have similar optical and electrical properties (although there are some notable differences) but are often processed differently. Small molecule devices are typically deposited by vacuum sublimation, which enables the controlled fabrication of complex multilayer structures but is not especially well suited to large area deposition. Polymer devices, by contrast, are typically fabricated using solution deposition techniques such as spin-coating or printing, which are better suited to large area deposition but are somewhat problematic for multilayer device fabrication (since deposition of successive layers tends to cause partial redissolution of already deposited layers).

As can be seen from Fig. 4.9, most organic semiconductor devices have simple multilayered structures, which may be fabricated in a straightforward

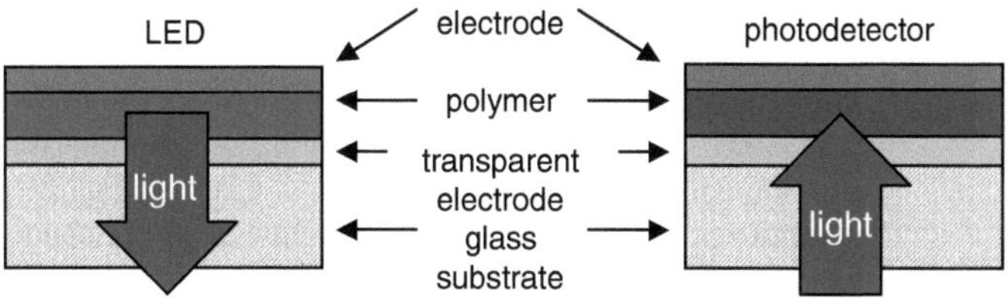

Fig. 4.9. Schematic of device structure of typical organic semiconductor-based light sources and detectors

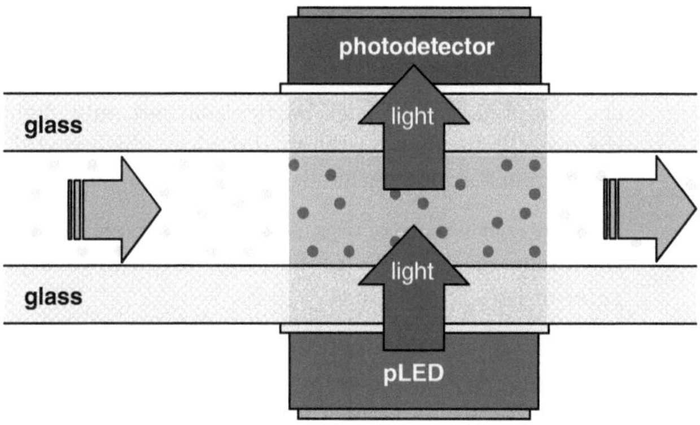

Fig. 4.10. Schematic of microfluidic detection set-up with polymer LED (pLED) light-source and organic photodetector

fashion by sequential deposition of appropriate semiconductors or electrodes. Typical thicknesses of the individual organic layers are 30–100 nm. For instance, a typical organic light-emitting diode (OLED) comprises one or more layers of semiconductor sandwiched between two electrodes, of which at least one must be transparent. The active layer emits light under electrical excitation and the devices may therefore be used as light sources. They may also be used in reverse as photodetectors by illuminating the active organic layer to generate a measurable electrical current. Accordingly, these two components may be used together as a simple detection system (Fig. 4.10). The LED and photodetector are arranged on either side (top and bottom surfaces) of a microfabricated channel containing an analyte under study. Analytes passing through the detection volume absorb photons from the OLED. Since many analytes of interest are fluorescent they may subsequently re-emit photons, which can be detected by the photodetector. Owing to the simple layer-by-layer deposition procedures for the polymer components and the planar structure of analytical microchips, such OLED/photodetector pairs may be easily integrated into existing chip structures at marginal additional cost.

There are a number of reasons why organic semiconductors are attractive as integrated sensors and some of the key advantages are discussed in Table 4.1. In short, they offer a unique means of integrating multiple light

Table 4.1. Features of organic semiconductor optics for microanalysis

Controllable properties	The properties of the LEDs and photodetectors may be systematically controlled by chemically modifying the photoactive molecules in a way that is not possible with conventional technologies. It is therefore possible to optimize light sources and photodetectors for specific analytes under investigation.
Small instrumental footprint	The active layers of organic LEDs and photodiodes are typically only a few hundred nanometers in thickness and thus add minimal weight and size to the microfluidic substrate. In most situations, the devices must be encapsulated against exposure to air and water. This can be achieved by sandwiching the organic device between substrates coated with multiple impermeable barrier layers. The inclusion of such barrier coatings only slightly increases the overall thickness of the final device, which is still significantly lower than for alternative light source and detector technologies.
Disposable use and custom fabrication	The low cost nature of organic devices and microfluidic chips make the integrated devices ideal for single-use disposable purposes, e.g., *point-of-care* applications.
Multipoint detection	As it is straightforward to fabricate arrays of closely spaced LEDs and photodetectors, it is possible to monitor the analyte at multiple locations along the flow path simultaneously. For some applications individual detectors need to be placed within 500 μm of one another to provide the spatial resolution required for useful analysis. This rules out many alternative detectors, such as photomultiplier tubes, whose physical bulk render them unsuitable for multipoint detection. Consequently, most detection methods for microanalysis have to date been limited to single-point schemes.
Solution processing and printing	In the case of polymers and solution processable molecules, high quality films may be deposited directly from solution using simple low-cost techniques such as spin-coating. This represents a major advantage over conventional semiconductors, and allows the fabrication of high-quality devices at far lower cost than using alternative technologies. The key advantage associated with soluble organic materials is the ability to use high-efficiency printing techniques (such as planographic or inkjet printing) to fabricate arrays of closely spaced OLED/photodetector pairs at precise locations on the chip substrate. The complexity of the achievable patterns is in principle limited only by the resolution of the printer, and thus enables the deposition of intricate arrays of sub-mm OLED/photodetector pairs with high precision and at low cost.

sources and photodetectors into analytical microchips without appreciably increasing the size, weight, or cost of the final device. Importantly, from the perspective of large-scale manufacturing, semiconducting polymers (and certain solution processable small molecules) are amenable to printing in much the same way as conventional inks. This offers a route for massively reducing the fabrication costs of light sources and photodetectors, and so opens up the possibility of creating disposable diagnostic devices.

The efficiencies of organic LEDs and photodetectors compare favorably with those based on conventional inorganic materials. For instance, external quantum efficiencies (defined as the ratio of photons emitted to electrons injected) as high as 29% [17] and 17% [18] have been reported for small molecule and polymer OLEDs, respectively, compared with values of 38.9% for InGaN-based devices [19]. Incident photon-to-conducted electron (IPCE) peak efficiencies of virtually 100% have been reported for both Si and organic photodiodes [20]. Importantly for sensor applications, organic photodiodes have an excellent dynamic range, and have been shown to exhibit linearity over four to six decades of light intensity (Fig. 4.11) [21].

However, there are also some respects in which organic semiconductor devices are less favorable than their inorganic counterparts. For example, it is difficult to obtain light emission below 450 nm from OLEDs and most organic photodiodes are only sensitive to light of wavelengths less than about 650 nm. The "operating window" for sensors based on pairs of organic light sources and photodetectors is therefore relatively narrow. This problem is further compounded by the broad emission from most OLEDs, which tends to span a spectral range of several hundred nanometers. The tail of the OLED emission

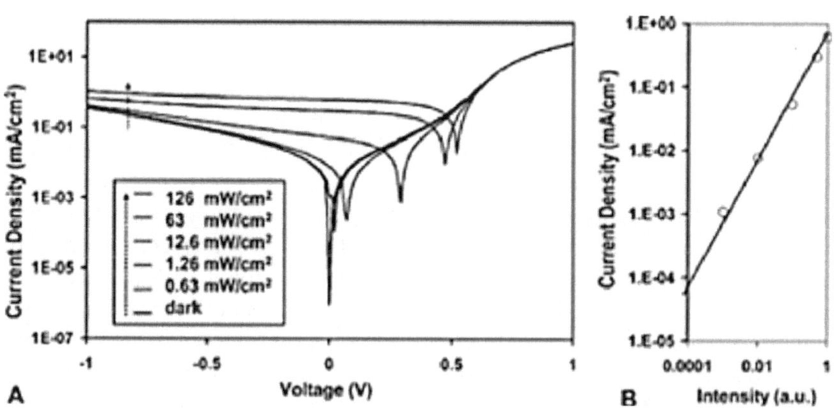

Fig. 4.11. (**a**) Current–voltage characteristics of P3HT:PCBM photodiodes under varying levels of 633 nm monochromatic illumination. See main text on pages 118 and 119 for further details about the devices. (**b**) The intensity dependence of the short-circuit photocurrent for the same device (From [21] – Reproduced by permission of The Royal Society of Chemistry)

therefore may overlap with the weaker emission from the analyte. In addition, because of the thin nature of the active layer, organic LEDs and photodiodes typically have large capacitances and so respond relatively slowly to changes in voltage or light levels, which precludes their use in certain time-resolved applications. Solutions exist to most of these issues but they introduce additional complexity into device fabrication or result in compromises elsewhere in device performance. The successful use of organic light sources and photodetectors as sensing elements is largely a question of engineering, in which one aspires to achieve sufficient detection sensitivity without making undue compromises elsewhere in the design process, e.g., cost, portability, power consumption, etc. In the following sections, we describe a number of ways in which organic light sources and photodiodes can be effectively integrated with microfluidic devices to create fully integrated self-contained sensors.

4.2.3 Towards Mass Manufacture

As noted above, semiconducting polymers (and certain solution processable small molecules) are amenable to printing in much the same way as conventional inks. There are two broad classes of printing techniques – contact methods in which the ink is physically pressed onto the surface and inkjet techniques in which the ink is squirted towards the surface. Inkjet techniques are currently the more advanced, and prototype inkjet-printed color displays as large as 40″ have been reported by companies including Sony and Samsung. Increasingly, however, attention is turning to contact-based methods such as gravure, flexographic, and offset printing, which are well suited to long print runs and historically have been applied to "low-tech" processes such as printing rolls of wallpaper. In gravure printing, for example, a recessed image is etched onto the surface of a rotating metal drum and the etched regions are filled with ink; excess ink is removed by a doctor blade and the remaining ink is transferred to a flexible film as it passes between the drum and a second roller (see Fig. 4.12). One of the key advantages of using printing techniques is the ability to switch from batch processing to *reel-to-reel* processing. The former approach suffers from considerable down-time due to the need to repeatedly load, set, and unload individual substrates whereas *reel-to-reel* processing is a continuous process that maximizes throughput. To exploit *reel-to-reel* processing it is necessary to use flexible substrate materials. This creates a number of difficulties from a technological perspective since indium tin oxide (the dominant transparent anode material for OLEDs) is typically deposited by thermal evaporation or sputtering at temperatures in excess of 350°C and, even when deposited at lower temperature, generally needs to be annealed in air or oxygen at those temperatures in order to achieve a suitably high conductivity. These issues, together with the tendency of ITO to crack when the substrate is flexed, have led researchers to seek alternative anode materials for flexible substrate applications (e.g., polymer–metal and polymer–fullerene composites and conducting polymers). Unfortunately, these materials have so far proved

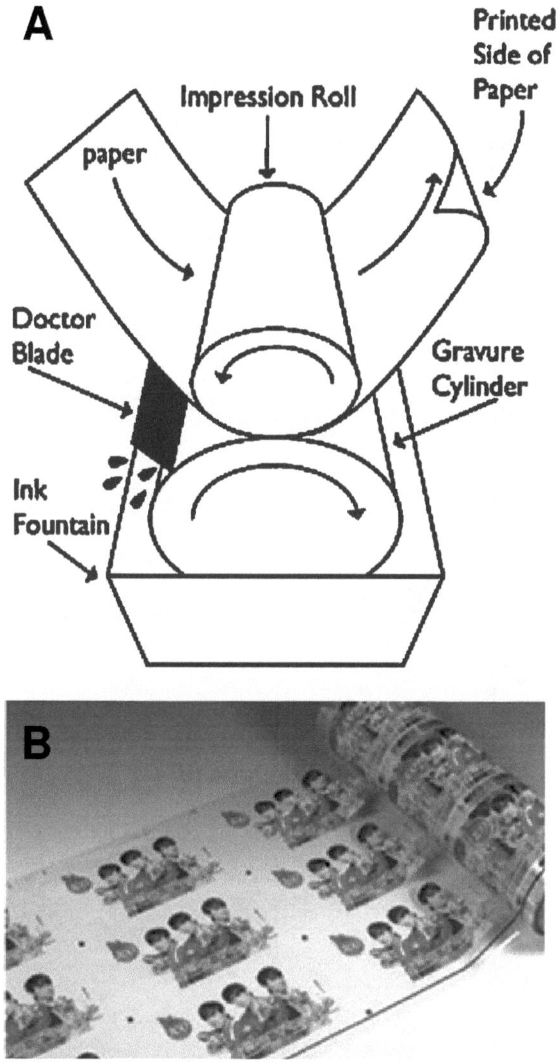

Fig. 4.12. (**a**) Gravure printing concept. (**b**) Images printed onto a flexible substrate (Pictures courtesy of Wikepedia)

inadequate for most display and lighting applications due to the difficulty of achieving a high conductivity without compromising transparency. The high sheet resistance of the resultant electrodes leads to excessive power consumption and significant heat generation, which accelerates device degradation, and considerable development is still needed before these materials are viable as electrode materials for display, lighting, and energy applications.

In the context of single-use sensors, where the devices are typically operated for just a few seconds or minutes before disposal, power consumption and extended operating stabilities are less of an issue, and in these circumstances it is possible to dispense with ITO altogether, using instead a simple layer of a conducting polymer as the anode material. The most widely used conducting polymer for device applications is poly(3,4-dioxythiophene):polystyrenesulphonate (PEDOT:PSS) due to its reasonable transparency and conductivity and its good chemical stability in the doped state [22]. PEDOT:PSS has a much higher resistivity than ITO (at least ten times higher), but it is nevertheless possible to make light sources and photodetectors using

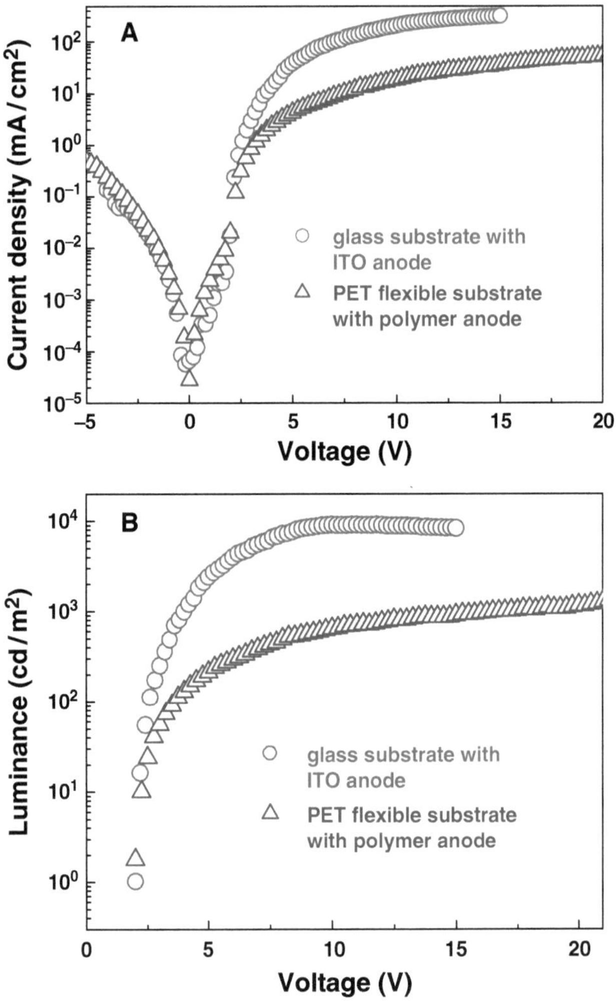

Fig. 4.13. Comparison of pLED on conventional ITO-coated glass substrate with pLED on ITO-free PET substrate. (**a**) Current density–voltage characteristics. (**b**) Luminance–voltage characteristics

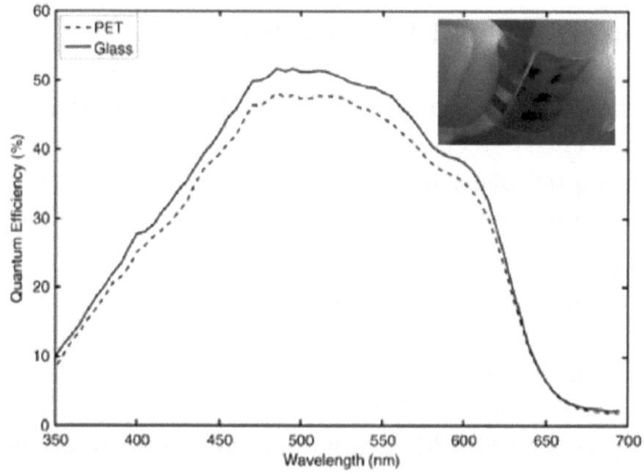

Fig. 4.14. Current–density characteristics of P3HT:PCBM photodetectors on ITO-free PET substrate. The Inset shows an image of the flexible photodetector (From [23] – Reproduced by permission of The Royal Society of Chemistry)

PEDOT:PSS anodes that are quite adequate for many sensor applications [23, 24]. In Fig. 4.13 we compare the current–voltage-luminance characteristics of a conventional LED on rigid ITO-coated glass with those of an otherwise identical device on flexible ITO-free polyethyleneterephalate (PET) coated with a PEDOT:PSS anode. The ITO free device compares reasonably well with the conventional device at moderate brightness ($<500\,\mathrm{cd\,m^{-2}}$), which is adequate for many diagnostic applications. In Fig. 4.14 we compare the spectral response of a conventional photodiode on rigid ITO-coated glass with that of an identical device on flexible ITO-free PET under (diagnostically relevant) low light levels. The quantum efficiency is virtually the same for the two devices, indicating that high quality organic photodiodes can be made on flexible substrates. It should be noted that the low conductivity of the PEDOT:PSS anode does not adversely affect the efficiency of the photodiode since the currents generated at low light levels are extremely small. The power dissipation in the anode is therefore also small and has a negligible influence on the overall efficiency of the device. At high illumination levels, as for example are incurred under solar illumination, the low conductivity of the PEDOT:PSS significantly reduces efficiencies compared with conventional ITO-containing devices. Hence, although the use of conducting polymer anodes is already more than sufficient for the diagnostic applications described in this chapter, it will be some years before they are suitable for solar energy applications.

In summary, the combination of flexible light sources and photodetectors with microfluidic chips fabricated from elastomers such as polydimethylsiloxane can enable the fabrication of low-cost all-plastic sensors that are ideally suited to low-cost *point-of-care* applications.

4.3 Functional Optical Components

In this section we describe in more detail the three key optical components for integrated fluorescence detection: (1) light sources, (2) detectors, and (3) optical filters. We also present proof-of-principle studies from our group describing the use of these components in chemical and biological analysis.

4.3.1 OLED Light Sources for Microchip Analysis

For microchip analysis, optical detection is often preferred due to the versatility and high sensitivities afforded by this technique. Although in the laboratory sophisticated benchtop mounted optics are commonly used for on-chip optical detection, this is clearly not an option for portable diagnostic applications. There exists a clear need for small low-cost optical components to replace the benchtop counterparts such as lasers, photomultiplier tubes (PMTs), or even sophisticated fluorescence microscopes. On the light source side organic light emitting diodes (OLEDs) present a viable alternative that can be integrated onto the microchip platform without significantly increasing the system complexity or power requirements. To this end we have pioneered the use of polymer light emitting diodes (pLED) as excitation sources for fluorescence detection in on-chip electrophoretic analysis (Fig. 4.15) and have also demonstrated their successful use in a diagnostic assay for kidney disease.

In our first study [25], we sought to assess whether polymer light-emitting diodes could be used as a "plug-in" replacement for mercury lamp excitation in chip-based capillary electrophoresis systems. The microfluidic device was fabricated in glass using standard lithographic techniques, and is shown

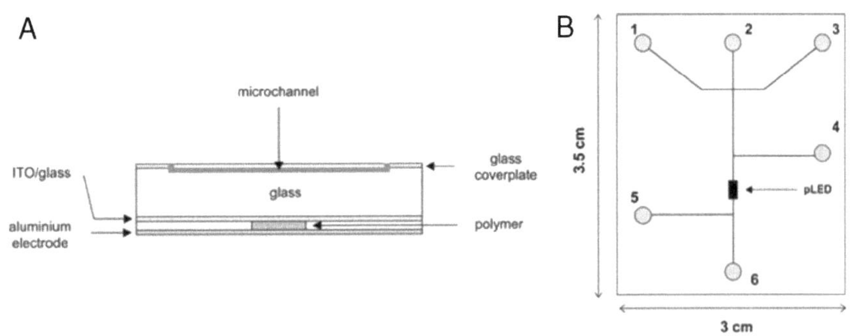

Fig. 4.15. (**a**) Side view (not to scale) of pLED integrated with a planar glass CE microdevice. (**b**) Layout of the CE microdevice comprising an injection channel and a variable length separation channel: 1, sample inlet; 2, buffer inlet; 3, sample outlet; 4–6, buffer inlets (4 and 5 unused). Microchannels are 50 μm wide and 40 μm deep. The pLED (active area: $40 \times 1{,}000\,\mu m^2$) is positioned below the separation channel between outlets 4 and 5 (From [25] – Reproduced by permission of The Royal Society of Chemistry)

schematically in Fig. 4.15a, b. Prior to the measurements, the buffer channel between ports 2 and 6 was prefilled with buffer solution. The injection channel was then filled with fluorescein solution at the sample inlet (1) by applying vacuum to the sample outlet (3), thereby displacing the buffer solution at the microchannel intersection. The sample volume in the intersection (~100 pL) was then injected into the separation channel by application of an electric field (~1 kV cm^{-1}) between the buffer inlet (2) and outlet reservoir (6). The arrival of the fluorophores at the pLED excitation zone was monitored using microscope optics and an external silicon diode. The pLED used for this work was a blue-emitting device based on a polyfluorene copolymer with an emission spectrum that matches well with the absorption spectrum of fluorescein (Fig. 4.16). The device was fabricated by sequentially depositing poly(3,4-ethylenedioxy-thiophene):polystyrene-sulfonate [PEDOT:PSS], 70 nm of the active layer, and a LiF/Al cathode onto an ITO-coated glass substrate. In use, signal-to-noise (S/N) ratios as high as 800 were obtained when the pLED was driven at 7 V, and separations of a variety of dye molecules including fluorescein and 5-carboxyfluorescein could easily be detected with mass detection limits of 50 fmol. The separation chip was then retested using a standard Hg lamp as the excitation source and broadly comparable results were obtained, indicating that the pLEDs can indeed be used as plug-in replacements for conventional light-sources. These studies demonstrated the first application of

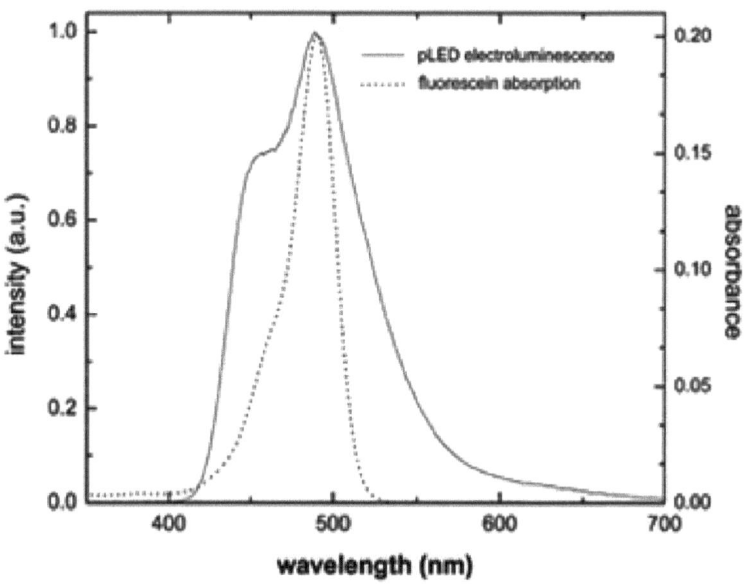

Fig. 4.16. Absorption and normalized electroluminescence emission spectra of 10 μM fluorescein and the polyfluorene pLED, respectively (From [25] – Reproduced by permission of The Royal Society of Chemistry)

thin-film polymer LEDs as integrated light sources for chemical analysis, and moreover afforded sensitivity and detection limits comparable to conventional mercury lamp excitation.

The variation of the total signal relative to the pLED signal and S/N as a function of the pLED applied bias is shown in Fig. 4.17a. It can be seen that at low drive voltages (4–5.5 V) an approximate linear increase in the ratio of total signal-to-pLED emission is observed. This ratio reaches a limiting value of ~8.5 for drive voltages above 7 V. It is interesting to note that the variation of S/N does not follow the same trend; in fact a continuous increase in S/N is observed as drive voltage is increased. We attribute this trend to the significantly lower noise levels encountered at higher pLED drive voltages. While these initial calibration studies clearly indicate that for sensitive detection high pLED drive voltages are beneficial, this can also severely compromise output stability and reduce the device lifetime. Consequently, for analytical applications we decided on a compromise driving voltage of 5.5 V.

Subsequent experiments were aimed at determining the sensitivity and linear range of our pLED based detection system. Figure 4.17b shows a calibration plot for fluorescein for a pLED driving voltage of 5.5 V. A detection limit of ~1 μM was established with a linear range spanning over four decades. These studies represent the first demonstration that pLEDs present a viable and low-cost alternative light source for high-sensitivity on-chip detection. In more recent studies we have also successfully used a poly(*p*-phenylene vinylene) (PPV) derivative OLED light source to monitor a diagnostically relevant fluorescence binding assay for renal disease. This work will be described in more detail in Sect. 4.4.1.

Current research efforts are focused on increasing the light output without compromising the lifetime of OLEDs. Further work is directed at extending the range of available OLED light sources to the red part of the light spectrum and to narrow the emission bandwidth, e.g., by using microcavity architectures.

4.3.2 Organic Photodetectors for Chemiluminescence Assays

One of the preferred methods for analyte detection in microfluidic devices is based on the phenomenon of chemiluminescence (CL), which offers a simple but sensitive means of monitoring low level analyte concentrations [26]. CL reactions typically involve the formation of a metastable reaction intermediate or product in an electronically excited state, which subsequently relaxes to the ground state with the emission of a photon. CL is particularly attractive for portable microfluidic assays, because the CL reaction acts as an internal light-source, thereby lowering instrumental requirements and significantly reducing power consumption and background interference compared to fluorescence assays. CL-based systems have been successfully applied to on-chip electrophoretic separation of metal ions, immunoassays, and enzyme assays [27], and consequently there is considerable interest in

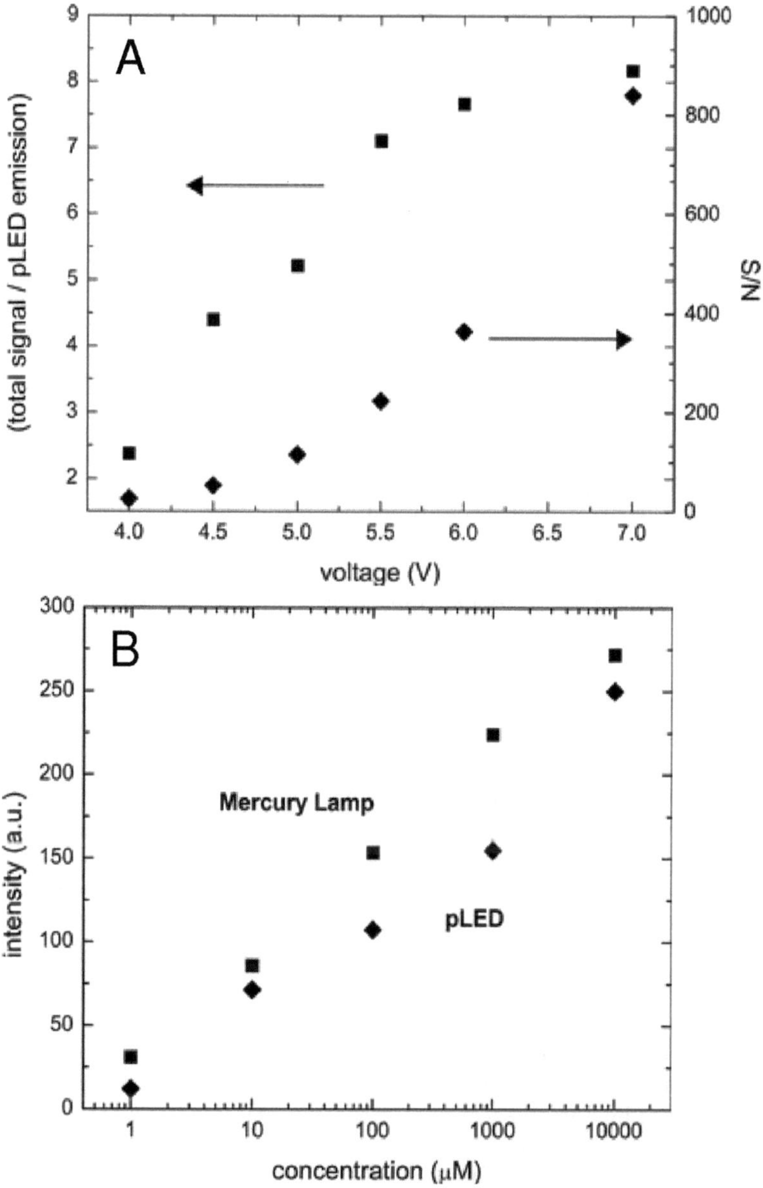

Fig. 4.17. (**a**) Variation of total signal to pLED emission ratio and signal-to-noise ratio as a function of pLED applied bias. (**b**) Calibration plot for fluorescein with on-chip mercury arc lamp and polyfluorene pLED excitation (drive voltage 5.5 V) (From [25] – Reproduced by permission of The Royal Society of Chemistry)

creating complete analytical devices that incorporate a CL assay and optical detector in a single integrated package.

In early reports of CL detection in microfluidic environments, the CL signal was generally detected and quantified using externally mounted photomultiplier tubes (PMTs) and/or microscope collection optics. Recently, however, Jorgensen et al. reported the use of integrated silicon photodiodes for monitoring peroxyoxolate-based chemiluminescence (PO-CL) reactions in micromachined silicon microfluidic chips [28]. They selected hydrogen peroxide as a model compound for quantitation, since it is produced by a number of enzymes in the presence of dissolved oxygen and certain analytes such as alcohol, glucose, and cholesterol. Using this approach they were able to attain measurable signals down to 10 µM, thus showing that high sensitivity chip-based CL detection could be implemented in a fully integrated microscale format.

The use of silicon photodiodes and micromachined silicon substrates, however, entails relatively expensive fabrication techniques that preclude the use of such devices in disposable *point-of-care* applications where low cost is of primary concern. In recent work, therefore, we investigated whether integrated microscale CL could be implemented in a lower cost format using polydimethylsiloxane (PDMS) instead of silicon as the substrate material and organic photodiodes in place of the Si detectors [28]. As previously discussed, PDMS has good biocompatibility and optical transparency over the visible range, and allows for rapid prototyping and scalable manufacturing at low cost and with high reliability. At the same time, organic devices may be fabricated at low temperature using simple layer-by-layer deposition procedures that are fully compatible with plastic substrates. The combination of PDMS microfluidic devices with organic photodiodes therefore offers an attractive route to fabricate low cost diagnostic devices, which incorporate the fluidic networks and the detectors in a single monolithic package.

In our first proof-of-principle studies, we used organic photodiodes based on vacuum-deposited bilayers of copper phthalocyanine (CuPc) and fullerene (C_{60}) to detect the emission signal from a PO-CL assay [29]. These measurements confirmed the feasibility of using organic devices for detection of the CL signal, but yielded relatively poor detection limits of 1 mM compared to 10 µM reported by Jorgensen et al. The low detection limits were attributable in part to a significant mismatch between the photodiode area ($\sim$16 mm^2) and the detection zone on the microfluidic chip ($\sim$2 mm^2), which resulted in a high background signal ($\sim$1 nA) due to the high photodiode dark current.

To improve the limit-of-detection, we replaced the 16 mm^2 vacuum deposited CuPc/C_{60} bilayer devices with 1 mm^2 solution-processed polymer devices based on 1:1 blends by weight of poly(3-hexylthiophene) [P3HT] and [6, 6]-phenyl-C61-butyric acid-methylester [PCBM] – a soluble derivative of C_{60}. The P3HT:PCBM devices fabricated in our laboratory typically have very low short-circuit dark current densities of $<10^{-6}$ mA cm^{-2}, and are consequently a good choice for high sensitivity detection [21]. The

$1 \times 1\,mm^2$ dimensions of the P3HT:PCBM devices were well matched to the $800\,\mu m \times 1\,mm$ detection zone of the microfluidic chip and thus minimized the background signal due to the dark current. The thin-film polymer photodiodes, when integrated with PDMS microfluidic chips, provide compact, sensitive, and potentially low-cost microscale CL devices with wide-ranging applications in chemical and biological analysis and clinical diagnostics.

The microscale CL devices consist of two parts: a Y-type micromixer fabricated in PDMS and a P3HT:PCBM-based photodiode supported on a glass substrate (Fig. 4.18). The channels of the micromixer were formed in a 2 mm layer of PDMS and were sealed with a 1 mm "lid" of unstructured PDMS. Finally, fluidic access holes were punched in the 2 mm PDMS layer to enable injection and extraction of the CL reagents. The depths of all channels in the micromixer were 800 µm, while the respective widths of the inlet and mixing channels were 400 and 800 µm. The inlet channels were 1 cm long and the length of the mixing channel was 5.2 cm (Fig. 4.18a). The organic photodiode comprised a 1 mm ITO-coated glass substrate that was successively coated with (1) a 60 nm layer of poly(3,4ethylenedioxythiophene):polystyrene-sulphonate (PEDOT:PSS) [Baytron P AI 4083]; (2) a 150 nm layer of 1:1 by weight regioregular poly(3-hexylthiophene) (P3HT) and 1-(3-methoxycarbonyl)-propyl-1-phenyl-(6,6)C61 (PCBM); and (3) a 200 nm Al cathode. The $1 \times 1\,mm^2$ active area of the pixel was defined by the spatial overlap of the ITO anode and the Al cathode. The entire device was encapsulated in an inert nitrogen atmosphere by securing a metal can (fitted with a desiccant patch) to the coated side of the glass substrate with a UV-cured adhesive.

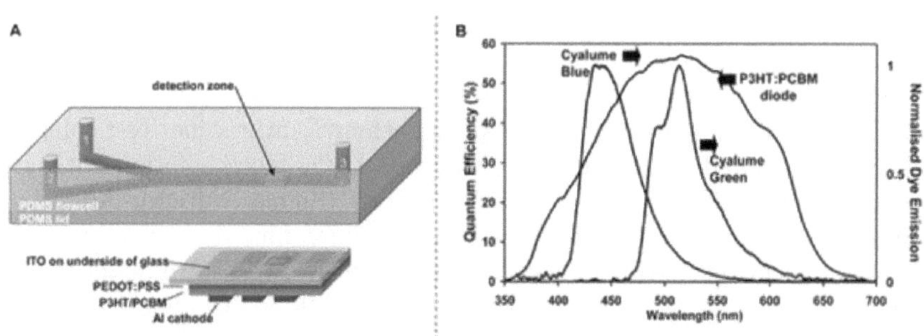

Fig. 4.18. (**a**) Schematic of PDMS microchip used for on-chip chemiluminescence assay. Reagents are loaded into inlets 1 and 2 and are hydrodynamically pumped towards outlet three. Light is generated when the two reagents mix. For detection a P3HT:PCBM photodetector is placed underneath the detection chamber of the microchip. (**b**) Responsivity of the organic photodetector and emission spectra of the chemiluminescence dyes (From [21] – Reproduced by permission of The Royal Society of Chemistry)

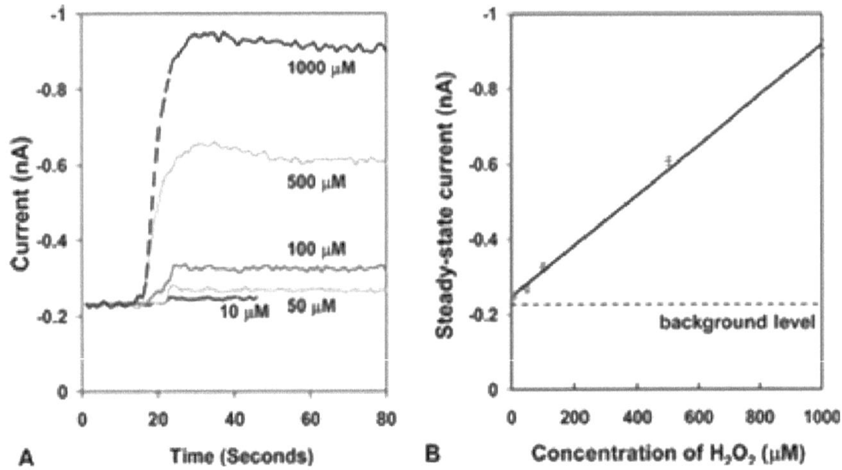

Fig. 4.19. (**a**) Transient chemiluminescence signal as recorded with P3HT:PCBM photodetector for different hydrogen peroxide concentrations (oxidation reagent). (**b**) Calibration plot for steady-state signal (From [21] – Reproduced by permission of The Royal Society of Chemistry)

The integrated CL device was completed by attaching the lid of the PDMS microchip to the glass side of the thin-film organic photodiode to form an integrated device in which the organic pixel was aligned with a detection zone on the fluidic chip and located 1 cm downstream of the point-of-confluence of the two inlet streams (see Fig. 4.18a). To perform the on-chip CL measurements, a stock solution (A) of CL reagent was first prepared by extracting the PO-CL reagents from Cyalume green light sticks (Omniglow, West Springfield, MA), which contain the active ingredients bis(2-carbopentyloxy-3,5,6-trichlorophenyl) oxalate (CPPO), dimethylaminopyridine (DMAP) catalyst, and the green dye molecule 9,10-bis(phenylethynyl)anthracene. Test solutions (B) were prepared by diluting 31% aqueous H_2O_2 stock solution with acetonitrile and adding 5 mM imidazole catalyst (Sigma-Aldrich). Additional measurements were undertaken using the Cyalume blue dye, which contains the dye molecule 9,10-diphenylanthracene. The two dyes have quantum efficiencies close to unity (0.91 ± 0.08 and 0.85 ± 0.03 for Cyalume green and Cyalume blue, respectively) and so are good choices for high sensitivity CL assays. The chemiluminescence spectra for the blue and green luminescent dyes are shown by the grey lines in Fig. 4.18b, and are essentially identical to the corresponding photoluminescence spectra for the dyes (not shown) in accordance with the indirect nature of the PO-CL emission mechanism. The emission spectra of the two dyes match reasonably well with the response of the photodiodes.

To initiate the CL reaction, the dye/catalyst mixture and the H_2O_2 were pumped hydrodynamically into inlets 1 and 2, respectively and, after waiting an appropriate time for the signal to settle, detection was performed 1 cm downstream of the point-confluence. The steady-state CL signal is plotted against hydrogen peroxide concentration in Fig. 4.19, and excellent linearity is obtained in the range 10 μM to 1 mM. The detection limit for the integrated chip/detector was <10 μM, which represents a 100-fold improvement compared to the data obtained using the CuPc/C_{60} devices described previously. Significantly, the sensitivity of the integrated CL devices described here is equivalent to that of the devices reported by Jorgensen et al. using integrated silicon photodiodes.

The limit-of-detection for our devices is determined primarily by the background signal (dark current) of the organic photodiodes, and current studies are aimed at reducing the dark current through improved device fabrication procedures. Changes to the CL chemistry may also enable appreciable increases in emission intensity since the Cyalume dyes used are primarily optimized for emission longevity and, in the current context, a short-lived high-intensity emission is preferable. We anticipate that taken together these changes may enable a further 100- to 1,000-fold improvement in detection limits. The current 10 μM limit-of-detection is already sufficient for many diagnostic applications, including the determination of alcohol, glucose, and cholesterol levels in blood. Further improvements to the 100 nM level will provide sufficient sensitivity for applications such as low level cancer marker detection and pharmacokinetics.

Finally, we note that these initial studies were conducted in a lab environment using standard instrumentation and equipment that is ill-suited to miniaturized low-cost applications. In ongoing studies, we are evaluating the use of on-chip reagent storage methods and passive capillarity-based fluid delivery schemes that remove the need for external fluid motivation. Such devices would offer a powerful low cost solution for chemical and biological analysis with potentially wide-ranging applications for *in-the-field* analysis and *point-of-care* diagnostics.

4.3.3 Optical Filters for Head-On Fluorescence Detection

In conventional fluorescence detection, the excitation source and detector are usually arranged orthogonally to one another to prevent direct illumination of the detector by the excitation source. This orthogonal geometry, however, is difficult to implement in a microfluidic environment since it requires the integration of optical components onto the side-surfaces of the microfluidic chip. Light sources and detectors are most conveniently located on the upper and lower faces of the microfluidic chip in a colinear geometry, but this often leads to detector saturation, with direct light from the excitation source masking the weaker fluorescence from the analyte. The key to achieving effective discrimination of the excitation and emission light in this "head-on" configuration is the use of a long-pass filter in front of the detector to block the excitation light and transmit the longer wavelength emission signal. The use of long-pass filters for this purpose is well established but has generally involved the use of discrete stand-alone filters – an approach that yields satisfactory optical performance but prevents monolithic integration and leads to inefficient collection of the fluorescence signal.

In this respect, it is preferable to use optical filters that are monolithically integrated with both the photodetector and the microfluidic chip substrate. Surprisingly, only a few monolithic approaches have been reported in the literature. One successful example was reported by Burns et al. who described the use of integrated filters in silicon-based microfluidic devices. In their work multilayer interference filters were fabricated on top of PIN silicon photodiodes [30]. The interference filters typically comprised up to 40 alternating layers of SiO_2/TiO_2 with ~5%-transmittance at 490 nm and ~90%-transmittance at 510 nm, and resulted in significant performance gains when incorporated into DNA analysis chips. An alternative lower cost approach was recently reported by Chediak et al., who developed integrated color filters for silicon microfluidic devices using thin layers of cadmium sulphide deposited directly on top of PIN silicon photodiodes [31]. The CdS filters exhibited strong blocking of the excitation light but relatively low transmission of the emission light (~40%). In addition thin-film CdS is known to exhibit appreciable fluorescence so, although not discussed by the authors, it is likely that these filters would exhibit significant autofluorescence – a serious issue for on-chip detection as discussed later.

The interference-filters and CdS-filters described above could in principle be straightforwardly integrated with glass microfluidic chips, but they are unsuitable for conformable elastomeric materials such as PDMS – a preferred substrate material for low-cost disposable applications – since polycrystalline materials such as CdS, TiO_2, and SiO_2 are typically deposited at relatively high temperatures (>300°C) and have a tendency to crack when the substrate is flexed.

All the above approaches also introduce additional steps of varying complexity into the fabrication process and, from a technological perspective, a preferable solution would be to use the microfluidic substrate itself as the color filter. The simplest way to implement this is to disperse appropriate dye molecules into the substrate material (plastic or glass) prior to chip fabrication. The colored substrate thereby obtained is able to serve concurrently as the microfluidic medium and optical filter, negating the need for an additional filter layer and allowing for improved collection of the fluorescence since the detector can be placed in closer proximity to the channel (Fig. 4.20). The dispersal of dye molecules in polymer matrices to produce color filters is well established, but the successful utilization of this approach to produce a dual-functioning microfluidic chip and color-filter presents a number of challenges: (1) the colinear detection geometry and typically weak intensity of the analyte emission necessitate a sharp filter cut-on with excellent blocking and transmission on either side; (2) because of the close proximity of the filter and photodetector, filter autofluorescence must be negligible

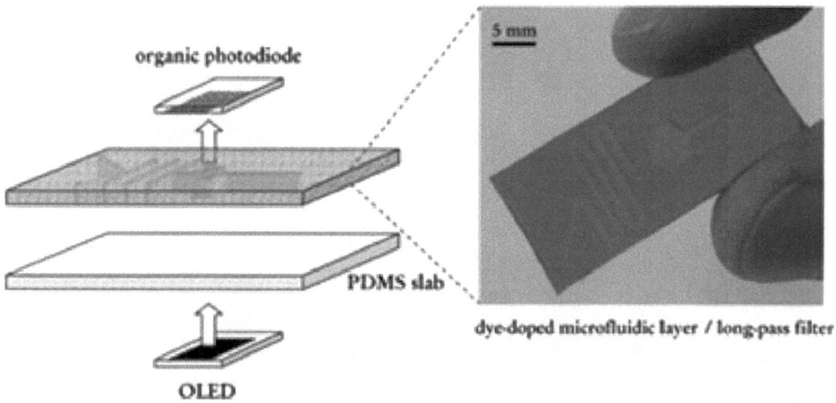

Fig. 4.20. Illustration of monolithically integrated optical filter concept. The dye-doped PDMS substrate serves concurrently as microfluidic conduit and optical longpass filter (From [32] – Reproduced by permission of The Royal Society of Chemistry)

since it is liable to raise the background signal from the photodiode and thus mask the analyte signal; (3) the incorporation of dye molecules into the host matrix must conserve the processability of the polymer, enabling high quality microfluidic chips to be fabricated from the dye-doped polymer; and (4) the resultant colored substrates must be stable (on the timescale of the experiment) against dye leaching due to solvent flow through the channel and photodegradation. The fabrication of high quality integrated microchip filters that retain the excellent processability of standard undoped PDMS but have comparable optical characteristics to state-of-the-art commercial filters is therefore a genuine challenge. Fortunately, we have found that the dispersion of lysochrome dyes into PDMS substrates offers a particularly effective solution (Fig. 4.21). In our method lysochrome dyes are dissolved in a small volume of apolar solvent and added to the PDMS monomer prior to polymerization over a master comprising the microfluidic layout. The resultant filters have excellent optical characteristics, e.g., red long-pass filters with $<0.01\%$ transmission at 500 nm and $>80\%$ transmission above 570 nm, which compares favorably with commercially available Schott glass filters. Importantly, such filters showed negligible autofluorescence, allowing them to be effectively employed in microchip-based fluorescence detection [32]. The filters proved robust in use, undergoing negligible leaching in aqueous solution and only marginal photodegradation under prolonged exposure to UV light. Patterning of the PDMS was unaffected by the dye doping, allowing for the fabrication of colored substrates that serve concurrently as channel medium and optical filter. In initial work, we were primarily interested in developing a range of long-pass filters. However, for diagnostic applications involving organic light-emitting diodes, short-pass filters are also of considerable importance. This is because organic semiconductors are broad-band fluorophores that typically

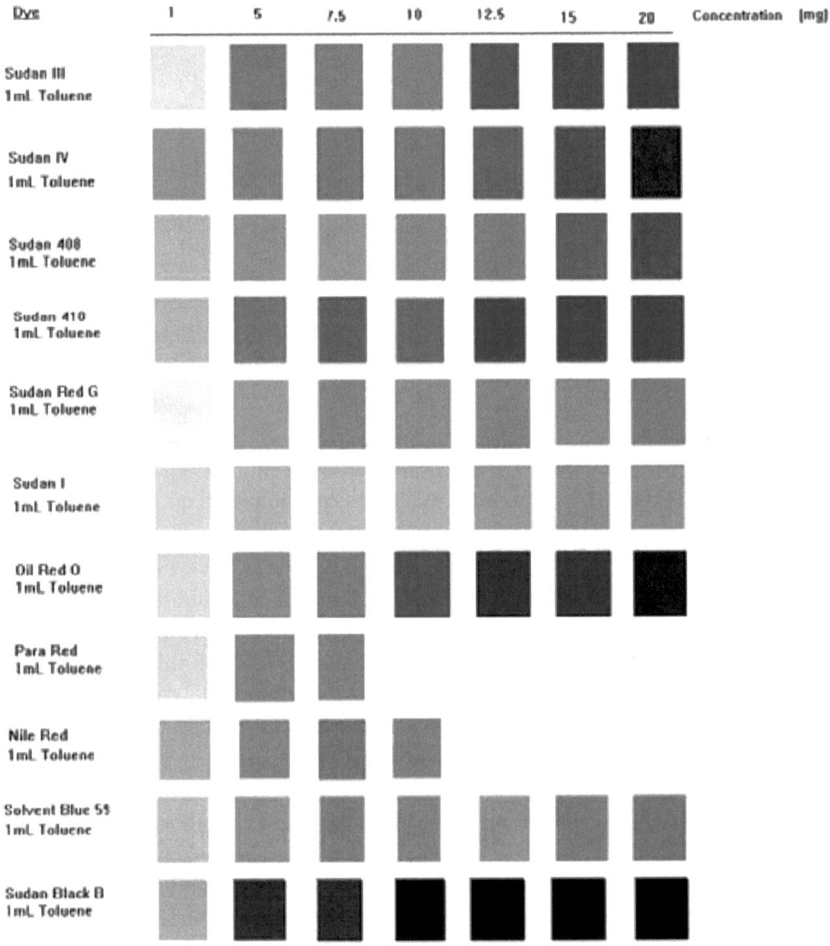

Fig. 4.21. Overview of available shortpass, longpass, and blocking filter based on dye-doping of the PDMS microfluidic substrate

have a long emission tail that extends several hundred of nanometers beyond the peak emission wavelength. This long tail tends to overlap and mask the analyte emission. To avoid this problem, short-pass filters are required between the LED and the channel to sharpen the emission from the LEDs and allow analyte emission to be detected. We have developed a broad range of short and long-pass filters for diagnostic applications, a selection of which can be seen in Fig. 4.21.

4.4 Applications

In this section we describe how the functional components described above have been applied to real-world analytical applications.

4.4.1 Microalbuminuria Determination On-Chip

As a first application for our integrated microfluidic platforms with organic semiconductor-based components we investigated the determination of microalbuminuria (MAU) in urine. MAU is defined as an increased urinary excretion of human serum albumin (HSA), which is indicative of renal problems in diabetic patients [33]. The American Diabetes Association thus recommends an annual MAU test for all diabetic patients to monitor the success of the diabetes treatment. Furthermore, MAU has recently been recognized as a potential risk factor for cardiac disease and is also known as an effective marker for battlefield trauma associated with blast damage. When MAU is detected at an early stage reno-protective and antihypertensive treatment can be applied to prolong patient's lives. Although semi-quantitative MAU dipstick tests are available for *point-of-care* use, unambiguous diagnosis can only be achieved through subsequent lab-based quantitative tests with sophisticated readers. Accordingly, the aim of our research was to enable quantitative MAU analysis at the *point-of-care* by developing a microfluidic chip-based analysis platform (with integrated OLED based detection) that could be used in the doctor's surgery or for home-testing by the patient itself.

An effective fluorescence binding assay originally developed by Kessler and coworkers was selected as a suitable assay format [34]. The assay is based on the Albumin Blue 580 (AB580) dye, which binds specifically to HSA, resulting in an enhancement in the fluorescence emission of two orders of magnitude. From a microfluidic point of view, MAU determination in urine is inherently simple in that no sample pretreatment is required (unlike blood-based diagnostics that often require sample filtration), with only two components being mixed during the assay (the albumin containing sample and AB580). From a detection point of view, conditions are also favorable in that there is a low background signal (AB580 on its own is only very weakly fluorescent) while the emission from the AB580-HSA complex is comparatively strong. Furthermore, the clinically relevant cut-off concentration for MAU of 15–40 mg L^{-1} HSA is high compared to other diagnostic markers such as Troponin, Myoglobin, and CK-MB in blood (typical cut-off limits <10 µg L^{-1}).

To achieve effective mixing of the sample and probe solutions we used a simple PDMS microfluidic device [35]. The device comprises two inlets, a meandering mixing channel, a widened detection chamber, and an outlet. The inlets were 400 µm wide, 800 µm deep, and 1 cm long, while the mixing channel was 800 µm wide, 800 µm deep, and 5.2 cm long (Fig. 4.22). The extended trapezoidal detection chamber was 5 mm long, 5 mm wide, and 1.6 mm deep. The total internal volume of the microchip was ∼27 µL. For initial testing we used standard syringe pumps to drive the HSA and AB580 solutions through the microchannel network. In our latest devices, however, as described in Sect. 4.4.3 we are using passive filling schemes. Optical excitation in the detection chamber was achieved through the use of a yellow thin-film OLED based on a light emitting poly(*p*-phenylene vinylene) (PPV) derivative. The device

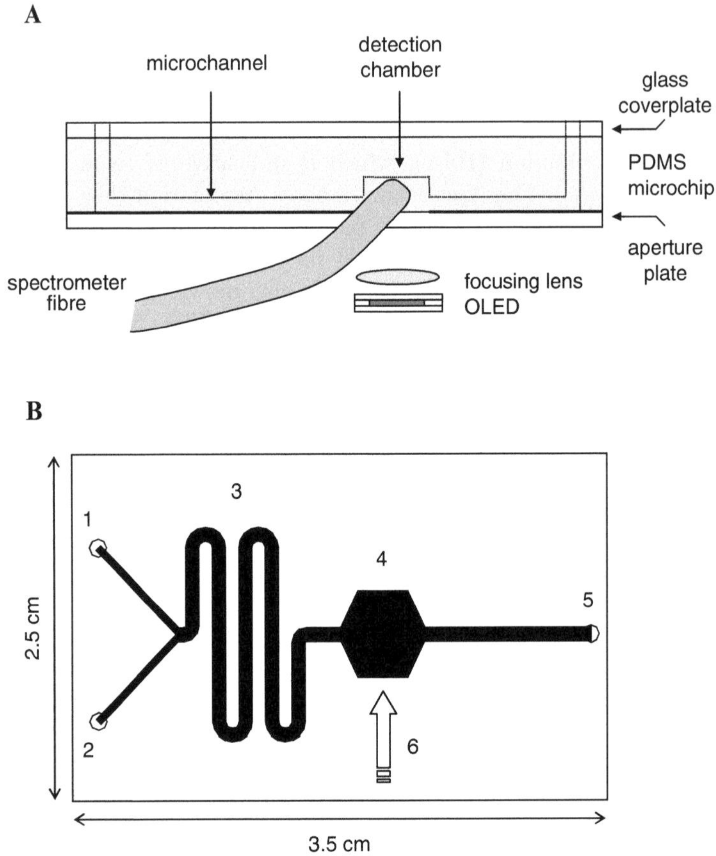

Fig. 4.22. (**a**) Schematic of experimental set-up for microalbuminuria determination. (**b**) Microchip layout with inlets for AB 580 (1) and HSA (2), mixing channel (3), detection chamber (4), outlet (5), and orthogonal detection point (6) (From [35] – Reproduced by permission of The Royal Society of Chemistry)

structure comprises a patterned indium tin oxide (ITO)-coated glass substrate, a hole-injecting layer of poly(3,4-ethlenedioxythiophene):polystyrene-sulfonate (PEDOT:PSS), the active PPV emission layer, and a thermally evaporated cathode layer with a lithium fluoride layer capped with aluminum. A metal encapsulation can comprising a desiccant patch was sealed against the glass substrate using epoxy adhesive.

Figure 4.23 shows the yellow OLED emission that spans from 500–700 nm with a maximum at 540 nm. As can be seen, the emission overlaps appreciably with the absorption spectrum of the HSA/AB580 complex rendering the OLED an efficient excitation source for this application. Emission from the HSA/AB580 complex spanned from ∼580–700 nm with a maximum at 610 nm. In a standard collinear "head-on" detection geometry with OLED

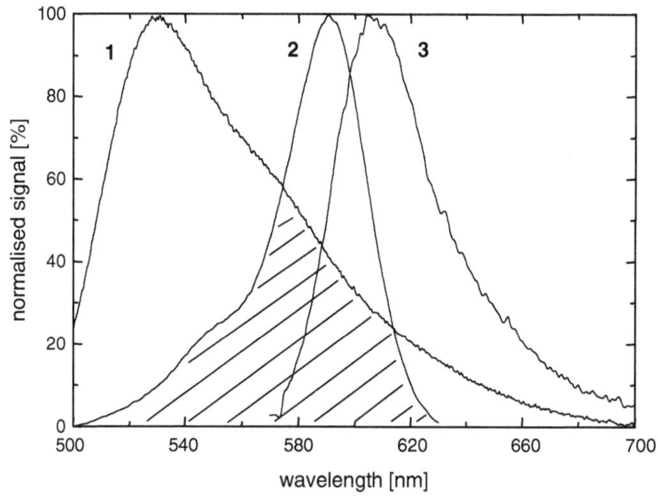

Fig. 4.23. OLED emission spectrum (1) and overlap with excitation (2) and emission spectra (3) of 1.2 μM AB 580 dye after addition of 100 mg L^{-1} HSA (From [35] – Reproduced by permission of The Royal Society of Chemistry)

and detector on either side of the microchannel, this overlap of complex emission with the tail of the OLED emission would necessitate optical filtering (e.g., the use of a short-pass filter on the light source side to cut off the tail of the OLED emission). The work described here predated our development of integrated short-pass and long-pass filters as described in Sect. 4.3.3. Hence, to circumvent this requirement we implemented an orthogonal detection geometry with the OLED light source positioned below the microfluidic chip and detection performed from the side, thereby minimizing the amount of OLED light hitting the detector.

To excite the HSA/AB580 complex on-chip, the yellow OLED was typically driven at 20–30 mA (7–8 V bias), yielding a brightness of 8,000–10,000 cd m^{-2}. To detect the complex emission a CCD spectrometer fiber was brought in close contact with the side surface of the microchip adjacent to the detection chamber. Initial results revealed an unusually high amount of detected excitation light, presumably due to light scatter of the diverging OLED light at the sidewalls of the detection chamber. The OLED contribution was greatly reduced through the use of an aperture between the OLED and the microchip, with the aperture size matched to the detection chamber. Typical complex spectra for HSA concentrations in the range 1–1,000 mg L^{-1} are depicted in Fig. 4.24a. To compensate for potential OLED intensity variations, the spectra were normalized to the 550 nm light source band. A general increase of the HSA/AB580 emission band at 620 nm can be observed for HSA concentrations above 10 mg L^{-1} with the pure AB580 dye serving as a negative control.

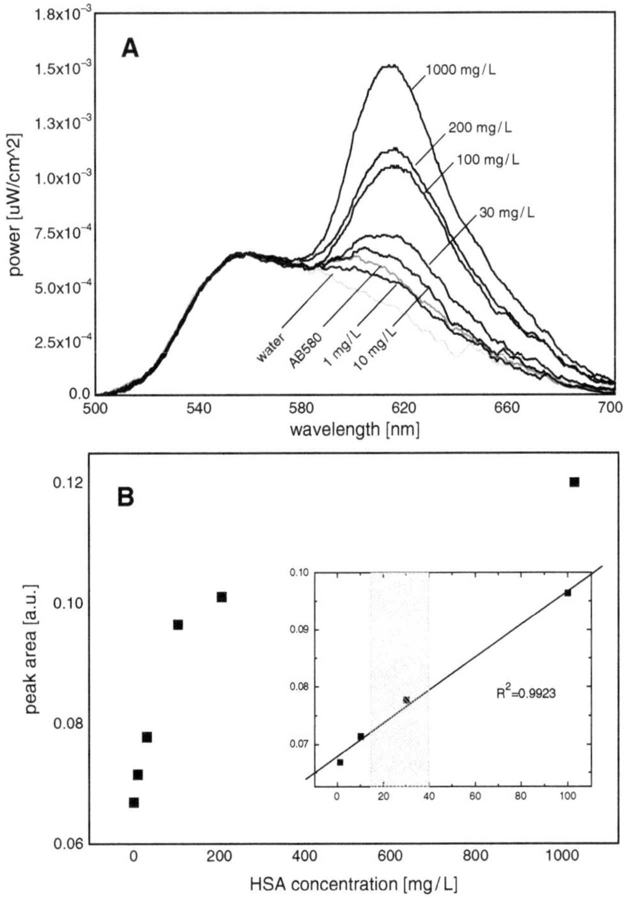

Fig. 4.24. Detection of different HSA concentrations after mixing with 1.2 µM AB 580 dye. (**a**) Spectra normalized on 550 nm peak. (**b**) HSA quantitation via peak area 500–700 nm. Inset shows linear range in diagnostically relevant HSA concentration range. Hatched area corresponds to 15–40 mg L^{-1} cut-off limit for MAU (From [35] – Reproduced by permission of The Royal Society of Chemistry)

Quantitation was subsequently performed by integration of the light intensity between 500 and 700 nm. The resulting calibration plot of peak area vs. HSA concentration is depicted in Fig. 4.24b. Although a strong signal increase is observed for lower HSA concentrations, the signal plateaus at very high concentrations due to the excess of AB580, which is consistent with previously published data. More importantly linearity between 1–100 mg L^{-1} HSA is obtained with an estimated limit-of-detection (LOD) of <10 mg L^{-1}. This is sufficient for the determination of MAU in a clinical setting with typical cut-off levels of 15–40 mg L^{-1} and compares to LODs of approximately 2 mg L^{-1} for cuvette-based measurements on conventional benchtop fluorometers.

While the successful application of OLED excitation sources for chip-based MAU determination clearly represents an important step towards the realization of portable diagnostic tests, engineering challenges still remain, especially the need to replace the expensive CCD with a low cost organic photodetector. To this end we are currently implementing organic photodetector to replace the employed CCD camera. This is most conveniently achieved in a colinear format rather than the orthogonal geometry employed here. To this end we are currently working on the integration of a short-pass filter on the OLED side (to cut off the emission tail) and long-pass filters on the detector side (to cut-off the OLED light while transmitting the longer wavelength analyte emission). As described in Sect. 4.4.3 we are also working on the implementation of passive microchip filling schemes to circumvent the use of bulky pumps for reagent delivery.

4.4.2 Chemiluminescence-Based Diagnostic Tests

As noted previously, CL offers a simple but sensitive means of monitoring low level marker concentrations in a microfluidic format. CL is particularly attractive for portable microfluidic assays because the CL reaction acts as an internal light source, thereby lowering instrumental requirements and significantly reducing power consumption while at the same time providing low background interference [36]. CL reactions typically involve the formation of a metastable product in an electronically excited state, which subsequently relaxes to the ground-state with the emission of light (direct CL). Indirect CL methods also exist where an electronically excited reaction intermediate is formed, followed by energy transfer to an emitting receptor dye. The most commonly used indirect method is PO-CL, where an oxalate ester is oxidized by hydrogen peroxide, resulting in the formation of metastable C_2O_4 and energy transfer to a variety of receptor dyes such as diphenylanthracene (blue), 9,10-Bis(phenylethynyl)anthracene (green), rubrene (yellow), rhodamine 6G (orange), and rhodamine B (red). The commercial availability of the reagents and the broad range of available emission colors render the PO-CL assay an ideal test platform for our chip-based portable diagnostic tests.

Potential analytes for PO-CL assays include oxidants, CL enhancers/quenchers, and fluorophore labels (e.g., the dyes listed above linked to a detection antibody, which is specific to the targeted marker). In chip-based diagnostic CL assays, however, specificity is most commonly afforded by upstream enzymatic assays that generate hydrogen peroxide only in the presence of a specific analyte (e.g., alcohol, glucose, and cholesterol) and dissolved oxygen. The generated hydrogen peroxide can then be detected through the generation of light when mixed with the peroxyoxalate reagent and a reporter dye. As discussed in Sect. 4.3.2, for our initial proof-of-concept tests we mixed varying concentrations of hydrogen peroxide with the PO-CL reagent and a green dye and recorded the resulting CL emission with our P3HT/PCBM organic photodetectors. We observed a linear dynamic range from 10 μM to 1 mM and a limit of detection <10 μM hydrogen peroxide, which is already sufficient for many diagnostic applications such as the determination of alcohol, glucose, and cholesterol in blood. However, further improvements to sub-100 nM levels are required to provide sufficient sensitivity for low-level cancer marker detection and pharmacokinetic applications.

For the first diagnostic application of our organic photodiode-based CL system we selected the determination of antioxidant capacity [37]. Antioxidants, either as food additives or as pharmaceutical supplements can terminate radical reactions in vivo, which can otherwise damage essential molecules such as nucleic acids and proteins. In a healthy person the production of reactive oxygen species (ROS) is balanced by the antioxidant defense system. However, in "oxidative stress" situations a serious imbalance between the production of ROS and antioxidant defense can occur. In such cases the intake of antioxidant is essential to prevent cell damage. The PO-CL assay described above provides a straightforward means of assessing the effectiveness of different antioxidants. In our initial experiments, antioxidant standards are injected into a hydrogen peroxide stream, which is subsequently reacted with the PO-CL reagent and an acceptor dye. The antioxidant standard essentially scavenges the hydrogen peroxide and hence reduces the CL emission that is generated.

Figure 4.25a depicts the recorded signal from the organic photodiode in response to the injection of varying concentrations of the known antioxidant Vitamin E [38]. As expected a decrease of CL emission is observed for the injection of each plug, which increases for higher antioxidant concentrations. Interestingly, some positive signal deviation before and after plug injection is also observed, which we assign to laminar flow-based stacking effects. Figure 4.25b shows similar signal profiles for unknown plant-based antioxidant extracts. Again a signal decrease is observed and an antioxidant content value can be calculated. The same experiment was repeated with conventional PMT-based detection and the comparison of the results is displayed in Fig. 4.26. It can be seen that the results are in good agreement, yielding a ranking order of the examined plant extracts.

While we have demonstrated the successful application of our integrated organic photodiode-based test platform to antioxidant capacity determination, we regard this as a first step towards a universal platform for chemiluminescence-based testing at the *point-of-care*. To this end we have investigated the use of passive fluid delivery schemes (to circumvent the use of valves and external pumps) and the implementation of low-cost electronic read-out circuits (for both size and cost reduction).

Figure 4.27 depicts a first demonstration system that comprises a microfluidic chip, an organic photodiode, and a low-cost printed circuit board (PCB) with amplifier. Powering and read-out is performed through a handheld computer. Future battery-driven devices are also envisaged to comprise an

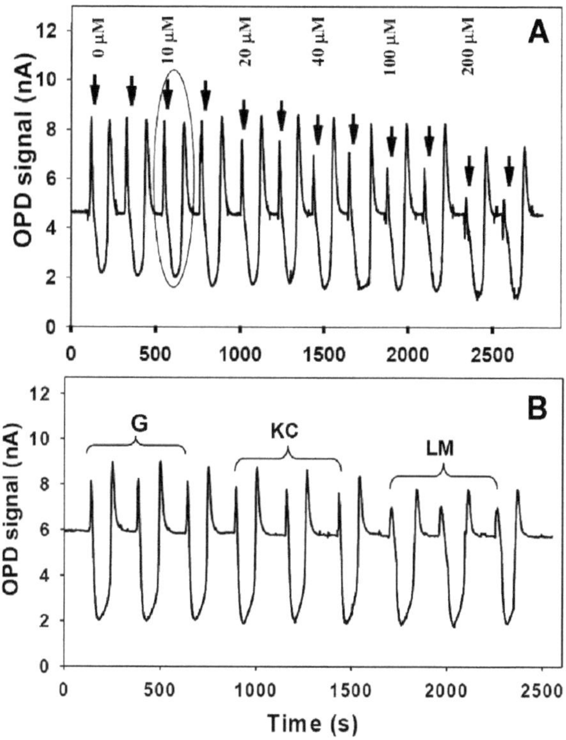

Fig. 4.25. PO-CL signal profiles after (**a**) duplicate injections of Vitamin E standards and (b) sequential injection of untested plant extracts Alpinia galanga – G, Kaempferia galanga – KC, and Cymbopogon citratus LM (**b**) (From [37] Copyright 2007 Springer)

integrated data display. In the current device assays are initiated by applying 30 µL of PO-CL reagent/dye and hydrogen peroxide, respectively, into the two inlet ports of the microchip. Filling of the microfluidic circuit is initiated by removing a lid from the microchip outlet, resulting in pressure equilibration and capillarity-driven filling of the microchannels. Diffusional mixing of the two reagent streams then occurs across the laminar flow interface. By the time the two reagent streams reach the detection chamber, which is positioned on top of the organic photodiode, the entire fluidic contents would have intermixed, resulting in CL emission. After closing the demonstrator lid, the photocurrent of the organic photodiode can be measured and subsequently quantitated against hydrogen peroxide concentration. Using this low-cost and passive technology, we have been able to achieve detection limits similar to our lab-based systems using syringe pumps and sensitive electrometer instrumentation.

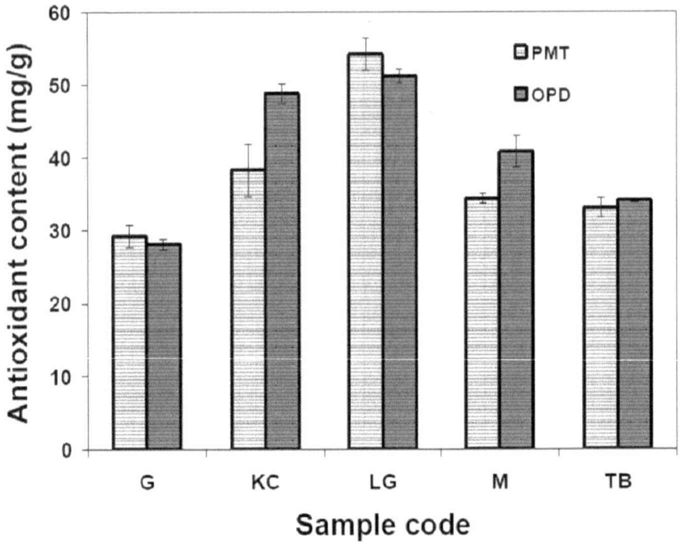

Fig. 4.26. Comparison of estimated antioxidant content of five herbal extracts as measured with P3HT:PCBM photodetector and PMT. G, Alpinia galanga; KC, Kaempferia galanga; LM, Cymbopogon citratus; M, Mentha piperita; TB, Ocimum spp (From [37] – Copyright 2007 Springer)

Fig. 4.27. Picture of chemiluminescence demonstrator with integrated organic photodiode and low-cost electronics

4.4.3 Towards Portable and Disposable Diagnostic Devices

Although the last two decades have seen significant breakthroughs in microfluidic technology, commercial success has been somewhat elusive. The few microfluidic-based analytical systems that have penetrated the market are primarily based on cartridge-reader systems where the microfluidic functionality is integrated in a disposable cartridge while read-out and data acquisition is accomplished in dedicated and often costly benchtop-sized reader systems. A typical example is Agilent's Bioanalyzer^{TM} system, which offers a variety of cartridges aimed at DNA sizing, and protein and cell binding assays (Fig. 4.28). Such benchtop-sized microfluidic devices have proven to be very powerful, e.g., DNA sizing can be accomplished in a few minutes compared to several hours or even days using standard gel electrophoresis techniques. However, there are significant drawbacks in that such systems remain bulky, require training or skilled staff, and their high cost is often prohibitive for use in the home or developing countries. The maturation of organic semiconductor-based detection technology provides a viable route towards the adaptation of powerful benchtop-sized microfluidic systems into portable, autonomous devices that can ultimately be made low-cost and disposable. When targeting the *in-the-field* or *point-of-care* diagnostic markets, additional design considerations have to be addressed [39]. First, there is a need for a user-friendly interface that allows simple sample application of the required bodily fluid. Sophisticated sample pretreatment methods as commonly used in clinical laboratories (e.g., centrifugation) are clearly not an

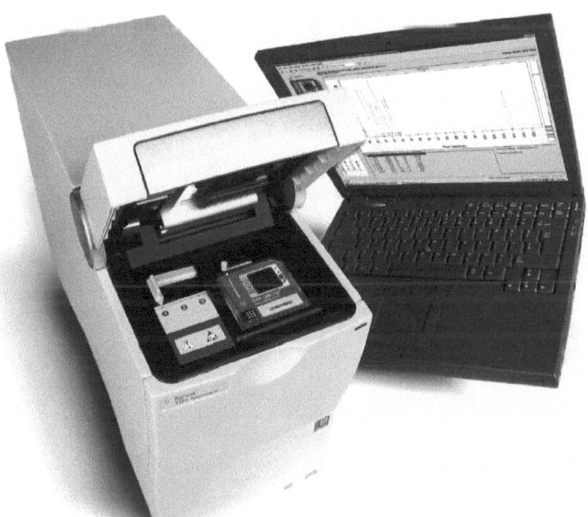

Fig. 4.28. Bench-top Agilent 2100 Bioanalyzer^{TM} with fluidic cartridge (Photograph courtesy of Agilent Technologies)

option for *point-of-care* use [40]. While urine samples can be directly assayed within microfluidic device, whole blood samples often require a filtration step to prevent blood cells from obstructing the microfluidic circuit [41]. Fortunately there exist a variety of methods for on-chip blood filtration, such as the use of microfabricated pillar arrays with pitches smaller than the diameter of blood cells. Secondly, once applied the sample should ideally be processed autonomously without requiring any further user intervention. This can be conveniently achieved using capillarity-based passive fluid delivery schemes, which circumvent the use of bulky pumps. Here, sample processing, such as the mixing with stored on-chip reagents, can be controlled via careful design of the microfluidic circuit. Read-out of the generated assay signal can then be achieved through integrated organic semiconductor-based detection systems with inherently low power requirements. For stand-alone devices (e.g., for home testing) low power requirements are essential in order to enable battery driven operation, but more sophisticated devices could draw power from widely available electronic consumer products such as mobile phones or PDAs. The main advantages of such configurations stem from the enhanced data processing power and added data handling capabilities, which for instance could enable wireless connections to a general practitioner (for test result evaluation) or connection to medical databases for data storage.

To this end we have developed an organic semiconductor based fluorescence demonstrator for *point-of-care* diagnostic testing (Fig. 4.29). The fluorescence demonstrator comprises an OLED-based light source in the lid, which is

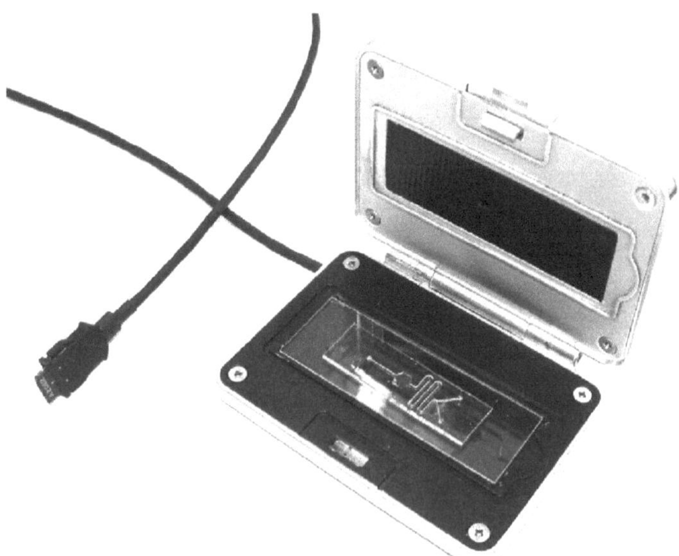

Fig. 4.29. Picture of fluorescence demonstrator with OLED light source (*in lid*), integrated organic photodetector (*in base*) and low-cost electronics

controlled via an integrated printed circuit board (PCB). The base of the demonstrator comprises an organic photodetector with integrated filtering and read-out electronics. A microfluidic chip is placed on top of the organic photodetector such that the detection chamber coincides with the detector pixel used for measuring the signal. After loading the sample into one of the two inlet ports of the microfluidic chip, a strip is removed from the outlet port to initiate autonomous filling and on-chip fluidic processing. When the lid of the device is in the closed position, the OLED light source is located on top of both the detection chamber and organic photodetector, resulting in a head-on or co-linear configuration. This allows for excitation of sample contained in the microfluidic detection chamber and measurement of any emitted fluorescence. To prevent excitation light from saturating the detector and thereby masking any analyte emission, optical filtering is required. In the described device an integrated optical long-pass filter is thus employed on the detector side to effectively block the light source contribution while transmitting the longer wavelength analyte fluorescence. Using fluorescent nanospheres as labels we can currently perform analyte detection in the nanomolar concentration range, and are working on lowering detection limits down to the picomolar range in order to widen the range of diagnostic assays that can be implemented in our low-cost detection platforms.

4.5 Conclusions and Outlook

The success of microfluidic systems as basic experimental tools in chemistry and biology has in large part been driven by a range of fundamental features that accompany system miniaturization. As has been shown, such features include the ability to process and handle small volumes of fluid, large improvements in analytical performance when compared to macroscale systems, reduced instrumental footprints, low unit cost, and facile integration of functional components. The last decade has seen a truly astonishing amount of research activity in the field of microfluidics. Significantly, the theoretical predictions (of increased efficiencies, speeds, throughput, and control) have in general been borne out by experiment, and integrated microfluidic systems are now being successfully used to solve a number of fundamental problems that have been inaccessible using conventional analytical instrumentation. Despite the obvious fact that functional integration of analytical components can be used to create microdevices that are physically small, the primary motivation for their development has been to perform chemistry and biology in a faster, more efficient, and more controlled fashion. Indeed, many early predictions of credit card-sized devices capable of performing complex analytical processes have been slow to materialize. As discussed, the lack of highly integrated microfluidic devices applicable to POC diagnostics has in large part been defined by the difficulties associated with integrating detectors that are small, sensitive, rapid, cheap, and applicable to the analysis of a wide range of biological analytes.

Fortunately, the technologies described in this chapter have begun to fulfill the above requirements and suggest a bright future for microfluidic-based POC diagnostics. Very simply, detectors based around organic semiconductors can in principle be mass-manufactured at extremely low unit cost using established processing methods, provide for sensitive optical detection in a range of formats and importantly can be tailored to fit many fluorescence-based biological assays. All these features are required for the mass deployment of POC diagnostic systems in a variety of environments. Although any POC diagnostic system will exhibit the core characteristics described previously, the application environment will dictate further specific qualities that must be realized. For example, a microfluidic POC diagnostic test for pathogen detection in Sub-Saharan Africa needs to be rugged, low-cost, suitable for operation by an unskilled user, have a long shelf-lifetime, be able to process samples containing contaminants and/or particulates, and operate within wide temperature and humidity regimes. On the other hand a microfluidic POC device for the detection of myocardial infarction by first responders in the US needs to provide a fast *time-to-test* result, but does not necessarily need to operate within wide temperature ranges or be able to function in dirty environments. Consequently, the acceptance, introduction, and sustainability of microfluidic POC diagnostic devices will ultimately be defined by the ability to lower unit costs to a level that is competitive with current test formats (such as lateral flow assays) while significantly improving performance markers (such a time-to-test result, accuracy, and quality control). This is by no means a simple task and commercial inertia in accepting these new formats is likely to be significant.

A number of key advances must be accomplished in the short-to-medium term if integrated microfluidic systems are to become established in the POC market. First, all the functional components within the device (i.e., microfluidic circuitry, detectors, and control circuitry) need to be fabricated using an integrated and continuous process to allow mass fabrication at low unit cost. Second, it is expected that most reagents required for a particular assay will be stored (in a solid form) on-chip and activated at the time of use. Accordingly, a key focus of future studies will address reagent capture, storage, and release procedures within microfluidic circuits. In a similar fashion, for many applications (e.g., in developing countries) it is expected that microfluidic POC devices may be stored for extended periods of time before use. Thus novel encapsulation approaches will need to be developed to ensure long shelf-lifetimes for both polymer LED and photodetector elements. Other areas of research focus will include the integration of multiple tests on single chip devices (panel testing) and the integration of components for data management and transferal to data bases.

In conclusion, it is clear that microfluidic formats offer immense, potentially revolutionary opportunities for POC diagnostics. Although in its infancy, the field of microfluidic technology has demonstrated a remarkable capability for analyzing, manipulating, and processing minute amounts of biological fluid. Nevertheless, the establishment and acceptance of robust *lab-on-a-chip* devices in medical diagnostics is still an aspiration, but one we hope will become a reality within the next five years.

References

1. A.J. deMello, Nature **442**, 394 (2006)
2. D.R. Reyes, D. Iossifidis, P.A. Auroux, A. Manz, Anal. Chem. **74**, 2623 (2002)
3. P.A. Auroux, D. Iossifidis, D.R. Reyes, A. Manz, Anal. Chem. **74**, 2637 (2002)
4. S.C. Terry, J.H. Jerman, J.B. Angell, IEEE Trans. Electron Devices **26**, 1880 (1979)
5. D.J. Harrison, K. Fluri, K. Seiler, Z.H. Fan, C.S. Effenhauser, A. Manz, Science **261**, 895 (1993)
6. R.G. Blazej, P. Kumaresan, R.A. Mathies, Proc. Natl. Acad. Sci. USA **103**, 7240 (2006)
7. P. Yager, T. Edwards, E. Fu, K. Helton, K. Nelson, M.R. Tam, B.H. Weigl, Nature **442**, 412 (2006)
8. A.J. de Mello, Lab Chip **3**, 29N (2003)
9. E. Verpoorte, Lab Chip **3**, 42N (2003)
10. K.B. Mogensen, H. Klank, J.P. Kutter, Electrophoresis **25**, 3498 (2004)
11. M.J. Madou, *Fundamentals in Microfabrication: The Science of Minaturization*, 2nd ed. (CRC Press, Boca Raton, 2002)
12. H. Becker, L.E. Locascio, Talanta **56**, 267 (2002)
13. J.C. McDonald, D.C. Duffy, J.R. Anderson, D.T. Chiu, H.K. Wu, O.J.A. Schueller, G.M. Whitesides, Electrophoresis **21**, 27 (2000)
14. D.C. Duffy, J.C. McDonald, O.J.A. Schueller, G.M. Whitesides, Anal. Chem. **70**, 4974 (1998)
15. T.U. Kampen, *Low Molecular Weight Organic Semiconductors* (Wiley-VCH, Weinheim, 2007)
16. G. Hadziioannou, G.G. Malliaras, *Semiconducting Polymers: Chemistry, Physics and Engineering*, 2nd ed. (Wiley-VCH, Weinheim, 2007)
17. D. Tanaka, H. Sasabe, Y.J. Li, S.J. Su, T. Takeda, J. Kido, Jpn. J. Appl. Phys. Part 2: Lett. Express Lett. **46**, L10 (2007)
18. Weblink (2007) http://www.japancorp.net/Article.Asp?Art_ID=14153
19. M.C. Schmidt, K.C. Kim, H. Sato, N. Fellows, H. Masui, S. Nakamura, S.P. DenBaars, J.S. Speck, Jpn. J. Appl. Phys. Part 2: Lett. Express Lett. **46**, L126 (2007)
20. J. Drechsel, B. Mannig, F. Kozlowski, D. Gebeyehu, A. Werner, M. Koch, K. Leo, M. Pfeiffer, Thin Solid Films **451–452**, 515 (2004)
21. X.H. Wang, O. Hofmann, R. Das, E.M. Barrett, A.J. Demello, J.C. Demello, D.D.C. Bradley, Lab Chip **7**, 58 (2007)
22. J. Huang, P.F. Miller, J.S. Wilson, A.J. de Mello, J.C. de Mello, D.D.C. Bradley, Adv. Funct. Mater. **15**, 290 (2005)
23. J. Huang, X. Wang, X.Y. Kim, A.J. deMello, D.D.C. Bradley, J.C. deMello, Phys. Chem. Chem. Phys. **8**, 3904 (2006)
24. J. Huang, R. Xia, Y. Kim, X. Wang, J. Dane, O. Hofmann, A. Mosley, A.J. de Mello, J.C. de Mello, D.D.C. Bradley, J. Mater. Chem. **17**, 1043 (2007)
25. J.B. Edel, N.P. Beard, O. Hofmann, J.C. deMello, D.D.C. Bradley, A.J. deMello, Lab Chip **4**, 136 (2004)

26. P. Fletcher, K.N. Andrew, A.C. Calokerinos, S. Forbes, P.J. Worsfold, Luminescence **16**, 1 (2001)
27. Z.Y. Zhang, S.C. Zhang, X.R. Zhang, Anal. Chim. Acta **541**, 37 (2005)
28. A.M. Jorgensen, K.B. Mogensen, J.P. Kutter, O. Geschke, Sens. Actuators B Chem. **90**, 15 (2003)
29. O. Hofmann, P. Miller, P. Sullivan, T.S. Jones, J.C. deMello, D.D.C. Bradley, A.J. deMello, Sens. Actuators B Chem. **106**, 878 (2005)
30. M.A. Burns, B.N. Johnson, S.N. Brahmasandra, K. Handique, J.R. Webster, M. Krishnan, T.S. Sammarco, P.M. Man, D. Jones, D. Heldsinger, C.H. Mastrangelo, D.T. Burke, Science **282**, 484 (1998)
31. J.A. Chediak, Z.S. Luo, J.G. Seo, N. Cheung, L.P. Lee, T.D. Sands, Sens. Actuators A Phys. **111**, 1 (2004)
32. O. Hofmann, X. Wang, J.C. deMello, D.D.C. Bradley, A.J. deMello, Lab Chip **6**, 981 (2006)
33. C.E. Mogensen, *Microalbuminuria, A Marker for Organ Damage* (London Science Press, London, 1993)
34. M.A. Kessler, A. Meinitzer, W. Petek, O.S. Wolfbeis, Clin. Chem. **43**, 996 (1997)
35. O. Hofmann, X. Wang, J.C. deMello, D.D.C. Bradley, A.J. deMello, Lab Chip **5**, 863 (2005)
36. X.R. Zhang, W.R.G. Baeyens, A.M. Garcia-Campana, J. Ouyang, Trends Anal. Chem. **18**, 384 (1999)
37. M. Amatatongchai, O. Hofmann, D. Nacapricha, O. Chailapakul, A.J. deMello, Anal. Bioanal. Chem. **387**, 277 (2007)
38. X. Wang, M. Amatatongchai, D. Nacapricha, O. Hofmann, J.C. deMello, A.J. deMello, D.D.C. Bradley, Lab Chip (submitted)
39. A.J. Tudos, G.A.J. Besselink, R.B.M. Schasfoort, Lab Chip **1**, 83 (2001)
40. C.D. Chin, V. Linder, S.K. Sia, Lab Chip **7**, 41 (2007)
41. E. Verpoorte, Electrophoresis **23**, 677 (2002)

5

Solid-State Chemosensitive Organic Devices for Vapor-Phase Detection

J. Ho, A. Rose, T. Swager, and V. Bulovič

5.1 Introduction

The need to extract increasing amounts of sensory information from our environment at faster rates fuels the need for smaller, cheaper, more sensitive, and more reliable chemical sensors. As chemical sensor research continues to mature and as new developments in sensor technology emerge, vapor-phase chemical sensors are finding applications in increasingly diverse areas. Vapor-phase sensors have the advantage of using a nondestructive, noninvasive method of detection with the potential for high degrees of sensitivity. Depending on the vapor pressure of the analyte, some vapor-phase sensors are capable of detecting as little as a femtogram of material (e.g., several thousand molecules) [1]. Chemosensitive organic materials have the advantages of being both synthetically flexible and amenable to room temperature processing, which allows for the development of a vast library of organic chemical sensors. While everyone can appreciate the use of sensors for the detection of explosives or for medical diagnosis, we must keep in mind that each application presents a unique set of challenges and specifications. What may work for the detection of landmines underfoot may not necessarily be useful for detecting disease within a human body. In fact, much of the bottleneck in commercializing chemical sensor research has been framing the right problem to be solved with the given sensor technology, all while minimizing costs. Despite decades of research that has produced many mature sensor technologies, recent advances in lasing to enhance the sensitivity of fluorescent conjugated polymers and in chemosensing heterojunction photoconductors continue to push the chemical sensing field in new directions as we search for better ways to give meaning to the odors in our environment.

5.1.1 Chemical Sensors and Electronic Noses

Before we survey the wide range of technologies available for chemical sensing, a distinction must be made between chemical sensors and electronic noses (e-noses). Most chemosensing applications favor one of two functionalities when sensing analytes: detection or distinction. For detection, the information regarding a specific chemical's presence in the environment is of utmost importance. For distinction, the relevant information is the identification of the chemical constituents that make up the odor. Typically, individual chemical sensors are implemented for chemical detection, while an array of such devices coupled with signal processing algorithms comprise an e-nose capable of distinction. Clearly, both types of devices will share some of the same technological challenges: reliability, portability, and sensitivity.

Chemical sensors used for detection are designed to be specific, with the ability to exclusively respond to one particular analyte. However, e-noses use arrays of sensors that are, ideally, orthogonal in odor space while having good coverage of the odor space. In odor space, one can imagine molecules being spatially located based on their similarity to other odors, as measured by human noses. In other words, two odors judged to have similar odor quality are positioned closer together in the multidimensional odor space than two different-smelling odors. The dimensions of odor space are related to the physical and chemical attributes of the molecules themselves: molecular weight, number of double bonds, number of carbon atoms, etc. Thus, the big challenge for e-noses is to make an array that incorporates enough sensors to cover the application-specific odor space while maintaining orthogonality among the various sensors to minimize the complexity of the data processing algorithms used. Another issue for sensor arrays is the variation in the rate of degradation between different sensors, necessitating complex calibration routines [2]. Algorithms and signal processing techniques are again the main approaches for solving this issue and can be difficult to implement depending on the sensor technology. Another challenge for e-noses is the creation of an accurate odor space itself. While there are numerous statistical odor classification schemes that weight each physicochemical attribute differently, to date there is no general relationship that can accurately predict the odor quality of molecules from their structure alone [3].

Because e-noses are composed of an array of chemical sensors, the performance of the sensor array is limited by the properties (sensitivity, selectivity, reproducibility, reversibility, etc.) of the individual sensors. Therefore, this chapter will focus on the individual chemical sensing approaches and their transduction mechanisms, rather than the aggregate performance of the array coupled with signal processing techniques.

5.2 Survey of State-of-the-Art Vapor-Phase Solid-State Chemosensing Organic Devices

The basic function of chemical sensors is to transduce a chemical signal into a measured entity. Within this broad class of devices are vapor-phase chemosensors, which convert the presence of specific volatile organic molecules in the gas phase into a measurable signal. Solid-state organic devices have the additional requirement of using carbon-based materials that are deposited as thin films. Despite these constraints, solid-state vapor-phase organic chemosensors encompass a diverse array of materials and device architectures. One simple sensor classification scheme uses the method of transduction to differentiate the technologies: electrical, optical, calorimetric (thermal), or gravimetric (mass). The electrical group can be further subdivided by device structure into chemiresistors, chemicapacitors, and organic thin-film transistors (OTFT). The optical group consists of chemoresponsive dyes, optical fiber sensors, surface plasmon resonance (SPR) sensors, and luminescent sensing schemes. The gravimetric group consists of surface acoustic wave (SAW) sensors, quartz crystal microbalances (QCM), and microcantilever sensors (Fig. 5.1). This survey will be limited in scope to cover the most recent advances in solid-state vapor-phase organic chemosensors, and as such, will not be a comprehensive survey of all existing vapor-phase chemosensors. The scope of this chapter restricts the discussion to research performed within the last 5 years and will reference older research, when necessary, to illustrate fundamental concepts. The focus of this survey also requires the organic (small-molecule, polymer, or carbon nanotube) materials to play a key role in the transduction mechanism by utilizing their electrical, optical, piezoelectric, or thermoelectric properties, rather than functionalizing an underlying inorganic sensor. Thus, the reader is encouraged to consult other sources to learn more about SAW [4], QCM [5,6], and microcantilever [7,8] sensing schemes that incorporate organic materials as functional groups to enhance selectivity.

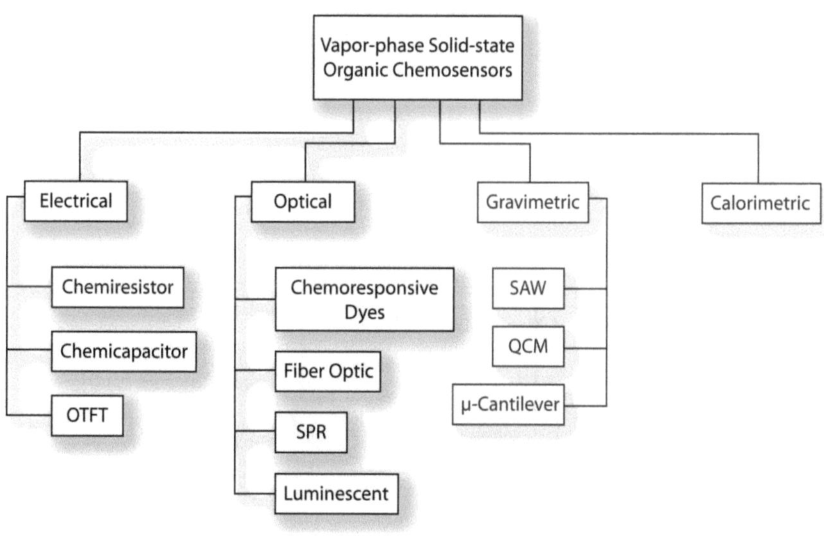

Fig. 5.1. Classification tree of different chemical sensor technologies. Lighter boxes will not be covered in this chapter

5.2.1 Electrical Odor Sensors

These chemosensors represent the simplest of gas sensors in that their chemical reactivity is directly transduced into an electrical signal. Changes in resistance, capacitance, voltage, or current indicate the presence of a particular analyte. These types of sensors are widely used for gas and odor measurements in commercial e-nose sensor arrays.

Chemiresistor

Chemiresistors represent the most common and simple type of sensor that incorporates organic materials. Generally speaking, the chemiresistor consists of a pair of electrodes that form contacts with the chemosensing material, which is deposited onto an insulating mechanical support (Fig. 5.2).

Normal operation involves applying a constant current through the sensing material while measuring the resulting potential difference at the electrodes. Thus, a change in resistivity would result in a change in the potential difference measured at the contacts. Inherently, this class of sensors requires less overhead electronics than more complex transduction schemes, which makes them suitable for miniaturization and portable applications. Some weaknesses include low signal-to-noise ratios and the variability in response repeatability

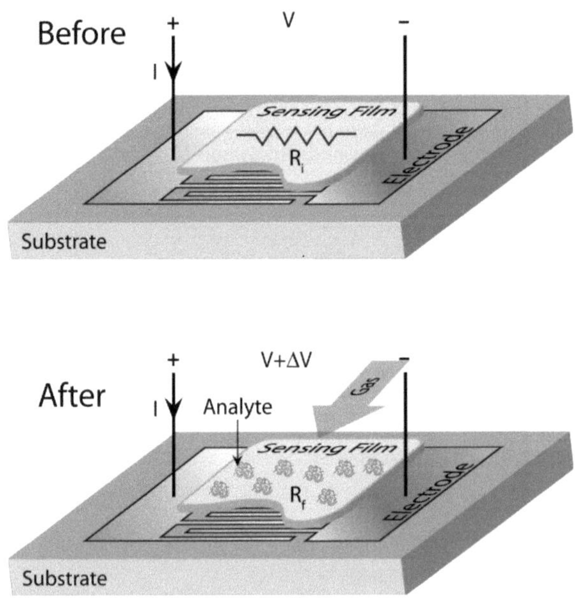

Fig. 5.2. Schematic of a typical chemiresistor device using interdigitated electrodes. There are two steady state conditions during operation: (1) no analyte present in the sampling environment and (2) analyte present in the sampling environment

to the same analyte, making them less useful when analyte concentrations need to be quantified. In addition, scalability is a major factor limiting the practical application of these sensors, because bulk resistance changes are less sensitive with decreasing device size. Within this class of sensors, there are several classes of materials that are used for the sensing layer: organic conjugated polymers and polymer composites, carbon black/organic composites, phthalocyanines, and carbon nanotubes. All of these materials exhibit an inherent change in conductivity due to the interaction with gases and can be used in different odor sensor schemes. For clarity, they are introduced here in the simplest sensor architecture.

Organic Conjugated Polymers and Polymer Composites

The alternating single and double bonds in organic conjugated polymers (CPs) allow for the delocalization of carriers along the polymer backbone, which gives the polymer the ability to conduct electricity. When this ability is coupled with the introduction of excess charges through doping, the polymer can then exhibit conductance from that of an insulator to that approaching a metallic conductor [9].

Recently, organic conducting polymers have become the focus of much of the materials research in chemosensing devices. Synthetic flexibility allows the chemical and physical properties of polymers to be tailored over a broad range of values for any given application. In addition, polymers exhibit tunable specificity to volatile organic compounds, which makes them ideal candidates for replacing canonical sensor materials such as metal oxide semiconductors.

In a conducting polymer film, when a gaseous analyte is introduced it acts as either an electron donor or an electron acceptor. Most organic conducting polymers are p-type materials with holes, as the majority charge carrier, determining the thin film conductance. If the conducting polymer donates an electron to the gas phase analyte, then its hole population and, consequently, its conductivity will increase. Alternatively, if the same conducting polymer receives an electron from the gas, then the hole population will decrease, reducing conductivity. Secondary effects include the slow diffusion of the gas into the conducting polymer itself, which can cause conformational changes in the polymer backbone [10]. Analyte gas diffusion generally results in a slow sensor response with more signal hysteresis.

Current trends in organic conducting polymer chemiresistor research involve the fabrication of nanostructures such as nanowires, nanotubes, or nanofibers made out of conducting polymers. The nanostructured conducting polymers have demonstrated improved sensor performance by increasing the exposed surface area as well as reducing the diffusion depth required for the penetration of gas molecules (due to high surface area to volume ratios) [11–14]. In addition, research on the incorporation of metal oxide nanostructures into the polymer films or incorporation of other polymers has yielded improvements in sensitivity, ageing, and the ability to detect nonpolar analytes [15–17].

The main advantages of using organic conducting polymer chemiresistors are their ease of fabrication, which allows the polymers to not only be selective and sensitive to a variety of analytes, but also to be deposited under room temperature conditions. In addition, their compatibility with microfabricated structures makes them particularly suitable for low-cost high throughput manufacturing. Another advantage the organic conducting polymer chemiresistors have is their ability to operate at room temperature. Metal oxide semiconductor chemiresistors are typically heated to between 200 and 500 °C for normal operation, which dissipates much more power. These polymers also typically have a sensor response time that is proportional to analyte concentration and the rapid adsorption/desorption kinetics as determined by van der Waals bonding at the surface, allowing the sensor to regenerate quickly. At the lowest analyte concentrations, the 10–100 s diffusion-limited response times [18] remain comparable to existing metal-oxide sensors that have response times in the range of a few seconds [19], depending on factors such as operating temperature and analyte concentration. Other weaknesses of organic conducting polymer chemiresistors include their variability from sensor to sensor and sensitivity to moisture.

Carbon Black Polymer Composite

Sensors using carbon black polymer composites utilize a polymeric insulating matrix to separate and fix conductive carbon black nanoparticles. The relative concentrations of the carbon black particles to the polymer and the form factor of the active sensor determine the conductivity of the composite and its signal-to-noise ratio [2]. With a high enough concentration of carbon black, the composite exhibits a simple, linear response to analyte concentration. Alternately, low concentrations of carbon black enable the composite to operate near the percolation threshold, yielding improvements in sensitivity with higher signal-to-noise ratios, at the cost of having less reproducibility, less stability, and a more complex, nonlinear response to analyte concentration [20].

The carbon black composite is usually spray-deposited on interdigitated electrodes, where vapors introduced to the composite interact with the polymer and cause it to swell, thereby inducing a conformational change in the distance between neighboring carbon black particles. Swelling results in a decrease in film conductivity. The sensitivity of sorption-based sensors is based on the interactions between the gaseous analyte and the polymer. By incorporating different functional groups into the polymer to target specific functional groups on the analyte, the selectivity of the sensor can be tuned to specific volatile organic compounds [21]. The sensitivity of the composite is determined by the amount of vapor the sensor can absorb. This interaction between gaseous analytes and the polymer matrix depends strongly on the density of functional groups within the matrix available to interact with the analyte.

Recent work done in this class of sensors has incorporated nonpolymeric materials to bind the carbon black particles, such as monomeric, low-vapor pressure organic molecules with functional groups attached [22]. The benefit of using smaller organic molecules in the matrix is an increased density of functional groups present in the composite, producing an increase in sensitivity. Also, the random orientation of the organic molecule matrix should create a more permeable structure for enhanced vapor diffusion. Finally, the use of nonpolymeric materials allows for new synthetic techniques to add chemical functionality and physical properties to the sensors that are not readily accessible in polymeric materials. Interestingly, these nonpolymeric carbon black composite sensors also exhibited a higher selectivity to the tested analytes, making them practical for array-based sensing.

In general, carbon black composites demonstrate fast response times, good reversibility, reproducibility, and stability. However, they lack the ability to react selectively to different gaseous analytes, making them better suited for a sensor array application where pattern recognition algorithms can be used to identify analytes.

Phthalocyanines

Phthalocyanines (PCs) are a robust class of semiconducting organic small molecules that are used in a wide range of applications: photovoltaics, nonlinear optics, xerography, and chemical sensing [23]. Chemical sensors based on metallophthalocyanine (MPC) thin films for the detection of donor and acceptor compounds (NH_3 [24,25], NO_x [26], Halogens) have been developed in the form of chemiresistors for over a decade. The bases for using these molecules as gas sensing materials are their ease of deposition as high quality, thin films (increasing the exposed surface area); their ability to achieve a desired molecular functionality by changing the central metal atom or adding substituents to the phthalocyanine ring; and their thermal and chemical stability.

The conductivity of MPC chemiresistors is sensitive to the presence of oxidizing and reducing gases, because a change in carrier concentration in the film occurs when charge transfer states arise from the adsorption of electron-withdrawing or donating gases. Previous studies have shown that there are many factors that can affect the MPC's ability to sense gases: film morphology, film thickness, operating temperature, and post-deposition annealing [27]. The PC films can be deposited on electrodes through a number of different methods such as Langmuir–Blodgett, spin-coating, thermal evaporation, or self-assembly.

Carbon Nanotubes

Discovered in 1991 by Sumio Iijima as a byproduct of an electrical discharge between graphite electrodes in an argon atmosphere [28], carbon nanotubes are chemical compounds that consist of concentric cylinders of covalently bonded hexagonal carbon rings. Carbon nanotubes range in size from a few nanometers in diameter to hundreds of micrometers in length. Single-walled carbon nanotubes (SWNT) consist of one concentric cylinder, while multiple-walled carbon nanotubes (MWNT) consist of several concentric cylinders. Carbon nanotubes have several unique mechanical and electrical properties that make them interesting from a sensing perspective: (1) carbon nanotubes have a large surface area to volume ratio [29], (2) they can be easily functionalized without disturbing their electronic structure, and (3) their electrical properties are extremely sensitive to the effects of charge transfer and chemical doping by various molecules.

Because of the difficulty in manipulating single nanotubes, carbon nanotube chemiresistors are based on films of spin-casted nanotubes on top of interdigitated electrodes. The interpenetrating network of carbon nanotubes has two sensing mechanisms. The first transduction mechanism is charge transfer from physically adsorbed analyte molecules, which modulate the Fermi level of the nanotube to cause a conductivity change. This intratube modulation is significant only if the nanotube is semiconducting in nature. The second transduction mechanism occurs when analyte molecules adsorb in the spaces between nanotubes, creating hopping conduction pathways between nanotubes that change the film's conductivity. This intertube modulation occurs for all types of molecules and for both metallic and semiconducting nanotubes, similar to the interaction between conjugated polymers and adsorbed molecules. The intratube modulation dominates the sensor response at lower concentrations with a superlinear dependence on concentration and a high sensitivity, while the intertube modulation shows a linear dependence and lower sensitivity over a broad range of concentrations [30].

In principle, carbon nanotubes are not sensitive to many organic vapors, especially nonpolar molecules that do not transfer charge. Thus, polymer coatings and sputtered Pd films have been used to functionalize the nanotubes to specifically detect Cl_2 and H_2, respectively [31]. Advantages of using carbon nanotube films include detection levels on the order of ppm at room temperature, good reversibility, as well as low variation between sensor response (less than 6%) making them suitable for use in arrays. However, nanosensors based on this mechanism face sizeable challenges for commercialization, such as the inability to selectively grow semiconducting nanotubes and the complex techniques required to handle single nanotubes.

Chemicapacitor

Chemicapacitors are structurally the same as chemiresistors, in that they consist of a pair of electrodes and a sensing material. The sensing layer can be either sandwiched between the two contacts [32] or deposited on top of a pair of planar interdigitated electrodes on an inert mechanical support [33, 34]. However, instead of relying on the changes in conductivity to detect the presence of analytes, this class of sensors measures the change in capacitance that develops across the chemosensitive film as the film absorbs the gaseous analyte. Absorption of analyte molecules changes the dielectric constant and

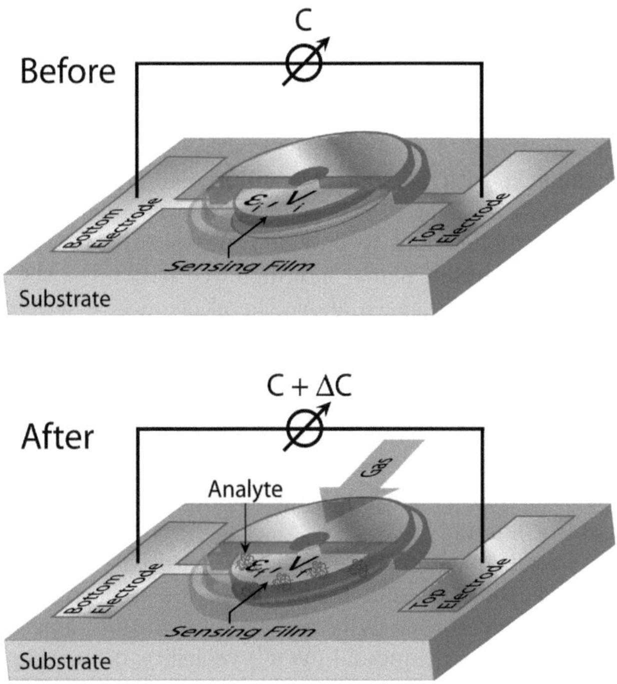

Fig. 5.3. Schematic of a typical chemicapacitor device using micromachined, parallel plate electrodes. In this scheme, both $\Delta\varepsilon$ and ΔV contribute to changes in capacitance

other physical properties (e.g., volume) to produce deviations from the baseline capacitance (Fig. 5.3). The chemicapacitors are usually operated with an AC voltage excitation (ranging from a few kilohertz up to 500 kHz), while the capacitances are measured using various readout circuits (frequency counters, output voltage, Sigma–Delta-modulators, etc.). Further details on the sensing mechanisms in chemicapacitors can be found in [35].

The simple device design and low power consumption makes this platform attractive for a wide range of gas sensing applications. In addition, the compatibility of the device structure with traditional semiconductor fabrication techniques allows for easy integration with complementary metal oxide semiconductor (CMOS) circuits and reduces the manufacturing costs needed to miniaturize the technology for portable gas sensing applications. While the sensor itself is rugged (no moving parts) and allows for a wide array of polymeric materials to be used as the sensor layer, the reliance on changes in the dielectric constant of the polymer makes these sensors susceptible to the high dielectric constant of moisture in the sampling environment. Humidity leads to large capacitive changes, which initially led to polyimide chemicapacitors being used as humidity sensors [36, 37].

More recent work in the chemicapacitor sensor field includes volatile organic compound detection using a variety of polymers [38] and liquid crystals [39]. However, most of the organic polymers that are used as sensitive coatings are only partially selective to most volatile organic compounds. To achieve finer selectivity of analytes, these sensors are usually placed in an array and combined with pattern recognition algorithms. Another method for enhancing the selectivity in chemicapacitive sensors is the use of configurable electrode configurations to provide two capacitive measurements with the same sensor. For simple, planar interdigitated structures, the electric field is mostly contained in the space above the electrodes to within a distance of half the electrode periodicity [40]. The interdigitated electrode periodicity can be changed by means of CMOS switches on the sensor chip; therefore, the effective ratio between sensing layer thickness and electrode periodicity changes. For thinner films, the capacitance change will depend more on the total polarizable material absorbed and less on the dielectric constant. The opposite is true for thicker films, making the differential measurement more robust to ambient humidity [41].

Typically, chemicapacitors have sufficient sensitivity to detect common solvents to below 100 ppm, and below 1 ppm for explosives and chemical warfare agents. However, sensitivity will be temperature dependent as heat increases volatile organic compound (VOC) volatility, decreasing the amount of absorbed analyte. The signal-to-noise ratio generally increases with an increase in polymer and analyte polarity. With dedicated signal conditioning circuits, the chemicapacitor is capable of resolving changes in capacitance as small as 0.07 fF [42]. The response times (time to reach 90% of full signal), which range from tens of seconds to minutes, are generally limited by the rates of diffusion of analyte molecules through the polymer [34].

Organic Thin-Film Transistor (OTFT)

Organic thin-film transistor (OTFT) technology is still considered a relatively new research area, with the first working OTFT fabricated out of polythiophene over two decades ago [43]. One argument in favor of using OTFTs as gas sensors is that they can offer more information about a particular analyte than an equivalent chemiresistor by monitoring changes in bulk conductivity, threshold voltage, field-induced conductivity, and field effect mobility [44]. The other advantage that OTFT sensors have over chemiresistors is the signal amplification inherent in transistor device structures, which produces gains in sensitivity and in signal-to-noise ratio [45].

The basic structure of an OTFT gas sensor is shown in Fig. 5.4. Typically, a conductive mechanical support is used as the gate electrode, upon which an insulating dielectric coating is deposited. Next, a thermally evaporated or spun-coated thin film of active organic semiconductor is added. Finally, the source and drain electrodes are deposited over the semiconductor and are patterned to allow exposure of the active region to the environment. The organic

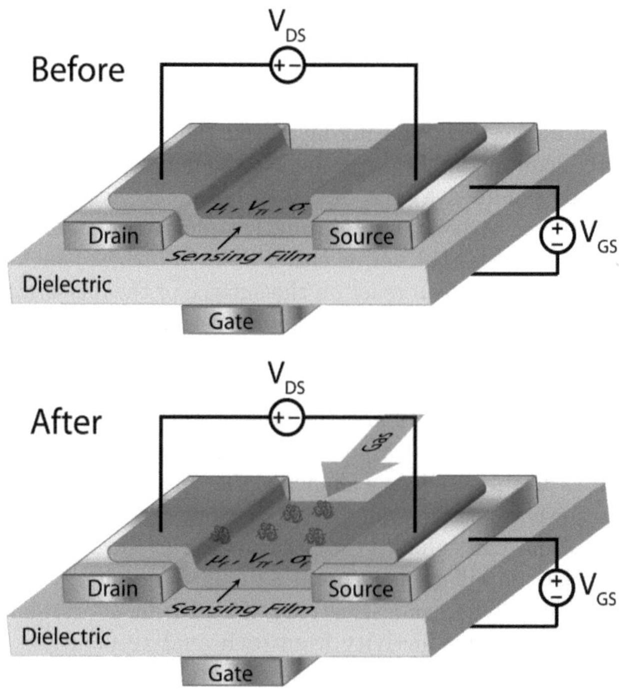

Fig. 5.4. Schematic of a typical OTFT gas sensor with a patterned gate electrode. The simultaneous measurement of multiple device properties yields more information and makes OTFTs a more attractive platform

field effect transistor is gate-biased so that a channel of field-induced charge is created at the interface between the organic layer and the dielectric. When a bias is applied across the source and drain contacts, current flows through the device. The transduction of the chemical signal into an electrical signal occurs when the introduction of an analyte changes the current flowing through the sensor. The major difference from a chemiresistor is that the gate voltage allows for modulation of the carrier density in the active sensing layer, which can enhance sensitivity. Because morphology and attached end/side groups on the organic semiconductor molecules provide sites for analyte interaction, both can be adjusted to provide additional improvements in sensitivity and selectivity [46]. There are several materials that are used for the sensing layer in this class of gas sensors: conducting polymers (CP) [47], phthalocyanines [48], and carbon nanotubes (CNT) [49]. The transduction mechanism for each of these materials remains unchanged from the case when the same materials are used for chemiresistors, as discussed earlier.

Generally, there are three different areas where an analyte can modify the properties of the OTFT sensor: the bulk of the organic semiconductor, the interface between the organic layer and the dielectric material, and the interface between the organic layer and the electrical contacts [50]. Just as with chemiresistors, the conductivity in the bulk organic semiconductor can change with the transfer of charge between the gaseous analyte and the sensing film. The other bulk effect is a change in mobility, which occurs when the interaction between the sensing layer and the analyte causes swelling of the organic semiconductor. Because the active channel of current flow in the OTFT is mainly confined to the first few molecular layers at the interface between the organic layer and the insulating dielectric material, amphiprotic molecules (molecules that act as either an acid or base) such as alcohols, organic acids, water, etc., can react on the surface of the dielectric to increase the device conductivity. This additional current is typically considered a parasitic conductive pathway and competes with the measured current signal from the organic semiconductor. Lastly, chemical modulation of the interface between the electrical contacts and the organic semiconductor can modulate the contact resistance and be responsible for the observed sensing effect.

Recent work has focused on enhancing the sensitivity of OTFT sensors. Crone et al. have reported that OTFT sensors are easily integrated into sensing circuits that can enhance the performance of the sensor. Ring oscillators and adaptive amplifiers, incorporating OTFT sensors, exhibit improved recovery time, sensitivity, and selectivity [51]. In an effort to increase the incremental change in conductivity from each analyte molecule by reducing the active area, Wang et al. have fabricated organic transistors with channel lengths on the nanometer scale. Although the results did not yield the expected gains in sensitivity, the studies have elucidated the dependence of the sensor response on the relative size of the channel length to the grain size [52, 53]. Because of their inherent simplicity and ease of integration with traditional silicon devices, OTFTs will continue to be an active area of vapor-phase chemical sensor research.

5.2.2 Optical Odor Sensor

For this class of chemosensors, changes in the optical properties of the organic sensing layer are used to detect gaseous analytes. Many organic materials have high absorption coefficients, high photoluminescent efficiencies, and absorption and emission profiles that span the visible spectrum (300–800 nm). For these reasons, organic materials inherently provide a straightforward transduction mechanism to convert chemical signals into visual outputs, which can then be detected by our eyes, photodetectors, or CCD arrays for further processing. Optical odor sensors are versatile, in that they can use a range of measurements (absorbance, reflectance, fluorescence, refractive index, polarization, colorimetry, interference, or scattering) to simultaneously collect information regarding intensity and wavelength.

Generally, optical sensors consist of four basic components: a mechanism or material that transduces a chemical signal into a change in optical properties, an excitation source to probe the sensor, optics for channeling light to and from the sensor, and a detector for converting optical signals into electrical signals. This class of sensors can be broadly organized into fluorescent and nonfluorescent technologies. Fluorescent technology is represented by amplified fluorescent polymers, nonfluorescent technology by chemoresponsive dyes and surface plasmon resonance (SPR), while optical fiber sensors encompass both types of sensor schemes.

Chemoresponsive Dye

Colorimetric sensors using chemoresponsive dyes are the simplest of the optical odor sensors and are analogous to simple pH indicator strips that change color in the presence of an acid or base. Colorimetric gas sensors work in much the same way by utilizing dyes that can change color in the presence of vapor-phase analytes. Not only do we acquire qualitative information about the presence of certain gaseous analytes through color changes that can be perceived by the human eye and brain, but if scanned with a CCD, that information can then be processed in a quantitative fashion to determine concentrations. While this type of sensor can use many different instruments to convert the optical signal into an electrical response (the human eye [54], flatbed scanner [55–57], photodetectors [58]), the sensors rely upon the same fundamental shifts in the absorption spectra of the chemoresponsive chromophore to indicate the presence of analyte.

When designing a chemoresponsive dye, two principles must be followed: (1) each dye must contain an interaction center that reacts strongly with analytes and (2) each interaction center must be strongly coupled to an efficient chromophore. The first principle requires that the analytes chemically interact with the dyes rather than merely physically adsorbing to the dye molecules. The importance of the strong dye–analyte interaction is highlighted by recent data that suggests that mammalian olfactory receptors are metalloproteins, and that chemical interactions with the metal center are intrinsic to the mechanism of smelling [59].

The metalloporphyrin family of organic molecules has found widespread use in colorimetric sensors due to their intense coloration, their large spectral shifts upon binding with foreign molecules, and the suite of organic chemistry techniques developed for these compounds. Many different chemoresponsive dyes can be synthesized by combining different metals with the porphyrin structure to form stable metalloporphyrin complexes. In these complexes, the central metal atom plays a primary role in the sensitivity and selectivity properties of the dye. Many volatile organic compounds are excellent ligands for metal ions and readily bind to the central metal atom to provide large color changes in the dye. In addition, the ability to add a wide range of side groups to peripheral positions on the porphyrin ring can enable enhanced selectivity.

Sensor arrays made from these chemoresponsive dyes are sensitive (110 ppb to 2 ppm) to volatile organic compounds and, with proper choice of dye and mechanical support, are not affected by ambient humidity. Also the response time of colorimetric sensors is relatively slow (2 min at 1 ppm analyte concentration), limited by the flow rate of the gas delivery to the sensor or equilibration of low vapor concentrations over large surface areas [60]. The simple, low-cost fabrication techniques for depositing chemoresponsive dye sensor arrays makes them ideal for general purpose vapor dosimeters or disposable analyte-specific detectors. However, implementation of a gas monitoring system using this technology becomes prohibitively expensive to install and maintain.

Surface Plasmon Resonance

Surface plasmon resonance (SPR) is an optical phenomenon where the conditions for total internal reflection at the interface between materials with different indices of refraction are satisfied, allowing an exponentially decaying evanescent wave to propagate in the medium with the lower index of refraction. When that interface is coated with a thin film of conductive metal (i.e., gold or silver), the evanescent wave can excite surface plasmon waves at the interface, which propagate parallel to the interface. The coupling between the incident light and the surface plasmon waves (SPW) in the thin film causes the reflected light to assume a particular wavelength, angle, intensity, or phase. Thus, any changes in the local refractive index at the interface will change the optical properties of the reflected light and produce a signal (see Fig. 5.5).

SPR chemical sensors consist of an optical system, a transducing medium, and an electronic system for data processing. The transducing medium

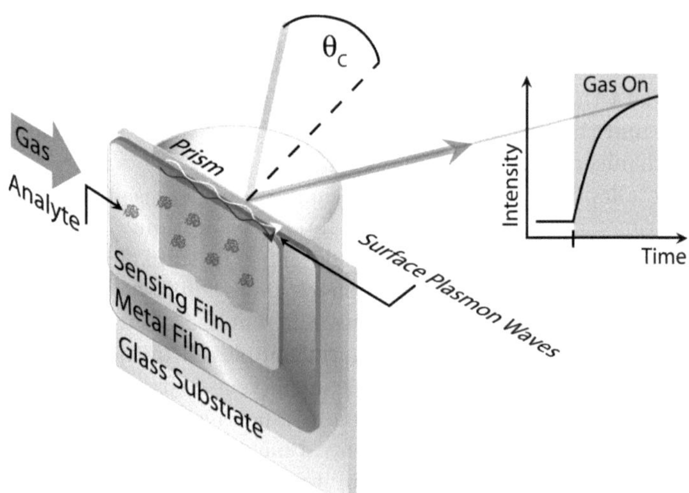

Fig. 5.5. Kretschmann configuration of SPR gas sensing method. The graph indicates a real-time measurement of intensity at a given wavelength and angle of reflected light

converts changes in the quantity of detected analyte into changes in refractive index, enabling the use of SPR to probe the variation. Typically, thin films of polymers [61] and small-molecule organics [62] are used as the transducing medium. As with most organic materials, the films can be deposited using spin-coating, dip-coating, or Langmuir–Blodgett techniques. SPR techniques can even be combined with optical fiber sensors to miniaturize the device design. While comparable sensitivities can be achieved, careful control over the measurement environment (mechanical, temperature, polarization) must be maintained [63, 64].

Sensor sensitivity, stability, and resolution depend upon the combined properties of the optical system and the sensing film. However, the selectivity, reversibility, stability, and response time are primarily determined by the properties of the sensing film. Because the SPR signals are strongly influenced by the optical properties of the sensing film, SPR sensors are extremely sensitive to small changes in the refractive index (down to 5×10^{-7} RIU) corresponding to better than 1 ppm levels of detection. Despite these high sensitivities, SPR will only detect adsorbates that have a molecular weight >500 Da to efficiently change refractive indices [65]. Some recent work has demonstrated the use of nematic liquid crystals [66] and cavitands (a class of supramolecule) [67] for SPR gas sensors. These materials exhibit relatively fast response times (<1 s), allowing this class of SPR sensors to be used for real-time monitoring of some volatile organic compounds.

Optical Fiber Sensor

Fiber optic technology first developed for the communications industry has been finding its way into chemical sensors. There are many technological reasons that make fiber optics sensors an attractive proposition. Optical fiber sensors are passive, requiring no electricity for operation. This can be important when trying to detect analytes in highly combustible environments or if the analyte itself is flammable at low vapor concentrations. For the same reason, fiber optic sensors are resistant to electromagnetic interference (EMI), making them useful for detecting analytes in environments with high and variable electric fields such as the power industry. In addition, fiber optics are capable of transmitting optical signals for long distances with low losses, enabling a receiver to be placed some distance away from the point of detection for remote sensing capabilities. By using existing optical instrumentation, optical sensing can be readily multiplexed to carry different measurement signals simultaneously. For most applications, however, fiber optic sensors are far more costly to implement than other gas sensing approaches [68].

Recent work has investigated conductive polymers [69, 70], sol–gels [71], and chemoresponsive dyes [72] for use as the gas-sensitive transduction medium in fiber optic sensors. In general, fiber optic sensors can be divided into two different sensing architectures: extrinsic and intrinsic. If the transduction of the chemical signal into an optical signal occurs within the fiber

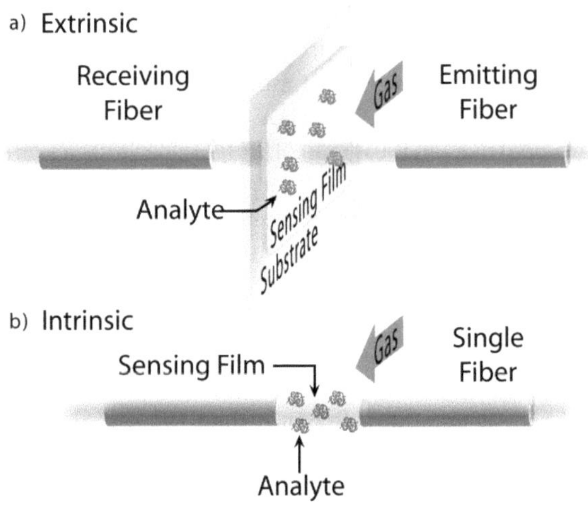

Fig. 5.6. (**a**) Extrinsic optical fiber sensor, where fibers are used to guide input and output signals. (**b**) Intrinsic optical fiber sensor, where sensor is incorporated into fiber structure

itself, it is known as an intrinsic sensor. Extrinsic sensors perform the transduction outside of the fiber (Fig. 5.6). In extrinsic sensors the fiber acts as a transmission line to bring light to a detection cell and to guide the light output to a photodetector or a spectrometer [73]. The detection cell contains the organic materials that provide selective gas detection. However, most fiber optic sensor research focuses on intrinsic sensors, with the most common being evanescent wave sensors.

In evanescent wave sensors, a small section of the passive cladding of the fiber is replaced with a chemically active sensing material. Thus, any interaction between a gaseous analyte and the sensitive material will manifest a change in the effective index of refraction in the fiber, as altering its transmission properties [70]. The chemically active material can be deposited on the fiber in a number of ways once the silica cladding is removed through polishing or etching: dip-coating, sol–gel deposition, or Langmuir–Blodgett [74]. For more information regarding the operation and theory of evanescent wave fiber optic sensors the reader is directed to literature [75].

A major limitation of the sensitivity of this type of sensor is the ability to transfer optical power into the evanescent field. Recent work has incorporated new types of fibers to increase the sensitivity of measurements. Newly developed hollow core fibers, which guide light mainly in the cladding, can be used to couple more power into the sensitive materials [76]. If the hollow core fiber is used as the active sensing area between two silica core fibers, then the transitions between the fibers will force the light to travel from the core into the cladding then back into the core. Another type of fiber is the microstructure fiber or photonic crystal fiber, which incorporates air holes into the silica cladding to enhance the interaction between the evanescent field and the gaseous analyte. By adjusting the size and placement of the air holes, light propagation in the fiber can be controlled [77].

Other advances in intrinsic fiber optic sensors include Bragg grating fiber sensors [78,79], active fiber core [71], optodes [80,81], fluorescent fibers [82,83], and some hybrid combination of the previous sensor schemes [64]. In general, the selectivity of the fiber optic sensor is governed by the light source wavelength and the chemical and physical properties of the sensing material. Special input signal modulation techniques can be used to enhance the selectivity of the sensors by reducing the cross-reactivity of other gases. Sensitivities are competitive with other sensor technologies (low ppm) and response times are relatively fast ($\sim$10 s), allowing for real-time monitoring of gaseous analytes such as ammonia [84]. However, prohibitive manufacturing costs, careful calibration requirements, and large amounts of overhead electronics and optics make fiber optic sensor systems less robust and more difficult to integrate into existing silicon circuit architectures. In addition, low dynamic range (as compared to other sensing schemes) and unsustainable reversibility limit the number of suitable applications.

Photoluminescent Sensor

Photoluminescence (PL) techniques are at the heart of many optical sensing schemes for a wide range of applications and can be extremely sensitive, enabling the detection of single molecules [85]. In general, PL is generated by fluorophores that absorb light at wavelengths within their absorption spectrum and emit light at longer wavelengths corresponding to the emission spectrum. The energy difference in emission and absorption spectra is referred to as a Stoke shift and represents vibrational relaxation and other energy loss mechanisms following absorption of a photon by a molecule.

The PL sensor scheme is surprisingly simple and can yield high sensitivities (Fig. 5.7). A thin film of the chemoresponsive material is excited by an excitation source (LED, laser, etc.) and a photodetector is used to collect the emission of the transducing material. The introduction of an analyte vapor will result in a change in optical properties, yielding a differential measured signal. An increase in emission results in a "turn-on" signal, while emission quenching results in a "turn-off" signal. "Turn-on" schemes require that the photoluminescent material be highly quenched. As a result, the distance between quenching sites will be less than the exciton diffusion length. In this situation, the sensitivity of the "turn-on" response will be decreased relative to the "turn-off" response, which relies on long exciton diffusion lengths to yield the high sensitivities necessary for ultra-trace detection. Additionally, the measured quantity in both cases is a differential intensity signal, making the lower limit of the signal-to-noise ratio (SNR) the determining factor of sensitivity. Thus, the only case where the "turn-on" response is easier to observe is if the new signal is produced against a completely dark background, so that

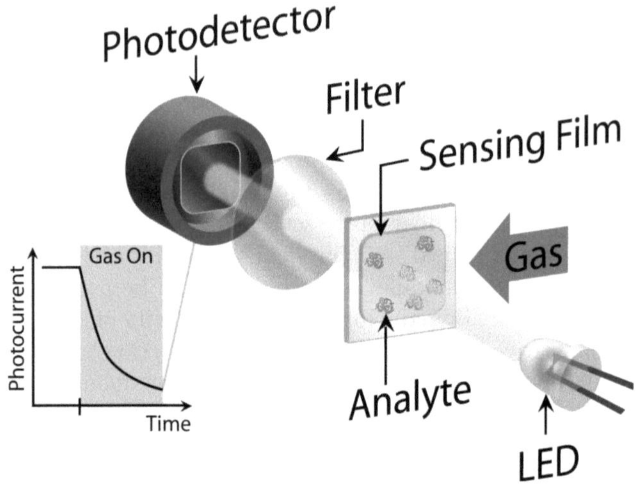

Fig. 5.7. Schematic of a typical configuration of a PL gas sensor. The LED is used to excite the sensing film, while the filter acts to prevent the excitation from reaching the photodetector. Upon introduction of a gaseous analyte, the sensing film's luminescence is quenched, producing a reduction in the PL intensity measured by the photodetector

the detection is no longer based on a ratio of signals. This is usually not the case, as background fluorescence often negates the sensitivity advantage of the "turn-on" response with regard to SNR. However, for trace detection in complex environments, "turn-on" detection schemes do have an advantage in selectivity as there are many more "false-positives" caused by adventitious quenching as opposed to fluorescence enhancement [86].

With the ability to measure a number of different parameters (decay time, energy transfer, polarization, quenching efficiency, intensity) and simple instrumentation, luminescent sensor techniques have become standard practice [87]. The ubiquity of luminescent sensing schemes and low-cost, high quality electronics, has even made possible the use of a computer screen as a programmable light source and a web camera as a photodetector in a chemical sensing technique known as the computer screen photo-assisted technique (CSPT) [88]. As luminescent sensing methods have matured, much of the ongoing research in this field focuses on the synthesis of new luminescent chemoresponsive materials to enhance analyte sensitivity and selectivity. Organic PL chemoresponsive materials span a wide range of compounds, including metalloporphyrin dye molecules [89], supramolecules, conjugated organic polymers [90], conjugated inorganic polymers [91], and phosphorescent molecules [92]. These luminescent compounds are either deposited as thin films [93], used as coatings for particles [94], or incorporated into inert, permeable matrices [95].

Organic Conjugated Polymers

The key advantage of organic CPs over small molecule or oligomeric materials is their ability to collectively respond to small perturbations. In particular, the CP's electrical and excitonic transport properties, as discussed above, can provide amplified sensitivity. While molecular sensing systems (one chemoresponsive molecule linked with up to three lumophores) have found use in practical applications, the luminescent properties of each molecule are only influenced by the local environment. Signal amplification comes from the ability to simultaneously change the properties of a large number of lumophore units with a single analyte molecule. The favorable exciton energy transport properties along their backbones allow excited states to efficiently migrate to many more attached lumophore units. Thus, any analyte molecule that is within the exciton diffusion length will quench the excited state, preventing it from emitting light. In other words, the exciton delocalization along the CP backbone effectively amplifies the transduction event.

Initial demonstrations of this approach used dilute solutions of conjugated polymers with integrated receptor units [96], restricting energy transport along a one-dimensional polymer chain. Thin films of conjugated polymers increase sensitivity by allowing excitons to sample receptor sites in three dimensions rather than redundantly sampling sites along the same backbone. In the solid state, however, conjugated polymers often form excimers or exhibit π-stacking, both of which can decrease photoluminescent (PL) efficiencies [97]. In one effective solution to this problem, rigid three-dimensional pentiptycene molecules were incorporated into the polymer backbone. This rigid scaffolding serves three important roles: (1) to prevent aggregation of the polymer, increasing PL efficiency; (2) to increase porosity of thin films for faster analyte diffusion; and (3) to create cavities that can selectively bind different sized analytes [98]. This approach has been used with great success for the selective detection of electron-deficient nitroaromatic compounds such as trinitrotoluene (TNT) [98]. By tuning the electrostatic interaction and film porosity/thickness, it is possible to detect TNT at sub-equilibrium vapor pressures (10–100 fg), which is comparable to canine olfactory performance [99].

To further enhance sensitivities, energy migration in three dimensions was explored with multilayer films deposited by the Langmuir–Blodgett technique. The construction of these layers created aligned polymer backbones that channeled exciton migration towards lower bandgap materials that were deposited on top [100]. This effectively created a staircase of energy levels, making it energetically favorable for the excitons to transfer their energy from layer to layer towards smaller bandgap materials. Thus, the energy migration efficiency is maximized by reducing the probability that excitons will revisit molecules. In addition, some research has centered on trying to extend excited state lifetimes to increase the probability of encountering a bound receptor site. This vein of research continues to yield promising materials for sensing applications [101].

Mathematical modeling of CP sensors has shown that the sensitivity of the CP to changes in analyte concentration has a nonmonotonic dependence on the number of receptors on each polymer backbone, increasing with the number of receptors up to a critical value at which point it decreases. In addition, the dynamic range of CP sensors is strongly dependent on the number of polymer backbones and the number of receptors on each backbone [102].

5.2.3 Summary

While much of the surveyed research exhibits promising vapor-phase sensing performance, many of the technologies remain experimental and bound to a laboratory setting. Most of the commercial gas sensors available today utilize older, more mature technologies such as electrochemical cells, catalytic beads, photoionization detectors (PID), SAW, metal oxide semiconductors (MOS), and QCM. The dearth of viable organic solid-state vapor-phase chemosensors indicates that there is much work still to be done (in terms of material stability, selectivity, etc.) before commercialization becomes commonplace for organic sensors.

5.3 Recent Advances

5.3.1 Chemosensing Lasing Action

As previously discussed, one approach for increasing the signal amplification in CP chemosensors is to change the energy transport properties by increasing excited-state lifetimes, or improving order in the polymer backbone to improve mobilities. Another, more recent, approach that can complement other sensitivity enhancement schemes is to use the lasing action of CPs or molecular organic thin films to generate greater differential intensity signals. Many organic materials, with high solid-state PL efficiencies, exhibit low lasing thresholds [103–105]. This is due mainly to the large Stokes shift that minimizes light reabsorption losses within the film. In addition, the molecular disorder inherent in organic materials creates a dispersion of energies through which excitons can diffuse downhill in energy. Exciton diffusion complements the energy relaxation due to molecular reconfiguration and increases the observed Stokes shift [106].

In general, the Franck–Condon principle is the approximation that electronic transitions occur much faster than nuclei can react. Following absorption and the nearly instantaneous promotion of an electron to an excited state, the surrounding nuclei must rearrange their energetic configurations to reach an equilibrated excited state. After emission, the nuclei are again in a nonequilibrium state and must change configuration to accommodate the relaxed electron. In all organic materials, the nuclear relaxation process causes the observed red shift in the emission profile relative to the absorption profile.

Table 5.1. Commercially available solid-state organic vapor-phase chemical sensors and e-nose instruments as of May 2007

Chemosensor type	Material classes	Analytes	Applications	Companies
Chemiresistor	OCP	VOCs	Medical diagnosis, food/beverage, process control, security	Scensive Technologies
	OCP	VOCs	Pharmaceutical, quality control	Alpha M.O.S.
	OCP	VOCs	Breath analysis	Abtech Scientific Inc.
	PC/MPC	NO_x, NH_3, O_2, Halogens	Environmental monitoring	(Research only)
	Carbon black composite	VOCs	Petrochemical, food/beverage, packaging, plastics, pet food	Smiths Detection (Cyranose)
	CNT	H_2, CO_2	Industrial gas detection, breath analysis	Nanomix
Chemicapacitor	OCP	VOCs, CWAs, TICs	Environmental monitoring, security, process control	Seacoast Science Inc.
	OCP	H_2O	Indoor air quality (relative humidity)	Jackson Systems
OTFT	OCP	VOCs	Environmental monitoring	(Research only)
	CNT	$H_2S, NO_x, CO_x, NH_3, N_2$	Environmental monitoring	(Research only)
Chemoresponsive dyes	Metalloporphyrins	Toxic Gas	Toxic gas dosimeter	Honeywell Analytics
	Metalloporphyrins	VOCs	Bacteria identification, breath analysis	ChemSensing Inc.
	Metalloporphyrins	Hydrocarbon	Meat spoilage	FQSI
	Metalloporphyrins	$H_2S, Cl_2, CO, N_2H_4, O_3$	Toxic gas dosimeter	American Gas and Chemical Co.

(*continued*)

Table 5.1. (continued)

Chemosensor type	Material classes	Analytes	Applications	Companies
SPR	Metalloporphyrins	O_2	Oxygen sensing	Sentronic
Optical fiber	Dye/Polymer matrix	Hydrocarbons	Environmental monitoring	PetroSense
	Dye/Polymer matrix	O_2	Oxygen sensing	Sentronic
	Dye/Polymer matrix	VOCs, O_2	Environmental monitoring, process control	Interlab (Optosen)
	Dye/Polymer matrix	O_2	Oxygen sensing	Ocean Optics
PL sensors	OCP	Explosives, CWAs	Security, life sciences	Nomadics/ICX
	OCP	VOCs	Air quality, security, process control	CogniScent
	OCP	VOCs	(In development)	Nanoident
	Metalloporphyrins	O_2	Oxygen sensing	Sentronic

Abbreviations: *OCP* organic conjugated polymer, *PC* phthalocyanine, *MPC* metallophthalocyanine, *CNT* carbon nanotube, *VOC* volatile organic compound, *CWA* chemical warfare agents, *TIC* toxic industrial compound, *OTFT* organic thin film transistor, *SPR* surface plasmon resonance, *PL* photoluminescent

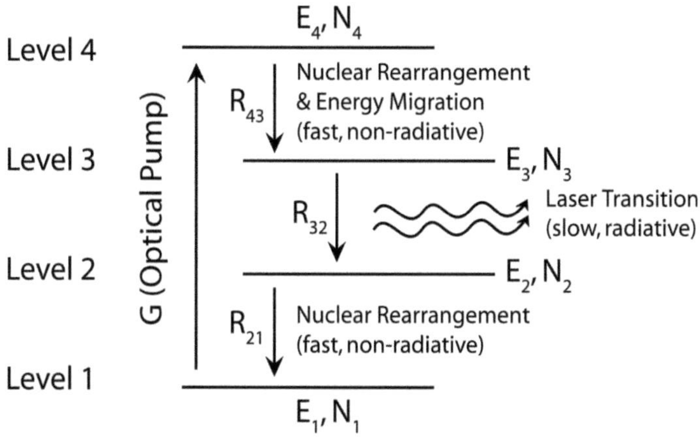

Fig. 5.8. Schematic of a four-level system representative of ASE of a conjugated polymer. E_x are the energies, N_x are the electron populations, and R_{ab} are the rates of transition between the different levels of the four-level system

These processes can be modeled using a four level system [107] (Fig. 5.8), where the different energy levels represent the following molecular configuration energies: (1) the electronic and nuclear ground state of the molecule, (2) the electronic ground state and the nuclear excited state, (3) the electronic excited state and the nuclear ground state, and (4) the electronic and nuclear excited state.

Modeling of Lasing Action

The model idealizes a CP thin film by disregarding energetic disorder, and places the film within an optical cavity that is excited by an ideal laser source (single wavelength, constant power). Following the treatment in [108], the following approximations can be made. First, the rates of transition from level 4 to level 3 and from level 2 to level 1 are approximated to be much faster than the transition from level 3 to level 2. This has two effects, first, electrons are approximated as being excited directly into level 3 and second, the electron population in level 2 is assumed to be negligible. The transition from level 1 to level 4 occurs when an exciton is generated in the CP and is characterized by the optical pumping rate G. The transition from level 3 to level 2 occurs through one of the four processes: spontaneous emission (with rate R_{sp}), stimulated emission (with rate R_{st}), spontaneous non-radiative decay (with rate R_{nr}), and quenching (with rate R_{q}).

Thus, the following rate equation governs the population in level 3 (N_3),

$$\frac{\mathrm{d}N}{\mathrm{d}t} = G - R_{32},$$

where

$$R_{32} = \begin{cases} R_{\text{st}} + R_{\text{e}} + R_{\text{nr}} + R_{\text{q}}, & \text{above lasing threshold} \\ R_{\text{e}} + R_{\text{nr}} + R_{\text{q}}, & \text{below lasing threshold} \end{cases}$$

In the absence of lasing with steady-state conditions, the exciton generation rate equals the sum of the relaxation rates and the population in level 2 is negligible ($N_2 \sim 0$), giving a constant population difference (N) of

$$N = G\tau_{32}, \quad \frac{1}{\tau_{32}} = \frac{R_{32}}{N} = \frac{1}{\tau_{\text{e}}} + \frac{1}{\tau_{\text{nr}}} + \frac{1}{\tau_{\text{q}}}.$$

To achieve lasing, N must surpass the required threshold population difference (N_{th}) by having the gain due to stimulated emission exceed the optical cavity losses.

The optical cavity losses can be evaluated by first noting that absorption of the lasing wavelength is negligible due to the large Stokes shift discussed earlier. Thus, only the mirror losses contribute. Assuming that both mirrors are identical (symmetric cavity), the distributed cavity loss coefficient (a) can be written as

$$\alpha = \frac{1}{d} \ln\left(\frac{1}{\Gamma}\right),$$

where d is the cavity length and Γ is the reflectivity of the mirrors.

Below threshold, the effect of stimulated emission on the population is small because the photon flux (φ) present in the cavity at the lasing mode is relatively low. So, the "small-signal" gain, $g(\lambda)$, is given by

$$g(\lambda) = N\sigma_{32}(\lambda) = G\tau_{32}\sigma_{32}(\lambda),$$

where $\sigma_{32}(\lambda)$ is the cross-section for stimulated emission at wavelength λ. It is important to note that the product of the photon flux with the cross-section for stimulated emission between level 3 and level 2 yields the stimulated emission rate, R_{st}:

$$R_{\text{st}} = \varphi\sigma_{32}(\lambda).$$

Thus, the lasing threshold should occur when

$$\begin{aligned} &\alpha = g(\lambda), \\ &\Rightarrow \frac{1}{d} \ln\left(\frac{1}{\Gamma}\right) = G\tau_{32}\sigma_{32}(\lambda), \\ &\Rightarrow G_{\text{th}} = \frac{1}{d\tau_{32}\sigma_{32}(\lambda)} \ln\left(\frac{1}{\Gamma}\right). \end{aligned}$$

From this expression, it is clear that the threshold pumping rate will be modified by the quenching rate through τ_{32} alone. Also, it has been shown that the relationship between the pumping rate and the incident photon flux and the optical absorption of the CP film will remain the same in the presence of an analyte.

When lasing, nearly every exciton generated above threshold contributes to optical power in the lasing mode, as can be seen by the following relation:

$$P_{\text{laser}} \propto P_{\text{pump}} - P_{\text{th}}.$$

P_{th} is the pumping optical power necessary to reach threshold and is linearly proportional to the threshold generation rate of excitons, G_{th}. Thus, if the introduction of an analyte causes the overall relaxation rate (τ_{32}) to change then the quenched laser power becomes

$$P'_{\text{laser}} \propto P_{\text{pump}} - P'_{\text{th}}, \quad P'_{\text{th}} = \frac{\tau_{32}}{\tau'_{32}} P_{\text{th}},$$

where τ'_{32} is the relation rate after quenching.

Now, it is possible to specify the fractional change in intensity of the lasing CP film in the presence of analyte using a fixed pump power,

$$\frac{\Delta I}{I_{0\ \text{lasing}}} = \frac{P'_{\text{laser}} - P_{\text{laser}}}{P_{\text{laser}}} = \frac{P_{\text{th}}\left(1 - \frac{\tau_{32}}{\tau'_{32}}\right)}{P_{\text{pump}} - P_{\text{th}}} = \frac{P_{\text{th}}\,(1-\beta)}{P_{\text{pump}} - P_{\text{th}}},$$

where

$$\beta = \frac{\tau_{32}}{\tau'_{32}}.$$

This expression can be compared to the fractional change in PL intensity of the CP film to specify the sensitivity enhancement due to lasing action. The PL efficiency (η) of the CP film can be written as

$$\eta = \frac{\text{Rate of photon emission}}{\text{Rate of exciton generation}} = \frac{R_{\text{e}}}{G} = \frac{\tau_{32}}{\tau_{\text{e}}}.$$

Considering that the PL intensity is directly proportional to the PL efficiency, the fractional change in PL intensity becomes

$$\frac{\Delta I}{I_{0\ \text{PL}}} = \frac{\eta' - \eta}{\eta} = \frac{\frac{\tau'_{32}}{\tau_e} - \frac{\tau_{32}}{\tau_e}}{\frac{\tau_{32}}{\tau_e}} = \frac{1-\beta}{\beta}.$$

From these expressions, it becomes clear that operating the chemosensing CP film near its lasing threshold will yield enhancements in fluorescence quenching, thereby enhancing sensitivity. Inspection of Fig. 5.9 shows that optimal sensitivity, at practical pump powers, is obtained when the pump power of the CP is at the onset of amplified stimulated emission (ASE), for the exposed film. Practical implementation of the concept would require rastering of the incident power intensity to locate the pump power that yields the maximum signal response.

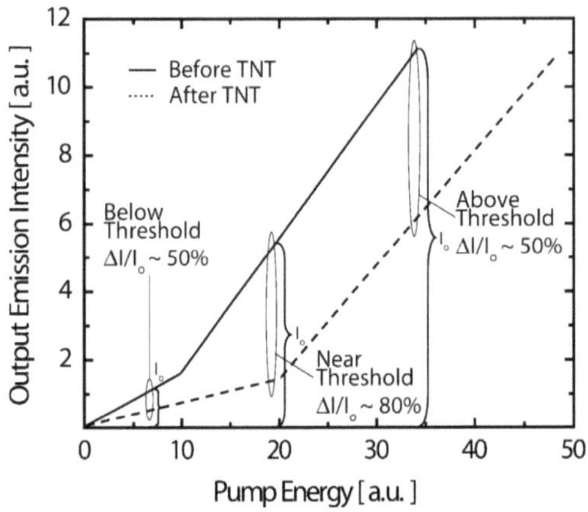

Fig. 5.9. Representative sketch of the ASE threshold increase after exposure of the CP film to a quencher such as TNT. The change in the output emission intensity divided by the initial output emission intensity is largest at pump energies just below the threshold of the analyte-exposed films

Experimental Demonstrations To-Date

Ongoing work in vapor-phase explosives detection in the Swager and Bulovic labs at MIT has demonstrated more than a 30-fold increase in detection sensitivity when operating polymer 1 (Fig. 5.10) near the lasing threshold. This polymer exhibits high durability, even operating at high incident pumping powers in ambient atmosphere. One design feature of the polymer is the incorporation of pendant aromatic rings that result in hydrocarbon side chains that run parallel to the polymer backbone. These side chains effectively encapsulate the backbone, preventing the aggregation that leads to self-quenching, and may be the source of the polymer's stability in thin film. Another feature of the polymer design is the extensive π-orbital interactions that promote longer exciton diffusion lengths (7–15 nm).

The sensitivity enhancement due to lasing action is a general principle that is independent of sensor architecture. Chemical sensitivity enhancements have been observed when the polymer is fabricated into a simple planar waveguide (Fig. 5.11a), deposited over a distributed feedback (DFB) grating (Fig. 5.11b) and coated on the exterior of an optical fiber (Fig. 5.11c).

Multimode lasing action in thin films of polymer 1 is generated by optically pumping the films with 4-ns nitrogen laser pulses (freq. = 30 Hz) at a wavelength of 337 nm in ambient conditions. The excitation source is focused into a $9 \times 0.09\,\text{mm}^2$ stripe, while the emission is collected at a 60° angle from the incident beam that is normal to the plane of the device. Initial experiments used simple asymmetric waveguides formed by spin-casting solutions of polymer 1 ($50\,\text{mg ml}^{-1}$ hexane) on glass slides. The film thickness was varied from 30 to 400 nm, and in films thicker than 50 nm, a multimode amplified

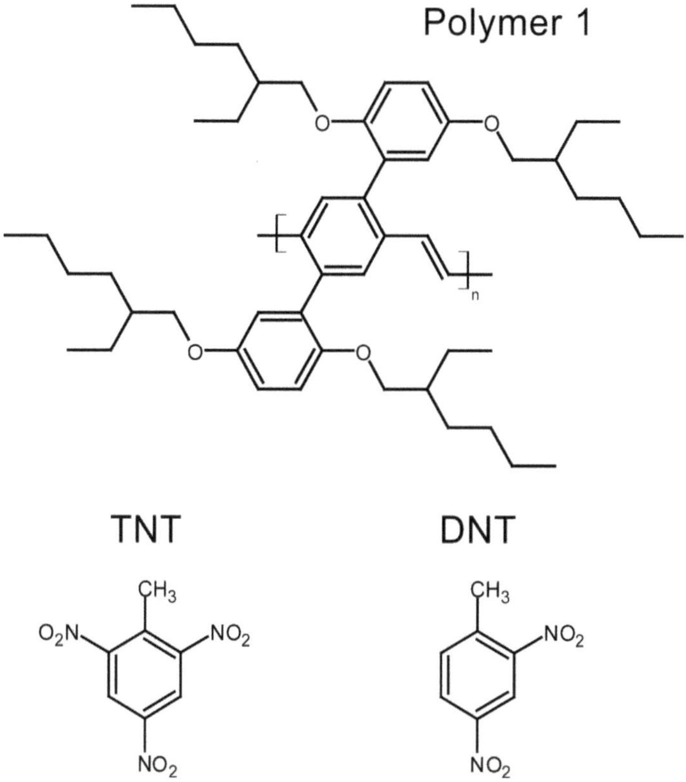

Fig. 5.10. Chemical structures of polymer 1, TNT, and DNT

spontaneous emission (ASE) peak at $\lambda = 535$ nm was observed. This wavelength coincides with the polymer's first vibronic transition (0,1). The (0,0) transition dominates the spontaneous emission due to its higher oscillator strength. The lack of ASE from the (0,0) mode is attributed to reabsorption losses that inhibit amplification.

Earlier studies cited above demonstrated that TNT and DNT bind strongly to electron-rich CPs due to the electron-deficient nature of nitro-aromatic compounds. With short exposures, these analytes are not able to diffuse effectively into the bulk of the film, remaining localized at the film's surface [97]. Thin CP films ensure that exciton quenching at the surface is not overshadowed by emission from the unquenched bulk, but very thin films will not sustain the waveguided modes needed for ASE and lasing action. Planar waveguides were formed by spin-coating a film of polymer 1 (refractive index, $n = 1.7$) onto an inert, transparent, index-matched thick film of parylene ($n = 1.67$) that guides most of the photoexcited radiation (Fig. 5.11a). With these structures, ASE is observed for layers as thin as 40 nm, with thinner films unable to absorb sufficient excitation light. Comparing the emission

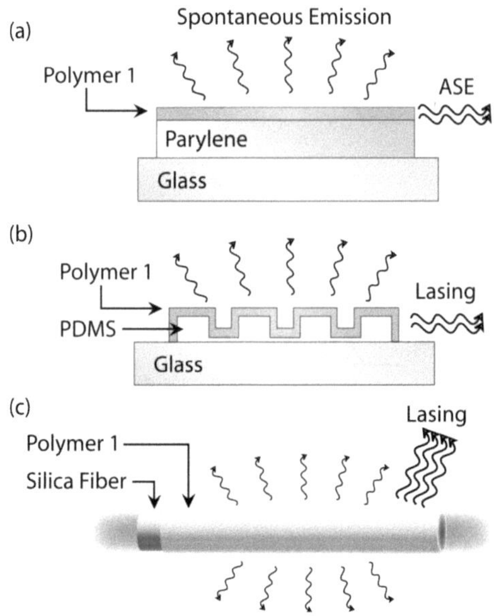

Fig. 5.11. (**a**) Spin-coated polymer 1 films (40–80 nm thick, with refractive index ($n \sim 1.70$) are coated on a transparent 200-nm-thick film of chemical vapor deposited parylene ($n \sim 1.67$), forming a two-layer index-matched waveguide on glass ($n \sim 1.45$). (**b**) Spin-coated polymer 1 films (40 nm) coated on DFB gratings fabricated from PDMS. (**c**) Ring-mode laser structure produced by dip-coating polymer 1 on a 25 mm diameter silica optical fiber

peaks at $\lambda = 500$ nm and $\lambda = 535$ nm in an 80 nm thick film reveals the onset of ASE at the integrated threshold power of 80 nW (peak pulse power, $P_{\text{th}} = 82\,\text{W cm}^{-2}$ and pulse energy, $E_{\text{th}} = 330\,\text{nJ cm}^{-2}$). At these low pumping powers, ASE signal attenuation due to photobleaching over a 20 s time period was negligible (~3% of total signal) (Fig. 5.12). As a comparison, the same thin film was subsequently exposed to the pump laser for 25 consecutive TNT quenching measurements (0.5 s each). As much greater quenching was measured due to TNT exposure, photobleaching contributions to fluorescence attenuation can be neglected. By reducing the thickness of the film to 40 nm and depositing polymer 1 on a polydimethylsiloxane (PDMS) DFB grating (Fig. 5.11b), the threshold energy can be further reduced to $E_{\text{th}} = 40\,\text{nJ cm}^{-2}$, which corresponds to a peak pump power of $P_{\text{th}} = 10\,\text{W cm}^{-2}$ (Fig. 5.13). Not only are these ASE threshold energies significantly lower than previously reported polyfluorene lasers, doped polymer, and doped molecular organic thin film lasers, but they also further reduce the associated photo-oxidation losses during quenching experiments. Again, the high performance of polymer 1 is associated with the improved polymer structure that maximizes PL efficiency in thin films, maximizing the optical gain.

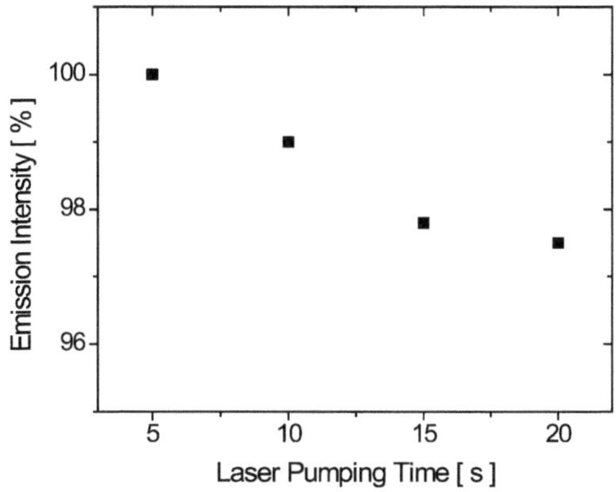

Fig. 5.12. Photostability of a thin film of polymer 1 coated on a silica optical fiber. A small amount of photobleaching is observed (~3% of total signal) over a 20 s time interval. During TNT measurements the sample was exposed to laser excitation for approximately 10 s (0.5 s for 25 measurements). Because the measured quenching due to TNT is much greater, the signal attenuation due to photobleaching can be neglected

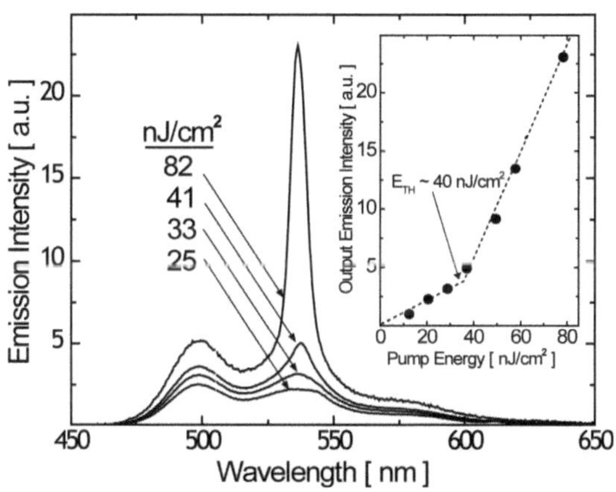

Fig. 5.13. Stimulated and spontaneous emission of polymer 1 spin-coated on top of a PDMS DFB structure (Fig. 5.10b) at different pumping powers. (*Inset*) Emission intensity of the $\lambda = 535$ nm peak as a function of input power

Experimental validation of the lasing enhanced sensitivity is demonstrated by comparing the sensory response of the spontaneous emission and ASE of polymer 1 to a saturated concentration of DNT vapor in air. After a 2 min. exposure, the spontaneous emission at $\lambda = 500$ nm displays an approximately two-fold quenching independent of pumping power. However, at pumping levels above threshold power ($P_{th} = 235$ W cm^{-2} corresponding to $E_{th} = 960$ nJ cm^{-2}), the ASE peak at $\lambda = 535$ nm exhibits a tenfold drop in intensity (Fig. 5.14). The differences between PL quenching and ASE

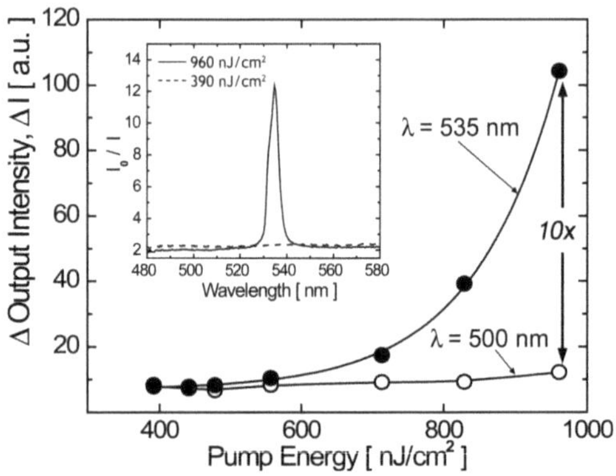

Fig. 5.14. Change in the output emission intensity, ΔI, at $\lambda = 500\,\text{nm}$ (spontaneous emission; open circles) and $\lambda = 535\,\text{nm}$ (ASE; filled circles) after 2 min. DNT exposure as a function of input power. ASE differential caused by quenching shows an increase at higher powers. (*Inset*) Emission intensity before exposure to DNT, I_0, divided by emission intensity after the exposure, I, plotted for $E = 390$ and $960\,\text{nJ cm}^{-2}$. The largest response occurs at ASE wavelengths ($\lambda = 535\,\text{nm}$)

quenching are more pronounced at 1 s exposure times of films to saturated DNT vapor, in which a significant attenuation in the ASE peak is reproducibly 30-fold more sensitive than the spontaneous emission peak.

While DNT (a byproduct of TNT) was readily detected with the DFB grating, it was more difficult to detect TNT molecules with their lower vapor pressure. In addition, the TNT molecules tend to bind tightly to the CP surface, making it ineffective at quenching excitons generated more than 15 nm from the surface. This property prompted the investigation of thin CP films in ring-mode laser cavities by dip-coating a 25 μm diameter silica fiber into a solution of polymer 1 ($10\,\text{mg ml}^{-1}$ in hexane). In Fig. 5.15, the attenuation of the lasing action on exposure to 5 ppb TNT is observed, while the spontaneous emission peak remains unchanged.

It is important to note that the quenching signals in the thin CP films, before and after TNT exposure, continue to increase with input pumping power. Unfortunately, the longer operating times (>1 min.) at the higher powers led to photobleaching of the thin films. Nevertheless, in applications that require the largest signal, increases in pump power result in increased quenching signals, which shorten the operating lifetime of the CP film.

While much of the chemosensing laser work has focused on using an amplified quenching mechanism to achieve sensitivity enhancements, ongoing work in turn-on lasing mechanisms has yielded promising results. As a proof of concept, a thin film of polymer 1 is mixed with the small molecule laser dye, 4-dicyanomethylene-2-methyl-6-*p*-dimethylaminostyryl-4H-pyran (DCM) (Fig. 5.10), in a 50:1 ratio by weight. When the film is exposed to excitation from a pulsed nitrogen laser ($\lambda = 337\,\text{nm}$), the DCM initially prevents the

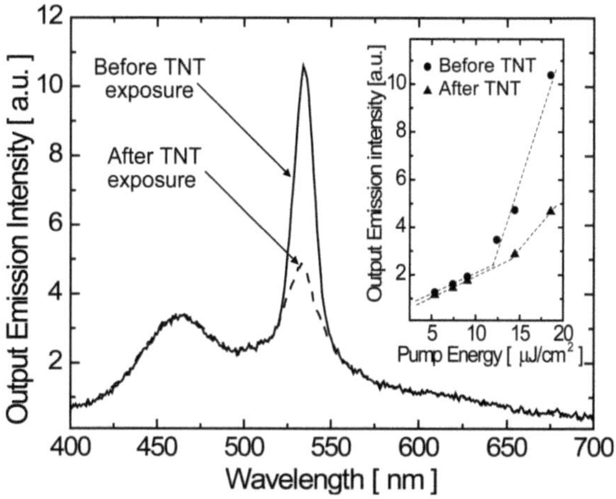

Fig. 5.15. Spectral response of a ring-mode structure consisting of a 25-mm diameter silica fiber dip-coated with a polymer 1 thin film. ASE attenuation in the absence of spontaneous emission attenuation occurs after 1.5 min exposure to saturated TNT vapor pressure. (*Inset*) Plots of ASE peak emission intensity ($\lambda = 535\,\text{nm}$) as a function of excitation power. TNT exposure increases the pump energy threshold

polymer 1 film from lasing. Two processes keep the lasing threshold beyond the limits of the pumping power; the DCM will absorb a small percentage of the incident optical power and, more significantly, the DCM will accept excitonic energy through Forster energy transfer from excited polymer 1 chains. If the excited DCM molecules are exposed to atmosphere, a photooxidation process occurs, rendering the DCM optically inactive. Thus, the competing loss mechanisms due to the DCM are removed under atmospheric conditions, allowing the film to lase (Fig. 5.16). This turn-on lasing mechanism is general and can be extended in the following fashion. If a chemosensing material that functions much the same way as DCM is incorporated into a polymer matrix, then the turn-on lasing can be modulated by the presence of a specific analyte.

Future Work

The devices used for the initial demonstrations of lasing-enhanced sensitivity were easily fabricated and convenient for evaluating and comparing the lasing characteristics of different CPs. However, these low quality laser cavities give high loss coefficients and higher lasing thresholds. Higher Q-factor laser cavities would lower device operating powers, lengthen CP film lifetimes, and reduce the lasing threshold, allowing for thinner CP active layers. These thinner active layers are expected to provide increased sensitivity and faster analyte response times. By incorporating CP thin films into high-Q cavities, improvements to sensitivity can be achieved, which could then be directly applied to the detection of explosives with low equilibrium vapor pressures

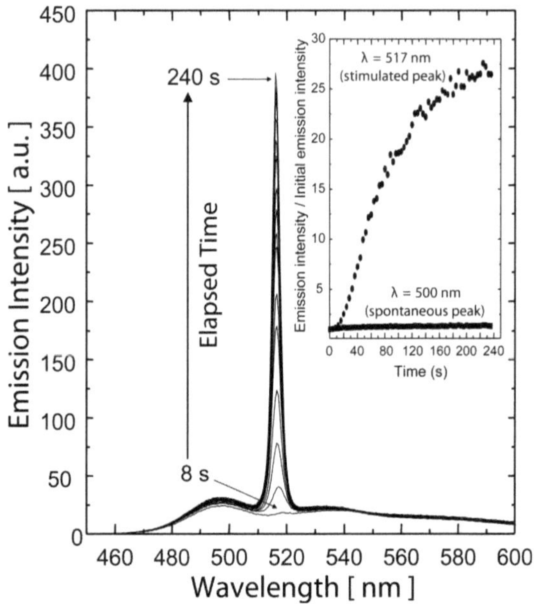

Fig. 5.16. Spectral response of a ring-mode structure consisting of a 25-mm diameter silica fiber dip-coated with a polymer 1 thin film. ASE attenuation in the absence of spontaneous emission attenuation occurs after 1.5 min exposure to saturated TNT vapor pressure. (*Inset*) Plots of ASE peak emission intensity ($\lambda = 535$ nm) as a function of excitation power. TNT exposure increases the pump energy threshold

such as RDX. Recent advances in nanofabrication technology allow the templating of large arrays of CP nanostructures in arbitrary shapes and sizes to make nanowire lasers [109]. This technique could also prove beneficial for the miniaturization and sensitivity enhancement of chemosensing lasers.

5.3.2 Chemical Sensing Heterojunction Photoconductors

Bi-layer heterojunctions play a key role in many organic optoelectronic devices such as photovoltaics [110], organic light emitting diodes (OLEDs) [111], and photoreceptors [112]. The main purpose of the bi-layer heterojunction is to present an energy band alignment useful for manipulating charge carrier dynamics. In the case of OLEDs, the interface provides a barrier for carrier transport such that exciton formation and electroluminescence occur at the interface. Alternately, in the case of photovoltaics and photoreceptors, the interface is responsible for the dissociation of photogenerated excitons into charge carriers that are swept out of the device by an applied bias.

Organic photoconductors are devices that consist of a pair of electrodes and a semiconducting material. Typically, interdigitated electrodes are used to maximize both the semiconductor area exposed to the light source and the device bandwidth [113]. As photons are absorbed by the semiconductor, excitons are generated in proportion to the intensity of the incident photon flux. There then needs to be a physical mechanism for the dissociation of excitons in order to generate charge carriers in the film. When a bias is applied across the electrodes, a field develops across the photoconductive film, which sweeps the charges out as photocurrent. In crystalline inorganic semiconductors, the exciton binding energy is small enough ($\sim$0.1 eV) that excitons readily separate into electrons and holes. However, in organic semiconductors, the photons will generate strongly bound (1 eV) Frenkel-type singlet excitons that are difficult to dissociate without the presence of carriers, dopants, or heterojunction interfaces [114].

Exciton dissociation in disordered organic materials has been attributed to both surface and bulk effects. Surface-enhanced exciton dissociation is based on the assumption that excitons diffuse to interfaces where they can either dissociate into a free electron–hole pair or they can interact with a donor or acceptor center to create a free carrier and a trapped carrier of opposite sign [112]. Exciton dissociation in the bulk film can occur when excitons encounter localized charges in the form of dopants or trapped carriers. This excitonic energy transfer can result in the liberation of a carrier from a deep-trap state, momentarily creating a nonequilibrium condition where the carrier must thermally hop amongst shallower states before becoming trapped again. While in this nonequilibrium state, the carrier will have a higher mobility than at equilibrium. The time it takes for the equilibration to be restored is dependent on the disorder in the film so much so that the carrier transit time across the film can be much shorter in strongly disordered materials. Thus, for every photoexcited carrier, many more injected carriers can be transported across the film before the excited carriers relax. This process is known as the disorder-controlled photomultiplication effect, and it is a complementary photogeneration process to surface-enhanced exciton dissociation in photoconductors [115].

Many studies [61] have examined photoconductivity in bulk organic heterojunctions, which are relevant to the fabrication of the most power-efficient organic photovoltaic devices [116]. The large interfacial area between materials is responsible for increasing the exciton dissociation efficiency by ensuring that interfaces are within exciton diffusion lengths. The tradeoff with increased exciton dissociation efficiency is reduced charge extraction efficiency. However, with the incorporation of ordered nanostructured materials, overall photogeneration efficiency can be increased [117]. In bulk heterojunction photoconductors, the inherent coupling of exciton dissociation properties with charge extraction complicates the study of intrinsic physical properties (mobility and carrier lifetime) and optimization of device performance. Therefore, significant simplification of device operation can be achieved through the design of a bi-layer heterojunction. By physically separating the device processes of exciton generation and charge transport into two different films, analysis of underlying physical processes can become simpler with a judicious choice of materials. While the bi-layer heterojunction structure has its roots in xerography, much of the research in this field has focused on device physics in the vertical direction (perpendicular to the plane of the films). The arrangement of electrodes in a lateral fashion (parallel to the plane of the films) creates a platform to study charge generation and recombination at the interface between different organic materials, which is crucial to understanding the physical mechanisms that dominate device performance in organic photovoltaics and organic light emitting devices.

Theory of Operation

The lateral bi-layer heterojunction photodetector consists of two layers deposited on interdigitated gold electrodes. Figure 5.17b, c shows two different views of the physical processes that operate in the device: Fig. 5.17b is an energy band diagram showing proposed HOMO and LUMO levels for both organic layers, while Fig. 5.17c depicts the cross section of a proposed device stack. We see that the charge transport layer (CTL) is in contact with the gold electrodes and the exciton generation layer (EGL) is on top of the CTL, exposed to the environment. Under steady-state bias conditions with no illumination,

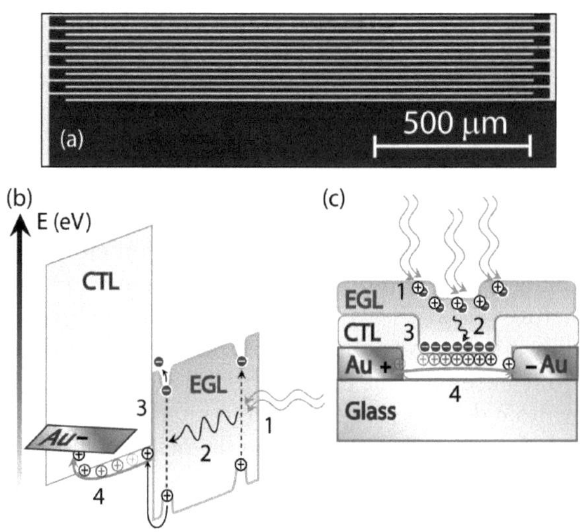

Fig. 5.17. (**a**) A digital image of a set of device electrodes under magnification (50×). The interdigitated gold fingers create 100 channels each 1,500 μm long by 10 μm wide, yielding an effective device area of approximately 1.5 mm^2. (**b**) Energy band diagram of a lateral bi-layer heterojunction photoconductor consisting of an exciton generation layer (EGL) and a charge transport layer (CTL). (**c**) A cross-sectional view of the same bi-layer device is also shown. Both illustrations depict the physical processes involved in steady-state device operation: (1) Absorption, (2) Exciton diffusion, (3) Exciton dissociation and charge transfer, and (4) Charge transport

conduction can occur through either the CTL, the charges at the CTL/EGL interface, or the EGL. Under illumination, light with an appropriate emission profile will be absorbed by the EGL and excitons will be generated. The generated excitons diffuse throughout the EGL, with some fraction ($\eta_{\mathrm{diffusion}}$) making their way to the interface. Once at the interface, it is energetically favorable for excitons to dissociate ($\eta_{\mathrm{diss}} \sim 1$) if the materials' energy bands are aligned such that $\mathrm{LUMO_{EGL}}-\mathrm{HOMO_{CTL}}$ are less than the exciton binding energy as shown in Fig. 5.17b. Since there are no electrodes in the vertical direction, the generated charge will continue to build up at the interface until equilibrium between generation and recombination and diffusion is reached. Because of charge neutrality, the holes accumulated in the CTL near the interface will balance the electron charge accumulated in the EGL. When a bias is applied to the lateral electrodes, the CTL's conductivity is enhanced by the population of light-generated charge carriers and yields a photoresponse.

Experimental Demonstrations

To test the device model we used two different small molecule organic materials: N,N'-bis(3-methylphenyl)-N,N'-diphenyl-1, 1′-biphenyl-4, 4′diamine (TPD) and 3,4,9,10-perylenetetracarboxylic bis-benzimidazole (PTCBI), whose absorption spectra and approximate energy band levels are depicted in Fig. 5.18. TPD (H.W. Sands) is a common hole transport material used in organic LEDs, making it suitable for use as the CTL. PTCBI (Sensient

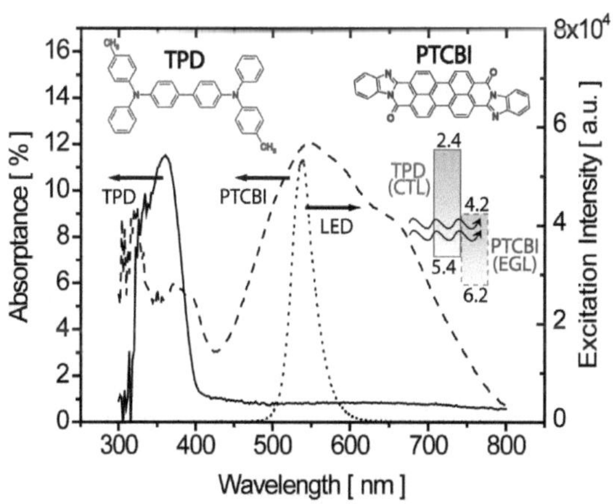

Fig. 5.18. Absorption spectra, chemical structure, and proposed energy band levels of the materials used in the bi-layer heterojunction photoconductor: N,N'-bis(3-methylphenyl)-N,N'-diphenyl-1,1′-biphenyl-4,4′-diamine (TPD, solid) and 3,4,9,10-perylenetetracarboxylic bis-benzimidazole (PTCBI, dashes). Also shown is the emission profile of the green LED (green, $\lambda_{\mathrm{peak}} \sim 532\,\mathrm{nm}$) operated with a forward current of 2 A and an intensity of approximately $40\,\mathrm{mW\ cm^{-2}}$ (dotted)

GmBh) is a perylene dye derivative that has found use as an absorber material in organic solar cells, making it suitable for use as an EGL. The rationale for choosing this combination is apparent from the absorption spectra (Fig. 5.18), which have little overlap for wavelengths longer than $\lambda = 400$ nm. This allows for preferential excitation of PTCBI by using a green LED centered at $\lambda = 532$ nm (Lamina Light Engine, BL-3000) in order to simplify analysis of the experimental results. The thermally evaporated films were deposited on a series of interdigitated gold fingers that have been photolithographically defined on glass (Fig. 5.17a). These fingers create one channel that is approximately 150 mm wide by 10 μm long, giving an effective device area of 1.5 mm^2.

Thin film absorption measurements were performed with an NKD-8000 (Aquila Instruments). The green LED light engine, powered by a sourcemeter (Keithley 2400), provides illumination with intensities of up to 40 mW cm^{-2} as measured with a calibrated photodetector (Newport, 818-UV). The emission profile of the green LED was measured using a spectrometer (Ocean Optics, USB4000). Current–voltage (I-V) characteristics were measured in a dark box using a picoammeter (Keithley 6487). Photocurrent spectra were measured using a digital lock-in amplifier (Stanford Research Systems, SR830) with monochromatic illumination provided by a 1,000 W Xenon arc lamp (Oriel) passed through a monochromator (Princeton Instruments/Acton Spectrapro 300i) chopped at a frequency of 40 Hz. Both I-V characteristics and photocurrent spectra were taken in a nitrogen glove box environment (MBraun) to avoid device degradation during measurements.

The current–voltage characteristics both with and without illumination from a series of devices is shown in Fig. 5.19. All of these devices involved either single films or bi-layer heterojunctions of PTCBI and TPD. The solid lines represent I-V sweeps in the light, while the dashed lines represent I-V sweeps in the dark. As shown, the device with a single 50 nm film of TPD generates no measurable photoresponse, which reflects the inability of TPD to absorb green light as its absorption band cuts off near 400 nm. PTCBI does generate a moderate photoresponse from the green light, which can be seen by comparing the gray solid line to the gray dashed line. This indicates that although there is no heterojunction, excitons are still able to dissociate (surface-enhanced and photomultiplication) in the single film of PTCBI and make their way to the electrodes. In addition, PTCBI exhibits lower currents due to its poor conductivity. However, the heterojunction devices indicate an improved photoresponse and a higher light-to-dark current ratio. This data suggests that the interface is dissociating excitons and generating charge under illumination, creating a more conductive channel in the CTL.

Photocurrent spectra from a series of lateral PTCBI/TPD devices are shown in Fig. 5.20. A single 5 nm film of TPD yields no measurable efficiency. The response of a single 10 nm film of PTCBI is represented by the dotted line. Responses of bi-layer heterojunctions of PTCBI/TPD are represented by the dashed and solid lines. In quantum efficiency measurements, the generated

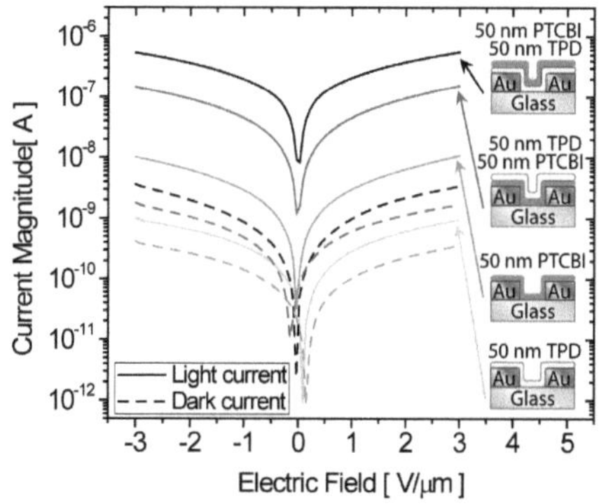

Fig. 5.19. Semi-logarithmic plot of the current–voltage characteristics from a series of lateral photoconductor devices using TPD and PTCBI. Current–voltage sweeps are taken in the dark (*dashed lines*) and under illumination (*solid lines*) through the glass support from a 40 mW cm^{-2} green LED ($\lambda_{\text{peak}} \sim 532$ nm). The inset cross-sectional diagrams describe the thicknesses and device structures. The Au/TPD device (*black*) exhibits no photoresponse to the green LED, consistent with its absorption spectrum. The Au/PTCBI device (*gray*) exhibits a steady-state photoresponse with lower currents due to PTCBI's low conductivity. Both heterojunction devices (Au/PTCBI/TPD [*black*] and Au/TPD/PTCBI [*gray*]) exhibit improvement in steady-state photoresponse and in light-to-dark current ratio

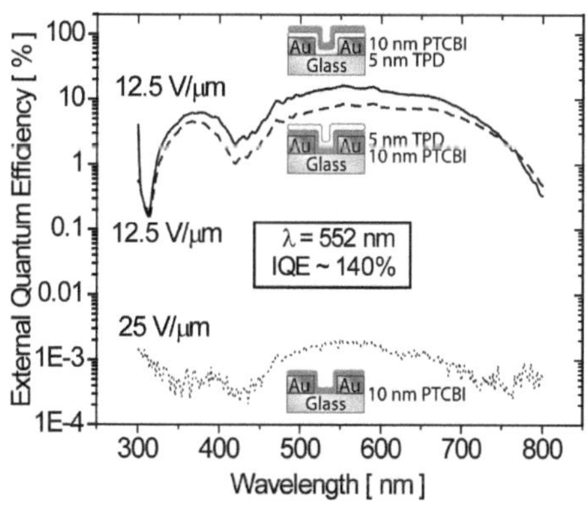

Fig. 5.20. Semi-logarithmic plot of external quantum efficiency (EQE) vs. wavelength. Inset illustrations depict device structures and film thicknesses. The Au/TPD device yields no measurable photoresponse and thus is not shown. Both heterojunction devices (*solid and dashed lines*) yield orders of magnitude improvement in the EQE over the Au/PTCBI device (*dotted line*) despite using lower bias voltages. At $\lambda = 552$ nm, the Au/PTCBI/TPD device (*dashed*) and Au/TPD/PTCBI device (*solid*) yield EQEs of approximately 8% and 16%, respectively. After accounting for the absorption in the PTCBI/TPD device, we calculate an internal quantum efficiency (IQE) of approximately 140%. This demonstrates that lateral bi-layer heterojunctions are capable of gain

photocurrent is picked up while the dark current is filtered out. A TPD thin film by itself, while being more conductive than PTCBI, is not able to dissociate enough excitons to generate a measurable quantum efficiency. A single film of PTCBI is capable of breaking up excitons and transporting them laterally through the film as previously shown in the *I*-*V* sweeps. However, the fields required to do this are much higher than the bi-layer heterojunctions because of PTCBI's low conductivity. Thus, this device yields low quantum efficiencies. Note that the introduction of an interface vastly improves external quantum efficiencies over single films, giving an external quantum efficiency on the order of 10% at $\lambda = 552\,\text{nm}$. Accounting for the number of photons absorbed in the thin films, at that wavelength, the internal quantum efficiency for Au/TPD(5 nm)/PTCBI(10 nm) is 140%. The significance of an internal quantum efficiency being larger than 100% is that there is gain present in the device. Another way of explaining this is to say that for every photon that is absorbed by this device, it is able to transport more than one charge carrier from one electrode to another.

Future Work

The lateral bi-layer photoconductor structure presented here exhibits gain and is shown to generate photocurrent exclusively from absorption in the PTCBI film. One benefit that this structure has over a single layer photoconductor is the separation of the charge generation mechanism from the charge transport mechanism. Doing so allows the engineering of each layer to optimize overall device performance. In addition, the physical processes that take place in device operation are not specific to the material set. The critical requirement is that the energy band alignment is favorable for exciton dissociation at the interface between the materials. Thus, both layers can be made of different materials to further optimize the light-to-dark current ratio or to add chemical sensing abilities, which is a natural application by virtue of having a lateral device with a top surface exposed to the environment.

The state-of-the-art fluorescent polymer sensor scheme illustrated in Fig. 5.7 is a commercially available solution for explosives detection (Nomadics/ICX). However, this transduction mechanism still has its drawbacks. Optical coupling losses are high due to the polymer's lambertian emitter profile. Typically, only a small fraction of the polymers' luminescence makes it through to the photodetector. This holds consequences for the signal-to-noise ratio because the detected fluorescence signal competes with the noise in the system from both optical and electronic sources. Despite having sensitivities on the order of parts per trillion in a fraction of a second, there are still compounds whose vapor pressures are beyond the detection limits of this sensing architecture, one example being RDX. Reducing coupling efficiency losses will yield an improvement in signal-to-noise ratio and ultimately in sensitivity. Additionally, the requirement of having a photodetector and a filter for detection add to the cost and the mechanical complexity of the sensor.

Thus, the challenge becomes one of simplifying the transduction of the change in luminescence while maintaining or even augmenting the sensitivity. It is intuitive to think that the most straightforward way of transducing a chemical detection event into an electrical signal is to use chemiresistors, where the polymer is acting as both the chemical sensing layer and the charge transport layer. Placing the film between two electrodes allows the conductivity in the film to be measured directly. However, the problem with these types of sensors is that charge transport and chemical specificity are physically coupled mechanisms. Functionalization of polymers to specific compounds usually degrades the charge transport and vice versa.

One solution is to separate the chemical sensing functionality from the charge transport by depositing an amplified fluorescent polymer on top of a CTL to create a chemosensing bi-layer heterojunction. For example, the TPD transport layer could be replaced by a metal oxide, while the PTCBI could be replaced with an amplified fluorescent polymer, which is sensitive to a specific analyte. If analyte binds to the surface of the polymer, a number of exciton quenching centers develop along the polymer backbone. Upon light excitation, the reduced exciton density in the EGL will manifest as a reduction in the population of charge carriers at the heterointerface. The reduction of charge carriers would then result in a smaller measured photocurrent, demonstrating chemosensitivity (Fig. 5.21).

A number of potential benefits arise from developing a chemosensing bi-layer heterojunction. First, by physically separating the sensing and transport functions in chemical sensors, optimization of the sensor's electrical properties can occur without affecting chemical specificity. Band energy differences, trapping distributions, interface properties, and bias voltages are just a few of the possible parameters that can be used to engineer better vapor-phase chemosensors. Another benefit is the development of a reusable device

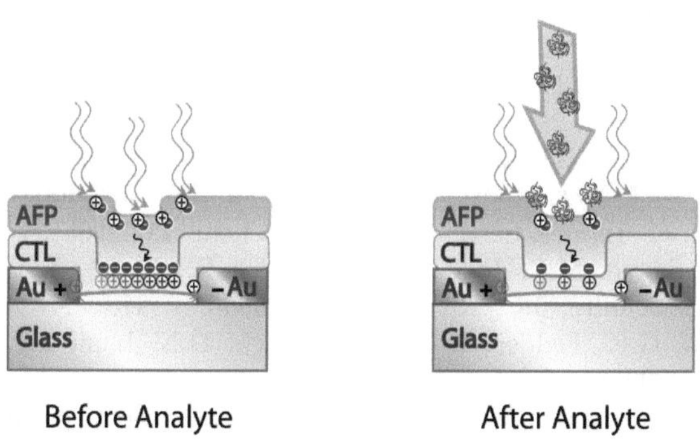

Fig. 5.21. Schematic diagram showing a cross-sectional view of a chemosensing bi-layer heterojunction operating both before and after the presence of analyte. Amplified fluorescent polymer (AFP) is deposited on top of a charge transport layer (CTL) to form the heterojunction

platform for a variety of chemosensing applications. Because the polymer's chemical specificity is no longer coupled to its conductivity, many existing chemosensing fluorescent polymers can be mated with the appropriate CTLs to yield vapor-phase sensors, as long as the polymers retain their photoluminescent efficiency in solid-state. Relaxing the constraint that the polymer must be conductive also simplifies the synthesis of new, chemically specific polymers for this sensing platform. Finally, the electrical amplification inherent in many photoconductive structures enables the use of much thinner active layers while maintaining a measurable signal. Thinning the chemosensing polymer layer will allow the analyte to diffuse throughout the entire film, reducing the background luminescence that is generated from unquenched polymer bulk.

References

1. A. Rose, Z. Zhu, C.F. Madigan, T.M. Swager, V. Bulovic, Proc. SPIE **6333**, 6330Y1–6330Y7 (2006)
2. B.C. Sisk, N.S. Lewis, Sens. Actuators B Chem. **104**, 249–268 (2005)
3. L. Turin, F. Yoshii, *Structure-Odor Relations: A Modern Perspective* (Marcel Dekker, New York, 2002)
4. M. Thompson, D.C. Stone, *Surface-Launched Acoustic Wave Sensors*, vol. 144 (Wiley, New York, 1997)
5. R. Lucklum, P. Hauptmann, Anal. Bioanal. Chem. **384**, 667–682 (2006)
6. C.K. O'Sullivan, G.G. Guilbault, Biosens. Bioelectron. **14**, 663–670 (1999)
7. A. Hierlemann, O. Brand, C. Hagleitner, H. Baltes, Proc. IEEE **91**, 839–863 (2003)
8. Y. Jin, K.M. Wang, R. Jin, Prog. Nat. Sci. **16**, 445–451 (2006)
9. A.J. Heeger, Angew. Chem. Int. Ed. Engl. **40**, 2591–2611 (2001)
10. W. Zheng, Y. Min, A.G. MacDiarmid, M. Angelopoulos, Y.H. Liao, A.J. Epstein, Synth. Met. **84**, 63–64 (1997)
11. J.X. Huang, S. Virji, B.H. Weiller, R.B. Kaner, J. Am. Chem. Soc. **125**, 314–315 (2003)
12. J. Huang, S. Virji, B.H. Weiller, R.B. Kaner, Chem. Eur. J. **10**, 1315–1319 (2004)
13. J. Liu, Y.H. Lin, L. Liang, J.A. Voigt, D.L. Huber, Z.R. Tian, E. Coker, B. McKenzie, M.J. McDermott, Chem. Eur. J. **9**, 605–611 (2003)
14. R. Kessick, G. Tepper, Sens. Actuators B Chem. **117**, 205–210 (2006)
15. B. Li, G. Sauve, M.C. Iovu, M. Jeffries-El, R. Zhang, J. Cooper, S. Santhanam, L. Schultz, J.C. Revelli, A.G. Kusne, T. Kowalewski, J.L. Snyder, L.E. Weiss, G.K. Fedder, R.D. McCullough, D.N. Lambeth, Nano Lett. **6**, 1598–1602 (2006)
16. M.K. Ram, O. Yavuz, V. Lahsangah, M. Aldissi, Sens. Actuators B Chem. **106**, 750–757 (2005)
17. L.M. Dai, P. Soundarrajan, T. Kim, Pure Appl. Chem. **74**, 1753–1772 (2002)
18. S.L. Tan, J.A. Covington, J.W. Gardner, IEEE Proc. Sci. Meas. Technol. **153**, 94–100 (2006)
19. T.C. Pearce, S.S. Schiffman, H.T. Nagle, J.W. Gardner, *Handbook of Machine Olfaction.* (Wiley, Weinheim, 2002)

20. B.C. Sisk, N.S. Lewis, Langmuir **22**, 7928—7935 (2006)
21. E.S. Tillman, M.E. Koscho, R.H. Grubbs, N.S. Lewis, Anal. Chem. **75**, 1748–1753 (2003)
22. T. Gao, M.D. Woodka, B.S. Brunschwig, N.S. Lewis, Chem. Mater. **18**, 5193–5202 (2006)
23. N. Kobayashi, Curr. Opin. Solid State Mater. Sci. **4**, 345–353 (1999)
24. J.P. Germain, A. Pauly, C. Maleysson, J.P. Blanc, B. Schollhorn, Thin Solid Films **333**, 235–239 (1998)
25. B. Schollhorn, J.P. Germain, A. Pauly, C. Maleysson, J.P. Blanc, Thin Solid Films **326**, 245–250 (1998)
26. W.F. Qiu, W.P. Hu, Y.Q. Liu, S.Q. Zhou, Y. Xu, D.B. Zhu, Sens. Actuators B Chem. **75**, 62–66 (2001)
27. Y.L. Lee, C.H. Chang, Sens. Actuators B Chem. **119**, 174–179 (2006)
28. S. Iijima, Nature **354**, 56–58 (1991)
29. J. Li, Y. Lu, M. Meyyappan, IEEE Sens. J. **6**, 1047–1051 (2006)
30. J. Li, Y. Lu, Q. Ye, M. Cinke, J. Han, M. Meyyappan, Nano Lett. **3**, 923–933 (2003)
31. I. Sayago, E. Terrado, M. Aleixandre, M.C. Horrillo, M.J. Fernandez, J. Lozano, E. Lafuente, W.K. Maser, A.M. Benito, M.T. Martinez, J. Gutierrez, E. Munoz, Sens. Actuators B Chem. **122**, 75–80 (2007)
32. H. Shibata, M. Ito, M. Asakursa, K. Watanabe, IEEE Trans. Instrum. Meas. **45**, 564–569 (1996)
33. R. Igreja, C.J. Dias, in *Advanced Materials Forum Ii*, vol. 455–456, Materials Science Forum (Trans Tech Publications Ltd, Zurich-Uetikon, 2004), pp. 420–424
34. S.V. Patel, T.E. Mlsna, B. Fruhberger, E. Klaassen, S. Cemalovic, D.R. Baselt, Sens. Actuators B Chem. **96**, 541–553 (2003)
35. A.M. Kummer, A. Hierlemann, H. Baltes, Anal. Chem. **76**, 2470–2477 (2004)
36. T. Boltshauser, H. Baltes, Sens. Actuators A Phys. **26**, 509–512 (1991)
37. T. Boltshauser, L. Chandran, H. Baltes, F. Bose, D. Steiner, Sens. Actuators B Chem. **5**, 161–164 (1991)
38. F. Josse, R. Lukas, R.N. Zhou, S. Schneider, D. Everhart, Sens. Actuators B Chem. **36**, 363–369 (1996)
39. F.L. Dickert, G.K. Zwissler, E. Obermeier, Berichte Der Bunsen-Gesellschaft-Phys. Chem. Chem. Phys. **97**, 184–188 (1993)
40. P. Van Gerwen, W. Laureyn, W. Laureys, G. Huyberechts, M.O. De Beeck, K. Baert, J. Suls, W. Sansen, P. Jacobs, L. Hermans, R. Mertens, Sens. Actuators B Chem. **49**, 73–80 (1998)
41. A.M. Kummer, A. Hierlemann, IEEE Sens. J. **6**, 3–10 (2006)
42. R. Igreja, C.J. Dias, in *Advanced Materials Forum Iii, Pts 1 and 2*, vol. 514–516, Materials Science Forum (Trans Tech Publications Ltd, Zurich-Uetikon, 2006), pp. 1064–1067
43. A. Tsumura, H. Koezuka, T. Ando, Appl. Phys. Lett. **49**, 1210–1212 (1986)
44. L. Torsi, A. Dodabalapur, L. Sabbatini, P.G. Zambonin, Sens. Actuators B Chem. **67**, 312–316 (2000)
45. J.T. Mabeck, G.G. Malliaras, Anal. Bioanal. Chem. **384**, 343–353 (2006)
46. J. Locklin, Z.N. Bao, Anal. Bioanal. Chem. **384**, 336–342 (2006)
47. J.B. Chang, V. Liu, V. Subramanian, K. Sivula, C. Luscombe, A. Murphy, J.S. Liu, J.M.J. Frechet, J. Appl. Phys. **100**, 7 (2006)

48. M. Bouvet, Anal. Bioanal. Chem. **384**, 366–373 (2006)
49. C.S. Huang, B.R. Huang, Y.H. Jang, M.S. Tsai, C.Y. Yeh, Diam. Relat. Mater. **14**, 1872–1875 (2005)
50. H. Chen, M. Josowicz, J. Janatax, Chem. Mater. **16**, 4728–4735 (2004)
51. B.K. Crone, A. Dodabalapur, R. Sarpeshkar, A. Gelperin, H.E. Katz, Z. Bao, J. Appl. Phys. **91**, 10140–10146 (2002)
52. L. Wang, D. Fine, A. Dodabalapur, Appl. Phys. Lett. **85**, 6386–6388 (2004)
53. L. Wang, D. Fine, S.I. Khondaker, T. Jung, A. Dodabalapur, Sens. Actuators B Chem. **113**, 539–544 (2006)
54. N.A. Rakow, K.S. Suslick, Nature **406**, 710–713 (2000)
55. N.A. Rakow, A. Sen, M.C. Janzen, J.B. Ponder, K.S. Suslick, Angew. Chem. Int. Ed. Engl. **44**, 4528–4532 (2005)
56. K.S. Suslick, MRS Bulletin **29**, 720–725 (2004)
57. K.S. Suslick, N.A. Rakow, A. Sen, Tetrahedron **60**, 11133–11138 (2004)
58. T. Nakamoto, M. Yosihioka, Y. Tanaka, K. Kobayashi, T. Moriizumi, S. Ueyama, W.S. Yerazunis, Sens. Actuators B Chem. **116**, 202–206 (2006)
59. J.Y. Wang, Z.A. Luthey-Schulten, K.S. Suslick, Proc. Natl. Acad. Sci. USA **100**, 3035–3039 (2003)
60. M.C. Janzen, J.B. Ponder, D.P. Bailey, C.K. Ingison, K.S. Suslick, Anal. Chem. **78**, 3591–3600 (2006)
61. A.V. Samoylov, V.M. Mirsky, Q. Hao, C. Swart, Y.M. Shirshov, O.S. Wolfbeis, Sens. Actuators B Chem. **106**, 369–372 (2005)
62. T. Basova, E. Kol'stov, A.K. Ray, A.K. Hassan, A.G. Gurek, V. Ahsen, Sens. Actuators B Chem. **113**, 127–134 (2006)
63. X. Bevenot, A. Trouillet, C. Veillas, H. Gagnaire, M. Clement, Meas. Sci. Technol. **13**, 118–124 (2002)
64. Y.C. Kim, W. Peng, S. Banerji, K.S. Booksh, Opt. Lett. **30**, 2218–2220 (2005)
65. J. Mavri, P. Raspor, M. Franko, Biosens. Bioelectron. **22**, 1163–1167 (2007)
66. E.B. Feresenbet, F. Taylor, T.M. Chinowsky, S.S. Yee, D.K. Shenoy, Sens. Lett. **2**, 145–152 (2004)
67. E.B. Feresenbet, E. Dalcanale, C. Dulcey, D.K. Shenoy, Sens. Actuators B Chem. **97**, 211–220 (2004)
68. B. Culshaw, J. Lightwave Technol. **22**, 39–50 (2004)
69. E. Scorsone, S. Christie, K.C. Persaud, F. Kvasnik, Sens. Actuators B Chem. **97**, 174–181 (2004)
70. S. Khalil, L. Bansal, M. El-Sherif, Opt. Eng. **43**, 2683–2688 (2004)
71. T. Shinquan, C.B. Winstead, R. Jindal, J.P. Singh, IEEE Sens. J. **4**, 322–328 (2004)
72. P.A.S. Jorge, M. Mayeh, R. Benrashid, P. Caldas, J.L. Santos, F. Farahi, Appl. Opt. **45**, 3760–3767 (2006)
73. S. Christie, E. Scorsone, K. Persaud, F. Kvasnik, Sens. Actuators B Chem. **90**, 163–169 (2003)
74. C. Elosua, I.R. Matias, C. Bariain, F.J. Arregui, Sensors **6**, 1440–1465 (2006)
75. A. Messica, A. Greenstein, A. Katzir, Appl. Opt. **35**, 2274–2284 (1996)
76. I.R. Matias, J. Bravo, F.J. Arregui, J.M. Corres, Opt. Eng. Lett. **45**, 1–3 (2006)
77. Y.L. Hoo, W. Jin, H.L. Ho, D.N. Wang, R.S. Windeler, Opt. Eng. **41**, 8–9 (2002)
78. S. Luo, Y. Liu, A. Sucheta, M. Evans, R.V. Tassell, Proc. SPIE **4920**, 193–204 (2002)

79. R. Falate, R.C. Kamikawachi, M. Muller, H.J. Kalinowski, J.L. Fabris, Sens. Actuators B Chem. **105**, 430–436 (2005)
80. M. Giordano, M. Russo, A. Cusano, G. Mensitieri, Sens. Actuators B Chem. **107**, 140–147 (2005)
81. C. Elosua, I.R. Matias, C. Bariain, F.J. Arregui, Sensors **6**, 578–592 (2006)
82. P.D. O'Neal, A. Meledeo, J.R. Davis, B.L. Ibey, V.A. Gant, M.V. Pishko, G.L. Cote, IEEE Sens. J. **4**, 728–734 (2004)
83. P.A.S. Jorge, P. Caldas, C.C. Rosa, A.G. Oliva, J.L. Santos, Sens. Actuators B Chem. **130**, 290–299 (2004)
84. W.Q. Cao, Y.X. Duan, Sens. Actuators B Chem. **110**, 252–259 (2005)
85. K.D. Weston, P.J. Carson, J.A. DeAro, S.K. Buratto, Chem. Phys. Lett. **308**, 58–64 (1999)
86. S.W. Thomas, G.D. Joly, T.M. Swager, Chem. Rev. **107**, 1339–1386 (2007)
87. J.F. Callan, A. Prasanna de Silva, D.C. Magri, Tetrahedron **61**, 8551–8588 (2005)
88. D. Filippini, A. Alimelli, C. Di Natale, R. Paolesse, A. D'Amico, I. Lundstrom, Angew. Chem. Int. Ed. Engl. **45**, 3800–3803 (2006)
89. C. Di Natale, R. Paolesse, A. D'Amico, Sens. Actuators B Chem. **121**, 238–246 (2007)
90. D.T. McQuade, A.E. Pullen, T.M. Swager, Chem. Rev. **100**, 2537–2574 (2000)
91. S.J. Toal, J.C. Sanchez, R.E. Dugan, W.C. Trogler, J. Forensic Sci. **52**, 79–83 (2007)
92. I. Sanchez-Barragan, J.M. Costa-Fernandez, M. Valledor, J.C. Campo, A. Sanz-Medel, Trends Anal. Chem. **25**, 958–967 (2006)
93. S.J. Zhang, F.T. Lu, L.N. Gao, L.P. Ding, Y. Fang, Langmuir **23**, 1584–1590 (2007)
94. S. Bencic-Nagale, T. Sternfeld, D.R. Walt, J. Am. Chem. Soc. **128**, 5041–5048 (2006)
95. S.Y. Tao, Z.Y. Shi, G.T. Li, P. Li, Chemphyschem **7**, 1902–1905 (2006)
96. Q. Zhou, T.M. Swager, J. Am. Chem. Soc. **117**, 12593–12602 (1995)
97. J.S. Yang, T.M. Swager, J. Am. Chem. Soc. **120**, 11864–11873 (1998)
98. J.S. Yang, T.M. Swager, J. Am. Chem. Soc. **120**, 5321–5322 (1998)
99. C.J. Cumming, C. Aker, M. Fisher, M. Fox, M.J. la Grone, D. Reust, M.G. Rockley, T.M. Swager, E. Towers, V. Williams, IEEE Trans. Geosci. Remote Sens. **39**, 1119–1128 (2001)
100. J.S. Kim, D.T. McQuade, A. Rose, Z.G. Zhu, T.M. Swager, J. Am. Chem. Soc. **123**, 11488–11489 (2001)
101. K. Kuroda, T.M. Swager, Macromol. Symp. **201**, 127–134 (2003)
102. J. Sung, R.J. Silbey, Anal. Chem. **77**, 6169–6173 (2005)
103. N. Tessler, G.J. Denton, R.H. Friend, Nature **382**, 695–697 (1996)
104. V.G. Kozlov, V. Bulovic, P.E. Burrows, S.R. Forrest, Nature **389**, 362–364 (1997)
105. I.D.W. Samuel, G.A. Turnbull, Chem. Rev. **107**, 1272–1295 (2007)
106. C.F. Madigan, V. Bulovic, Phys. Rev. Lett. **96** (2006)
107. V.G. Kozlov, V. Bulovic, P.E. Burrows, M. Baldo, V.B. Khalfin, G. Parthasarathy, S.R. Forrest, J. Appl. Phys. **84**, 4096–4108 (1998)
108. A. Rose, Z.G. Zhu, C.F. Madigan, T.M. Swager, V. Bulovic, Nature **434**, 876–879 (2005)
109. D. O'Carroll, I. Lieberwirth, G. Redmond, Nature Nanotechnol. **2**, 180–184 (2007)

110. P. Peumans, A. Yakimov, S.R. Forrest, J. Appl. Phys. **93**, 3693–3723 (2003)
111. C.W. Tang, S.A. Vanslyke, Appl. Phys. Lett. **51**, 913–915 (1987)
112. P.M. Borsenberger, D.S. Weiss, *Organic Photoreceptors for Xerography* (Marcel Dekker, New York, 1998)
113. B.E.A. Saleh, M.C. Teich, *Fundamentals of Photonics* (Wiley-Interscience, New York, 1991)
114. M. Pope, C.E. Swenberg, *Electronic processes in organic crystals and polymers*, 2nd ed. (Oxford University Press, New York, 1999)
115. J. Reynaert, V.I. Arkhipov, P. Heremans, J. Poortmans, Adv. Funct. Mater. **16**, 784–790 (2006)
116. P. Peumans, S. Uchida, S.R. Forrest, Nature **425**, 158–162 (2003)
117. J. Peet, C. Soci, R.C. Coffin, T.Q. Nguyen, A. Mikhailovsky, D. Moses, G.C. Bazana, Appl. Phys. Lett. **89**, 3 (2006)

6

Detection of Chemical and Physical Parameters by Means of Organic Field-Effect Transistors

A. Bonfiglio, I. Manunza, P. Cosseddu, and E. Orgiu

6.1 Introduction

Organic semiconductors have been studied so far mainly for their optoelectronic properties, and in fact applications based on their light emitting and absorbing properties (OLED, organic light emitting diodes, and OSC, organic solar cells) are at a mature level of development [1, 2]. This interest has been mainly due to the fact that this class of materials allows devices to obtain a unique trade-off between electrical conductivity, light emission/absorption, mechanical flexibility, and, last but not least, low cost of deposition and patterning techniques. In particular, the possibility of exploiting "easy" techniques such as printing for realizing devices on unusual surfaces, like paper, or plastic, and over large areas, is a reason for further exploring new applications.

On the other hand, there are several drawbacks associated with these materials: among them, low-carrier mobility that severely limits the possibility of applications in electronic circuits, and poor environmental stability upon exposure to external agents, which has so far limited the lifetime of devices in all practical applications. Therefore, in order to fully exploit the great potential of these materials without being limited by these drawbacks, it is advisable to focus on those applications where high performance in terms of switching speed are not required and environmental sensitivity could be a plus, rather than a minus.

Sensors seem to be the optimal candidate for fully profiting from the unique properties of these materials. For instance, slow electronic responses are generally not a problem for (bio)chemosensors because the global speed of the device is already limited by the dynamics of the (bio)chemoreaction occurring at the sensing area. Environmental sensitivity can also be exploited in order to achieve specific functionality in gas sensing devices. Furthermore, the ability to obtain large sensing areas is certainly a benefit for a wide set of applications (as it is for solar cells) and using printing techniques for creating sensing devices on unusual surfaces could certainly widen the set of possible applications where sensing is required. Sensors include the broad class of (bio)chemosensing devices but also detectors for physical parameters such as pressure, strain, temperature, etc. The integration of different devices endowed with the properties of organic materials on the same mechanical support is especially interesting as it allows obtaining "smart," flexible surfaces that can be employed in emerging applications such as the recently proposed electronic skin (for robotic applications) and smart textiles.

In the following sections, we will focus our attention on two particular classes of such devices, namely (bio)chemodetectors for ionic solutions and strain/pressure sensors. We will present the current state-of-the-art technology and applications, and will show how to take advantage of flexible, free-standing dielectric films for obtaining different sensing devices on conformable, large surfaces.

6.2 An Overview of Organic Field-Effect Sensors

An organic field-effect sensor is, first of all, a transistor, i.e., an active device, able to produce an output that is the amplified copy of the input. Alternatively, the device can behave as a bistable component, able to switch between the ON and OFF states depending on the input. In a field-effect device, the input signal is a voltage, applied through a capacitive structure to the device channel. A field-effect sensor is based on the idea that a signal induced by the parameter to be sensed alters the input voltage or the semiconductor mobility and therefore modifies the output current. This reversible modification may then be exploited to measure the signal amplitude.

An Organic semiconductor-based Field Effect transistor (OFET) is usually realized in a thin film configuration which is a structure that was developed for the first time for amorphous silicon devices [3]. An OFET is comprised of a multilayered structure where the gate capacitor is formed by a metal, an insulator, and a thin organic semiconductor layer. On the semiconductor side, two metal contacts, source and drain, are used for extracting a current that depends both on the drain–source voltage and on the gate–source voltage. A thin organic semiconductor layer is needed because this device does not work in inversion mode, i.e., the carriers that accumulate in the channel are the same that normally flow in the bulk in the OFF state. Therefore, if the semiconductor is too thick, the OFF-current is too high and the switching ability of the device is compromised. For the same reason, this structure is used only for low-mobility semiconductors, as high mobility semiconductors would give rise, even in thin film devices, to high OFF-currents.

To obtain a sensor from an OFET, it is necessary to achieve, in one of the layers that form the device, a specific sensitivity/selectivity toward an external stimulus applied to the device. If this is a chemical stimulus, often a specific chemical functionalization of one of the device layers is required. Organic semiconductors, though intrinsically affected by many, still unsolved, problems, such as low mobility and poor stability [4], have a very unique property that makes them interesting for chemosensing: they can be chemically tailored in order to obtain specific properties. This aspect, together with the low cost of deposition and fabrication techniques, makes them really attractive for future electronic concepts. Organic semiconductors are not meant to compete with inorganic materials in terms of "classical" electronic properties (like mobility, ON/OFF ratio, etc.) but rather in terms of new properties, unachievable by traditional electronic materials. Chemical tailoring is used, for instance, to tune the energy bandgap in order to obtain a large variety of emission/absorption colors in optoelectronic devices. Similarly, this feature can be extremely interesting in the field of chemical sensing. In fact, the synthesis of semiconductors with specific chemical functionalities can result in specific chemical sensitivities in the active layer of the device. Chemical tailoring is in fact the key for obtaining chemical selectivity in any sensing technique but it has not been exploited much with inorganic semiconductors devices (because in this case it is intrinsically difficult to obtain).

In an active device, like an organic field-effect transistor, chemical tailoring can be applied not only to the semiconductor but also to metallic and insulating layers, thus allowing different localizations of the device-sensing area. This possibility broadens the set of sensing principles exploitable for these devices. In addition, from the electrical characterization of an active device it is possible to simultaneously extract different parameters, which correlate to the identification of a chemical species in a mixture. In this way it is possible to obtain a sort of "fingerprint" of a compound [5].

Chemical field-effect sensors in which the organic semiconductor comes in direct contact with the analyte and acts as the sensitive layer, are the subject of intense activity [6] and are described in detail in Chap. 5. In this chapter we will focus our attention on those field-effect devices in which the sensing area is localized on the gate side of the device and is used to achieve chemical detection in a solution brought in contact with it. We will refer to this class of devices as charge-modulated (CM) field-effect devices, because they exploit the effect of a charge variation at the gate level to induce a conductance variation in the device channel. In our lab, we have developed this concept first using complimentary metal–oxide–semiconductor (CMOS) technology [7, 8], for detecting DNA hybridization, and with organics on a fully flexible plastic film [9].

To obtain detectors for physical parameters, again the external stimulus must reversibly affect one of the different layers of the device and result in variation of one or more of the electronic parameters (mobility, threshold voltage, etc.) that can be extracted from the output curves. In this chapter, we will describe pressure/strain sensors that can be obtained starting from an organic field-effect transistor. The integration of different detection principles on the same flexible mechanical support is an interesting example of what can be obtained thanks to the intriguing properties of organic sensors. It is likely that this possibility will become progressively more attractive as demonstrated by the interest gained by innovative applications such as the electronic skin, first proposed by Someya et al. [10].

6.3 (Bio)chemosensing in Solution

6.3.1 Ion Sensitive Organic Field-Effect Transistors (ISOFETs)

Ion Sensitive Field-Effect Transistors (ISFETs) developed to measure pH and to sense a variety of analytes in solution are well known in the silicon technology field [11]. In these devices the gate insulator is in direct contact with the electrolyte solution (see Fig. 6.1).

By coating the gate insulator with, for instance, enzymes as the selecting agent, highly selective sensors have been developed. Such enzyme-modified ISFETs can in principle be constructed with any enzyme that upon reaction with the analyte induces a local change of pH. Silicon-based, ISFET biosensors are used in medical diagnostics, environmental monitoring, and food quality control [12].

Conventional ISFET sensors are manufactured by CMOS technology on silicon; long-term stability problems correlated with the fabrication and packaging costs are major limiting factors for practical applications. Disposable microsensors for health-related applications are highly desirable due to the safety requirements and limited lifetime of the biological components involved. The discovery of semiconducting polymers and the ability to dope these polymers over the full range from insulator to metal has resulted in a class of materials that combine the electronic and optical properties of semiconductors and metals with the attractive mechanical properties and processing advantages of polymers. For such applications in sensors, where disposable devices are desirable, organic semiconductor materials present several advantages such as simpler fabrication techniques compared to silicon, compatibility with plastic mechanical supports and biocompatibility. In 2000, Bartic et al. [13, 14] proposed the first example of an ion-sensitive organic field-effect transistor (ISOFET), whose structure is shown in Fig. 6.2.

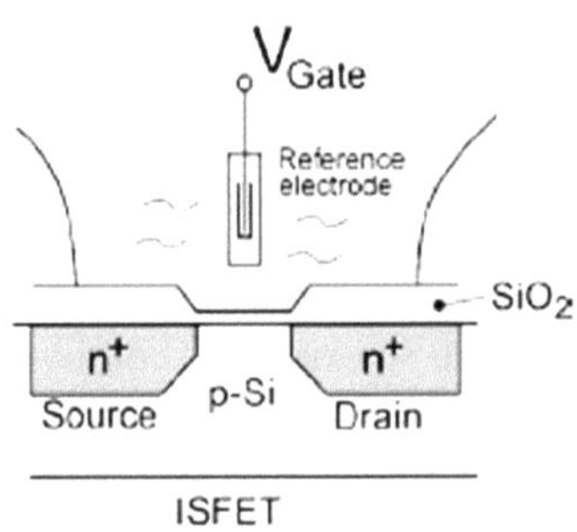

Fig. 6.1. Schematic representation of an ISFET [10]. Copyright 2003 IEEE

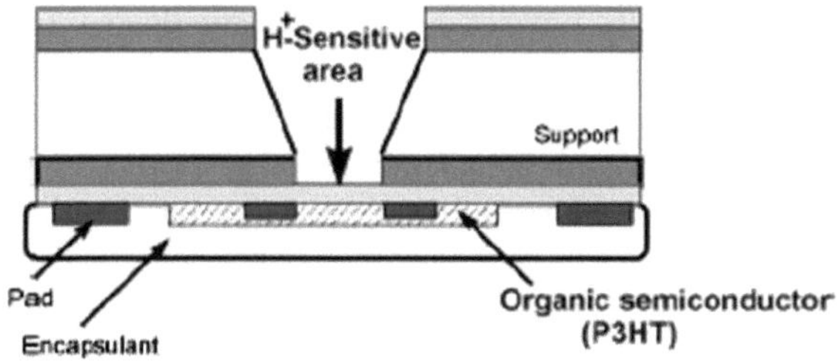

Fig. 6.2. Cross section of an ion sensitive organic field-effect transistor (ISOFET) [14]. Copyright 2001 IEEE

In ISOFETs as well as in ISFETs, the metallic gate electrode deposited on the dielectric layer is omitted and the dielectric layer is exposed to an aqueous solution where an Ag/AgCl electrode is used as the gate. The use of a non-polarizable electrode guarantees a constant potential drop at this interface. In other words, the potential difference across it is virtually fixed and consequently acts as a "reference." The drain current is modulated by field-effect doping as in OFETs but the electric field across the insulating gate dielectric is controlled by ions at the electrolyte/insulator interface. The detection mechanism in ISOFETs is similar to that of ISFETs as pH variations modify the voltage drop across the dielectric/semiconductor interface, resulting in a variation in the drain current.

With the aim of developing low-cost transducers for bioanalytical applications, Bartic et al. [14] have fabricated and characterized this preliminary prototype device in order to investigate the ability of an organic transistor to detect charge. In this structure, a regioregular poly(3-hexylthiophene) (P3HT) layer acts as a p-type semiconductor. The choice of the material was based on its good solubility, processability and high environmental and thermal stability. The device chemical sensitivity toward protons is achieved by exposing a silicon nitride (Si_3N_4) layer to the aqueous solution under investigation. A silicon wafer with a thickness of 650 μm was used as the mechanical support for the organic transistor. To expose the gate insulator to the solution, a window was anisotropically etched in the silicon. After etching, the resulting device structure is very frail even if the thickness of the sensitive layer has been chosen in order to ensure low-mechanical stress. However, the fact that the sensitive area and the electrical contacts are on opposite sides of the insulating layer is very convenient from the packaging point of view. In fact, the device can be encapsulated with a room temperature curable epoxy resin and the semiconductor is therefore protected from ambient and the solution to be monitored.

To investigate the pH response, the device was immersed in the buffer solution together with the reference electrode and was biased as a typical FET in a common source configuration. The authors explain the working mechanism by coupling the site-binding theory developed for electrolyte/dielectric systems with the solid state physics of the electronic device. In fact, while the electrochemical phenomena at the electrolyte/dielectric interface are exactly the same as in silicon-based devices, the effect of pH variation on the conduction mechanism of organic semiconductors is more complex: in organic materials, a large density of trapping levels is present and the charge carrier mobility is gate-field and therefore pH dependent. P3HT behaves as a lightly doped, p-type semiconductor; therefore, when a small bias, V_{DS}, is applied between the source and drain electrodes, with $V_{REF} = 0\,V$, a drain current is measured, corresponding to the bulk conductivity of the film. It is known that in p-type organic transistors, the semiconductor contains positive charge carriers induced by the unintentional doping in the bulk of the polymer as well as by the field-effect that creates the accumulation layer. The bulk charge is constant, depending only on the density of the free carriers at equilibrium, while the accumulated charge is gate-voltage dependent. The drain current can therefore be considered to have a field-effect component and a bulk component.

In a given pH buffer, the potential drop at the electrolyte/dielectric interface occurs if an external voltage, V_{REF}, is applied on the reference electrode with respect to the source electrode of the transistor. When the pH of the solution changes, a bias variation, $\delta\psi$, occurs at the electrolyte/dielectric interface. This voltage drop will be reflected in the band bending at the dielectric/semiconductor interface, leading to a charge variation $\delta Q = C_{ox} \cdot \delta\psi$, where C_{ox} is the total series capacitance of the insulating material and the double layer capacitance (Gouy–Chapman theory). Depending on the direction of the pH change, the charge in the channel will be incremented (accumulation) or decreased (depletion) by this amount. The excess charge is provided by the source/drain contacts [15].

A particularity of ISOFETs as compared to their inorganic counterparts (i.e., silicon ISFETs) is the pH dependence of the charge mobility, especially at high biases applied on the reference electrode. The mobility in some organic semiconductors increases with charge density (due to filling of low-lying states). Since charge accumulated at the semiconductor/insulator interface is a function of V_{GS} and ψ_0, with ψ_0 dependent on both pH and V_{GS}, it is expected that the mobility will be also pH-dependent. This is an interesting aspect that can be used to improve the performances of these devices by using organic semiconductors with intrinsically higher mobility values rather than polymers like P3HT [11].

The chemical sensitivity of these organic devices has been demonstrated for both protons and glucose by Bartic et al. [17]. Again P3HT is used as the active semiconducting layer but in this case tantalium oxide has been used as the dielectric layer owing to its larger number of proton-sensitive surface sites, which improves the pH sensitivity and because it is well known that the use of high dielectric constant insulators enables the reduction of the operation voltage of organic transistors. A simple glucose biosensor has been realized by anchoring an enzymatic layer (glucose oxidase) onto the gate

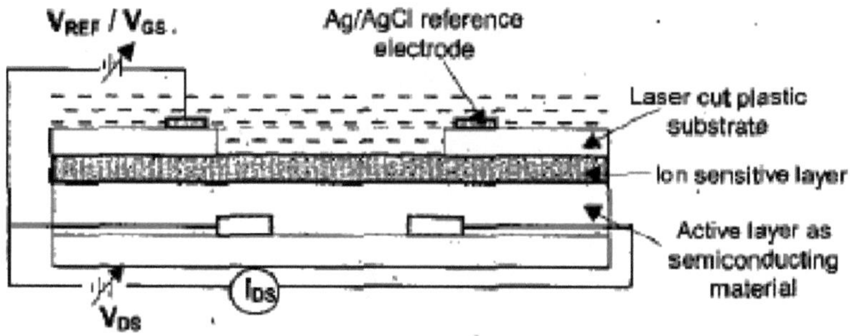

Fig. 6.3. Schematic of the device structure proposed in [18]. Copyright 2003 IEEE

insulator. The hydrolysis of the glucose catalyzed by the enzyme, increases the proton concentration at the insulating surface and consequently generates a current variation proportional to the glucose concentration. The effect of the biocatalytic conversion of the glucose to gluconic acid is reflected by the decrease of the drain current when the glucose is present in the solution. The increase of the acid concentration (pH decrease) in the enzymatic layer is responsible for this effect and this is in perfect agreement with the behavior of pH sensors where the drain current increases with the pH.

Bringing this technology to another level, Gao et al. [18] have proposed a "full polymer" disposable FET on a polymer mechanical support, which is integrated with a microfluidic system for pH measurement. While a conventional ISOFET sensor uses an external Ag/AgCl reference electrode as the gate electrode, they developed and integrated the gate electrode on the device structure by using laser micromachining technology. The device structure is shown in Fig. 6.3.

The reference electrode and the microfluidic channel are directly assembled on top of the gate insulating layer. As a result, the polymer FET reported in this work can be integrated with disposable lab-on-a-chip platforms, which is very attractive for in vivo applications. Also in this case the application of the device can be extended to the detection of biomolecules by modifying the gate layer with a proper functionalization membrane.

We have proposed still another approach [9]: A 900-nm thick Mylar® foil is used as H^+-sensitive layer and also as mechanical support of the whole device. The main feature of the proposed device is that it is a mechanical support-free, lightweight and totally flexible structure (see Fig. 6.4).

The working principle of this device is similar to other ISOFETs described here, with some differences arising due to the different materials employed. In particular, while the electrochemical phenomena occurring at the electrolyte/inorganic dielectric interface are well known, a site-binding model that explains in detail the mechanism at the surface for the Mylar® foils has not yet been developed. Nevertheless, the sensitivity of the device to the pH variation has been demonstrated (as shown in Fig. 6.5)

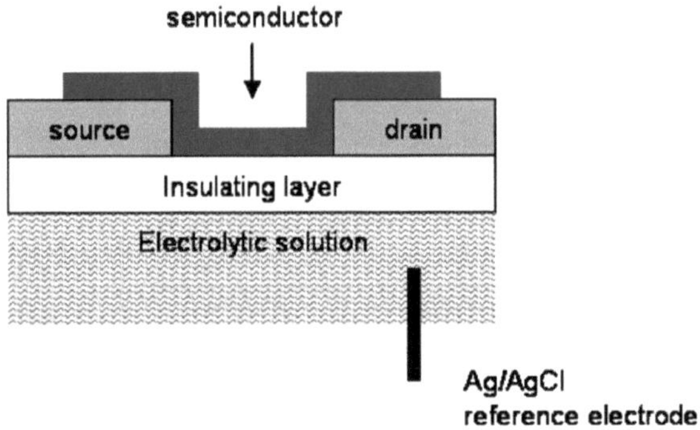

Fig. 6.4. Structure of the ion-sensitive organic field-effect device proposed in [9]. Reprinted with permission. Copyright 2005, American Institute of Physics

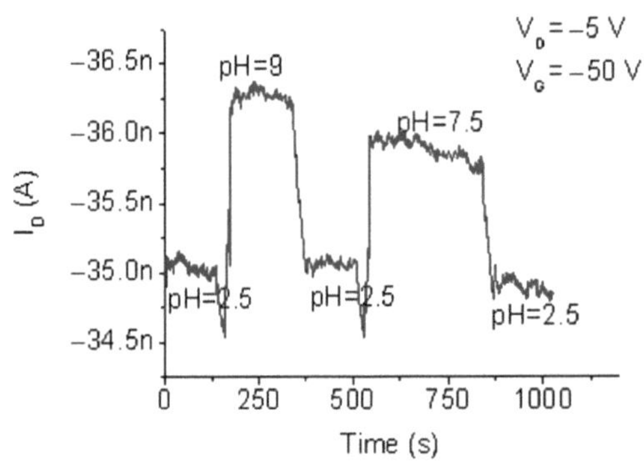

Fig. 6.5. I_D vs. time curve taken with different pH values of the electrolytic solution (linear regime)

In addition, Mylar® (and PET in general) is a widely used biocompatible material. For this reason many approaches to the modification and functionalization of the polymer surface by wet chemistry, plasma processes, or UV treatment have been reported in the literature [19–22]. These surface modification approaches demonstrate that it is possible to improve the reactivity of the PET surface in order to generate specific groups on the surface, or to immobilize biomolecules. Therefore, possibilities for (bio)chemosensing on a fully flexible mechanical support can be envisioned and are very interesting for innovative applications such as smart packaging and biotechnology.

However, chemical functionalization of metal is a much easier task than grafting molecules on polymeric surfaces. As an example, well-known processes such as thiol self-assembly could be very easy to apply to this kind of device structure. Therefore, one of the possible developments of the ion-sensitive strategy concerns the development of floating gate devices similar to those cited previously [7,8], first realized in our lab using CMOS technology. This structure can be employed to detect changes of electric charge without the need for any external components (e.g., an external reference electrode).

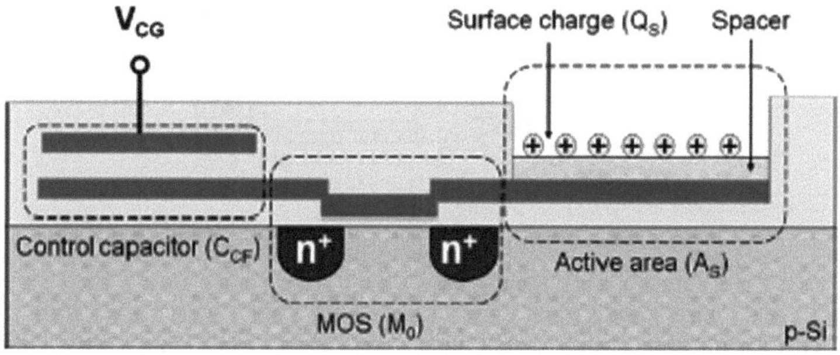

Fig. 6.6. Structure of an ion-sensitive organic field-effect floating gate device for chemical sensing [7]. Copyright 2006 IEEE

The schematic of this device is shown in Fig. 6.6. Both the control-capacitor and the charge immobilized on the active area will determine the actual gate voltage drop of the transistor. The final voltage will be due to the combined action of these two parameters and will determine the drain current flowing in the transistor channel. Besides the elimination of the external electrode, another interesting feature of this structure is that the control gate allows the ability to address a single device in an array rather than switching-on all devices with a common reference electrode. This is very important for practical purposes, for example in the case of sensor arrays for the detection of DNA hybridization or any other biomolecular process involving a change in electric charge (antigen/antibody interaction, proteomics, etc.) immobilized on the sensing gate of the device [7, 8]. The organic version of this device is currently under investigation in our laboratory.

6.4 Strain and Pressure Sensors

Despite the rather large range of gas and chemical sensors based on organic field-effect transistors reported in the literature, only a few reports of mechanical sensors have been published so far. Relatively little progress has been made in the field of pressure or bending recognition compared to the areas of gas and chemical sensing, mainly because mechanical sensing requires attributes of conformability and flexibility as well as three-dimensional large area shaping that in many cases are difficult to achieve even for organic devices. On the other hand, in terms of applications for these devices, artificial sense of touch is considered an essential feature for future generations of robots and wearable electronics have become one of the hottest themes in electronics aimed at the design and production of a new generation of garments with distributed sensors and electronic functions. For both of these fields, conformable strain/pressure sensors are highly desirable. In the following paragraphs an overview of this field is presented.

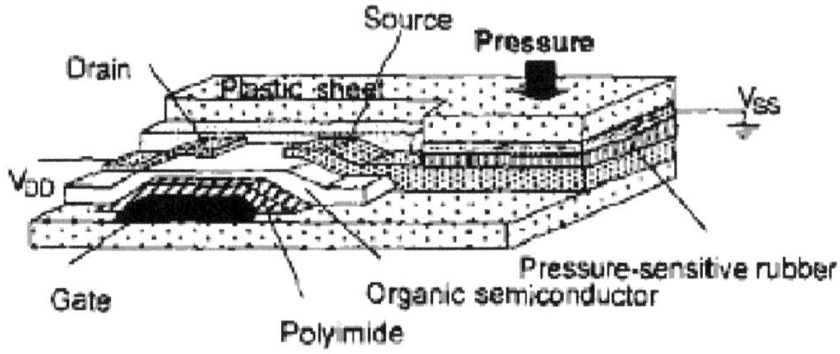

Fig. 6.7. Schematic of the device structure reported in [10]. Copyright 2003 IEEE

6.4.1 State of the Art of Mechanical Sensors Including OFETs

Silicon technology is not suitable for manufacturing low-cost, large-area sensor devices, because these devices need to be light, flexible, and even disposable (for some biomedical applications). The inherent high-temperature fabrication processes associated with silicon devices make it very difficult to use inexpensive flexible mechanical support materials, resulting in high fabrication costs. It is well known that the mobility of organic materials is about three order of magnitude lower than that of crystalline semiconductors [23], however in the case of sensors for large areas, for electronic skin in particular, the lower speed is tolerable and the use of organic materials therefore allows mechanical flexibility, large-area, low-cost, and relative ease of fabrication without suffering from their drawbacks.

The first example of a large area pressure network fabricated on a plastic sheet by means of integration of organic transistors and rubbery pressure-sensitive elements was reported by Someya et al. [10] in 2003. The device structure is shown in Fig. 6.7.

In order to fabricate the organic transistor, first a glass resin was spin-coated and cured onto a 50-μm thick polyimide film with an 8-μm thick copper film acting as the gate electrode. Then a pentacene layer (semiconductor) of nominal thickness (30 nm) was vacuum sublimed onto these films at ambient temperature. Finally, source and drain gold electrodes were deposited by means of vacuum evaporation through shadow masks. Pressure sensors may also be made using pressure-sensitive conducting rubbery sheets sandwiched between two 100 μm width metal lines that cross at right angles. One of the metal lines is connected to an organic transistor while the other line is connected to the ground. The pressure-sensitive sheet is a 0.5 mm thick silicone rubber containing graphite. In this work organic transistors are used to address the rubber pressure-sensitive elements in a sensor array. The equivalent circuit diagram is shown in Fig. 6.8.

In 2004, Someya et al. [24] improved the fabrication technique for these devices and developed an electronic artificial "skin." In this work, once again, organic transistors were not used as sensors in themselves but as addressing elements of a flexible matrix which was used to read out pressure maps

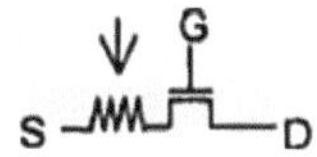

Fig. 6.8. Equivalent circuit diagram of the device [10]. Copyright 2003 IEEE

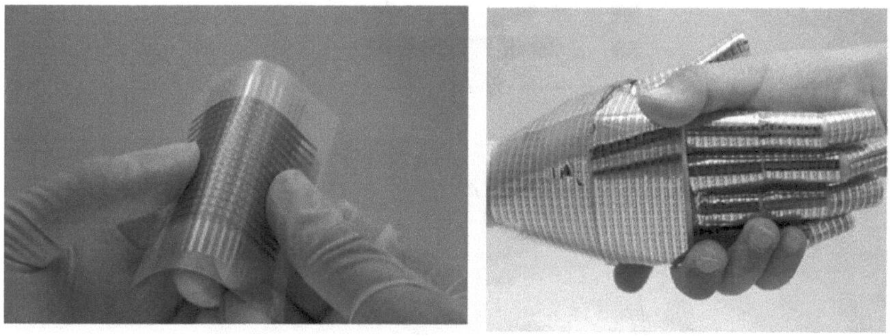

Fig. 6.9. Electronic artificial "skin" [24]. Copyright 2004 National Academy of Sciences, U.S.A.

from pressure-sensitive rubber elements containing graphite. The resulting electronic artificial "skin" is shown in Fig. 6.9.

The mobility of organic transistors at −100 V is comparable to that of amorphous silicon but this operating voltage is not realistic for artificial skin applications. At −20 V the mobility is still large ($0.3\,cm^2\,V\,s^{-1}$) and the device still functions. In the active driving method presented, only one transistor needs to be in the on-state for each cell where pressure is applied so this design is suitable for low-power applications where a high number of cells is required over large areas, such as electronic skin. The device can detect a few tens of kilopascals, which is comparable to the sensitivity of discrete pressure sensors and the time response of the pressure-sensitive rubber is typically of the order of hundreds of milliseconds.

Recently, low-cost manufacturing processes have been further optimized for the development of flexible active matrices using ink-jet printed electrodes and gate dielectric layers [25]. This work demonstrates the feasibility of a printed organic FET active matrix as a read out circuit for sensor applications. Figure 6.10 shows a cross section of this device structure (Fig. 6.10a), an image of the large area pressure sensor (Fig. 6.10b), and an image of the stand-alone organic transistor and the circuit diagram (Fig. 6.10c).

Other groups have also attempted to use organic semiconductors as sensing elements. For instance, Rang et al. [26] have investigated the hydrostatic-pressure dependence of *I–V* curves in organic transistors. This device was realized on a heavily doped silicon wafer and measured in a hydrostatic pressure apparatus (a hydraulic press). The authors found a large and reversible dependence of drain current and hole mobility on hydrostatic pressure, suggesting that this kind of device could be suitable for sensor applications.

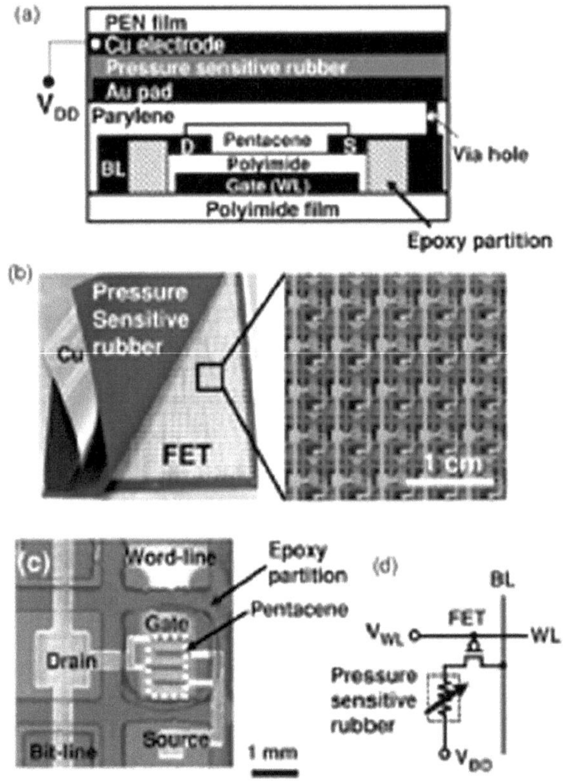

Fig. 6.10. (**a**) Cross section of a pressure sensor; (**b**) pressure sensor comprising an organic FET active matrix, a pressure-sensitive rubber, and a PEN film with a Cu electrode. A magnified image of the active matrix is also shown; (**c**) micrograph of stand-alone pentacene FETs; (**d**) circuit diagram of a stand-alone pressure sensor cell. Reprinted with permission from [25]. Copyright 2006, American Institute of Physics

However, the proposed device was not flexible, therefore not suitable for applications like robot skin, e-textiles, etc.

Darlinski et al. [27] have studied the possibility of developing pressure sensors based solely on organic transistors, without the need for any additional sensing element. In this way the organic device itself acts as the sensing element. To study the pressure dependence of the electrical performance of these devices, the authors applied mechanical force directly on the transistors using a tungsten microneedle moved by a step motor as shown in Fig. 6.11. During measurements, the device is placed on a balance in order to measure the applied pressure. The authors explain the force-induced change in drain current in terms of variation of the distribution and activity of trap states at or near the semiconductor/dielectric interface. Again, a major drawback of

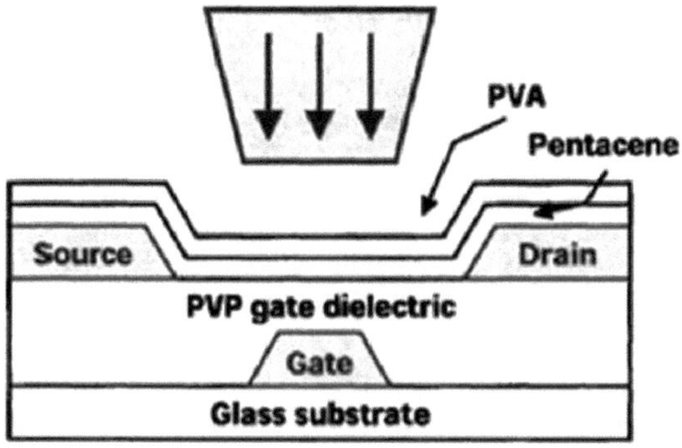

Fig. 6.11. Schematic of the device structure reported in [27]. Reprinted with permission. Copyright 2005, American Institute of Physics

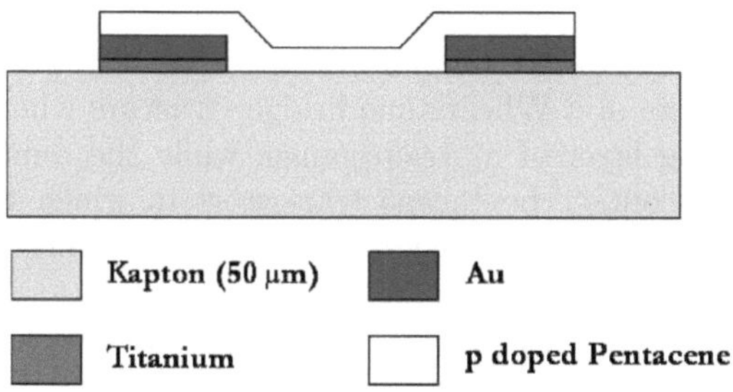

Fig. 6.12. Cross section of the organic semiconductor strain sensor [28]. Copyright 2005 IEEE

the reported device is that, because it was built on a glass slide, it was not mechanically flexible.

Strain sensors using an organic semiconductor as the sensor (resistive) element of a strain gauge have been also reported by Jung and Jackson [28]. In conventional strain gauges, the large stiffness mismatch generated between the inorganic semiconductor element and the flexible (polymeric) mechanical support may lead to irreversible plastic deformations, which can be problematic. The stress in the sensor element (i.e., the inorganic semiconductor or the metal, with a high Young modulus) is not representative of the stress present in the mechanical support (i.e., the polymeric material with low Young modulus). As a result, the sensor performance is reduced in terms of reliability and reproducibility. On the opposite, it is expected that the use of organic semiconductors with a low Young modulus as the sensing elements would minimize the induced stress concentration. A cross section of the sensor is shown in Fig. 6.12.

For these sensors, 2-nm thick Ti and 20-nm thick Au layers were deposited on 50-μm thick polyimide films by vacuum evaporation. Next, a 50 nm pentacene layer was deposited, again by vacuum sublimation. The pentacene layer was then doped p-type by exposure to a 1% solution of ferric chloride in water.

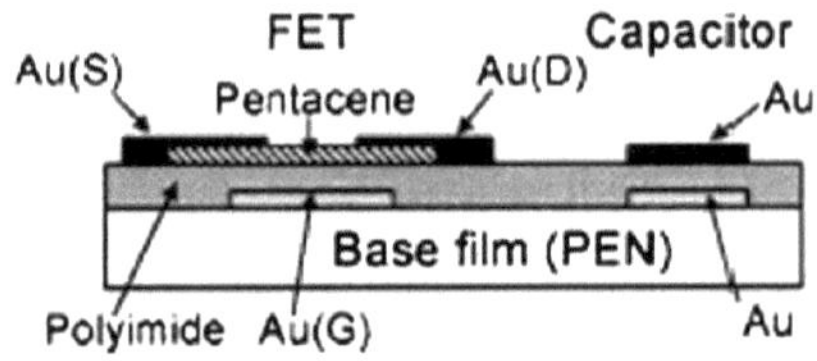

Fig. 6.13. Cross section of the organic FET and capacitor reported in [30]. Reprinted with permission. Copyright 2005, American Institute of Physics

The maximum process temperature used to fabricate the organic strain sensors is 110°C. The devices were tested using a Wheatstone bridge configuration and the results indicate that it is possible to fabricate a strain sensor at low temperatures with mechanical characteristics matched to low Young modulus supports using organic semiconductors.

Joung et al. [29] have also demonstrated the possibility of combining these sensors with pentacene-based thin film transistors as temperature sensors. The strain sensor consists of a Wheatstone bridge structure where the pentacene film acts as sensing layer of a strain gauge, while the temperature sensors consist of bottom-contact pentacene transistors in which the variations of the drain currents in the subthreshold regime are measured with respect to temperature.

The effects of strain on pentacene transistor characteristics while changing the bending radius of the structure have been investigated by Sekitani et al. [30]. A cross section of their device structure is shown in Fig. 6.13.

For the development of this device, first, a gate electrode consisting of 5 nm Cr and 100 nm Au was vacuum evaporated onto a 125-μm thick polyethylenenaphthalate (PEN) film. Polyimide precursors were then spin-coated and cured at 180°C to form 900-nm thick gate dielectric layers. A 50-nm thick pentacene film was vacuum sublimed and finally, 60-nm thick Au drain and source electrodes were vacuum evaporated through a shadow mask. For comparison, a capacitor was also manufactured simultaneously using the same base film. The authors observed large changes in the drain current that cannot be explained by the deformation of the device structure alone. In the analysis of the electrical characteristics, changes in the structural parameters of the transistors (namely the channel width and length, and the thickness of the gate insulator) were taken into account in order to evaluate possible variations in the mobility of the transistor.

The transfer characteristics of the transistor and the mobility variations observed are shown in Fig. 6.14 as a function of strain or bending radius. The mobility increases monotonically when the strain is changed from tension to compression passing through the flat state. The observed strain effect on the mobility of organic FETs is quite reasonable since the transport properties of organic molecular systems follow the hopping model rather than the band model. Under compressive strain, as described in a theoretical study [31], the hopping rate increases due to the smaller average distance between molecules,

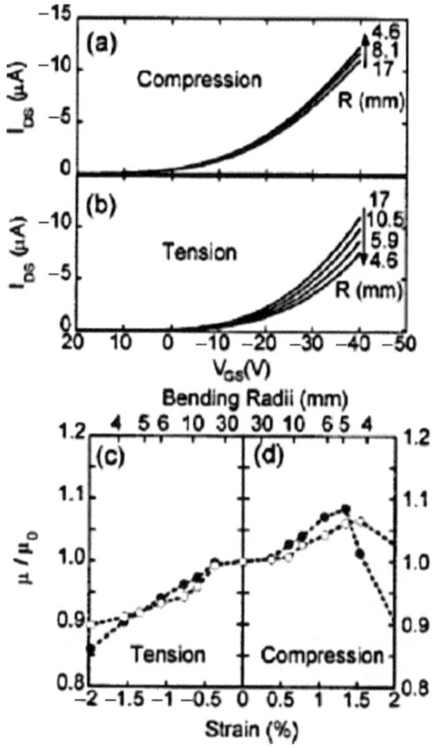

Fig. 6.14. Transfer curves on (**a**) compressive and (**b**) tensile strains. Mobility as function of strain or bending radius in (**a**) tensile strain and (**b**) compressive strain [30]. Reprinted with permission Copyright 2005, American Institute of Physics

resulting in an increase in the mobility of the FETs. In contrast, tensile strain causes a larger spacing, resulting in a decrease in mobility. Interestingly, strain dependence of mobility is identical in transistors arrayed perpendicular or parallel to the direction of strain. Such an isotropic electrical property indicates that transport is also consistent with the transport in polycrystalline thin films with randomly oriented crystalline grains.

6.4.2 Flexible Structures for Mechanical Sensors

Organic transistors should be inherently mechanically flexible, but many of the devices reported in the literature are fabricated on rigid mechanical supports such as silicon or glass. Often, even in devices assembled on plastic films, the presence of a mechanical support results in a reduction of flexibility. In addition, if the whole device is flexible, the mechanical stimulus can be applied to the semiconductor through the support itself (thus avoiding the risk of damaging the semiconductor layer). We have recently proposed a support-free and totally flexible structure for mechanical sensors [32].

An organic transistor was assembled starting from a 1.6-μm thick polyethylenetherephthalate foil (Mylar®, Du Pont, dielectric constant of 3.3, dielectric rigidity of $10^5\,\mathrm{V\,cm^{-1}}$) that acts as gate insulator and also as mechanical support of the whole device. Vacuum evaporated bottom-contact gold electrodes were patterned on the upper side of the organic insulating layer while the gold gate electrode was deposited on the opposite side. A vacuum sublimed pentacene layer was used as the active layer. The device structure is shown in Fig. 6.15. The proposed device is completely flexible (given that the total thickness of the device is less than 2 μm) and combines the ability to carry out both switching and sensing functions. Furthermore, thanks to the transparency of the dielectric film [33], it is possible to obtain self-aligned contacts with a consequent strong reduction of parasitic capacitance effects that could limit the dynamic performance of these devices. The marked sensitivity of the drain current to an elastic deformation induced by a mechanical stimulus on the device channel has been exploited for detecting a pressure applied by means of air flow on the gate side of the free-standing device. In Fig. 6.16 the current variation in response to different increasing values of applied pressure is displayed.

Figure 6.17 shows what happens when a constant value of pressure is repeatedly applied and removed over time. The drain current variation is reproducible and reversible, despite the presence of a hysteresis. The sensor

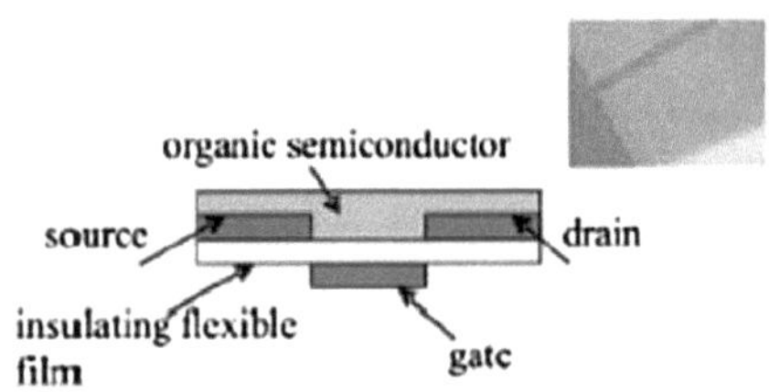

Fig. 6.15. Cross section of the mechanical sensor reported in [32]. Reprinted with permission. Copyright 2005, American Institute of Physics

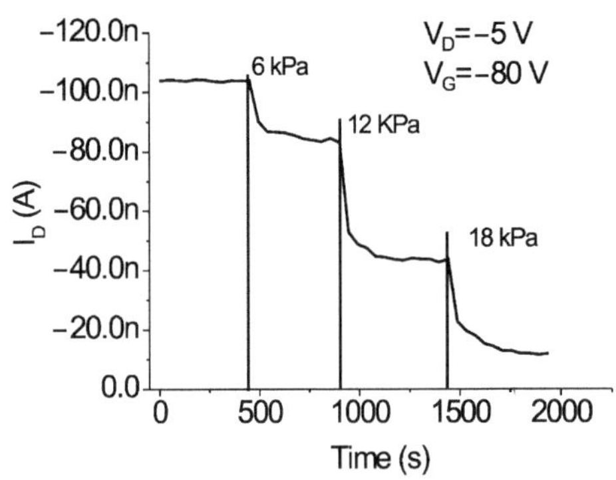

Fig. 6.16. I_D vs. time curve with different applied pressures. Reprinted with permission from [32]. Copyright 2005, American Institute of Physics

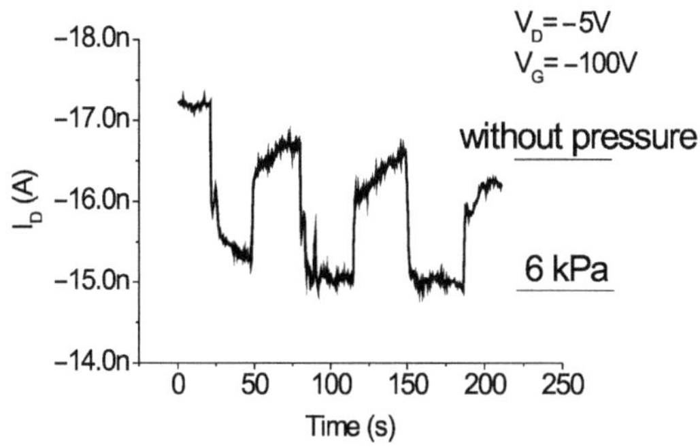

Fig. 6.17. Reversibility of I_{D} vs. time [34]

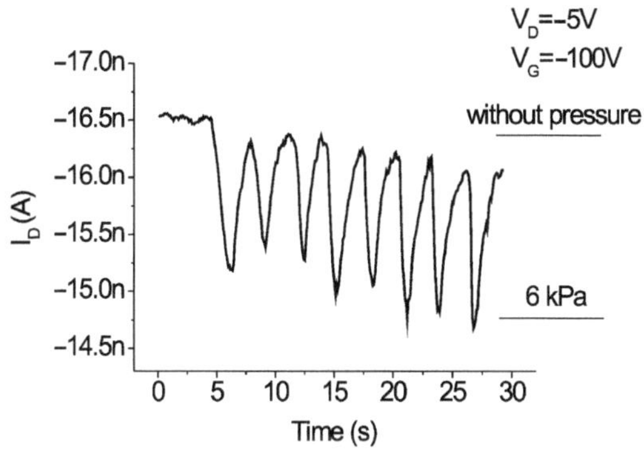

Fig. 6.18. I_{D} vs. time

responds very fast to the mechanical stimulus (i.e., within tens to hundreds of milliseconds) but the time required to reach the steady state is much higher (tens to hundreds of seconds) with a significant variability from sample to sample.

Figure 6.18 shows the sensor response to a "breathing-like" signal. The pressurised air flow is applied and removed according to a natural human breathing rhythm. As shown, despite the hysteresis, the sensor response is reversible, reproducible, and fast enough to monitor physiological parameters in wearable sensor applications.

A careful analysis of the pressure dependence of the current [34] shows that this dependence can be explained in terms of variation in the mobility, the threshold voltage and in the contact resistance of the transistor. The variation of the mobility may be attributed to a direct dependence of the semiconductor conductivity on the pressure applied to the device while the distribution and activity of trap states are typically responsible for variations in contact resistances and threshold voltages.

In order to clarify the influence of structural effects (in particular at the metal/semiconductor interfaces) on the pressure sensitivity, we have also developed, on the same insulating layer, couples of bottom-contact and top-contact devices with the same active layer as described in [34]. The different metal–semiconductor interface is expected to affect the electrical characteristics of the transistors even if no pressure is applied. In fact, the morphology of the pentacene film in the device channel region close to the electrode edges is different in top-contact and bottom-contact devices. In the bottom-contact structure near to the edge of the electrodes there is an area with a large number of grain boundaries that can act as charge carrier traps and are believed to be responsible for the reduced performance of bottom-contact devices as compared to top-contact devices. The results show that the threshold voltage decreases with respect to pressure in bottom-contact devices while in top-contact devices it is pressure-insensitive. This observation confirms the role of insulator/semiconductor and metal/semiconductor interfaces in determining the pressure sensitivity of these devices.

On the other hand, the mobility had a very similar behavior in top-contact and in bottom-contact devices, indicating also a direct contribution of the semiconductor mobility to the observed sensitivity. Despite the fact that the underlying mechanism of the observed pressure sensitivity is not completely clarified yet, pressure sensitivity seems to result from a combination of mobility variations in the channel and interface effects in the source/drain surrounding areas, likely due to morphological modifications of the pentacene layer under stress.

The main drawbacks of these devices seem to be the high operating voltage and a limited stability that could be overcome by using a proper flexible encapsulation layer in order to protect the semiconductor layer from exposure to ambient (humidity, light, etc.) as these factors are known to cause drift and in general to negatively affect the device performance. Nevertheless, our results [34] suggest that, despite the current degradation occurring in a nonencapsulated device with time, the pressure sensitivity of the device is weakly affected by degradation.

6.5 Design and Technology of Organic Field-Effect Sensors

In principle, the choice of materials used to develop both the electrodes and the insulator of an OFET-based sensor should be basically the same as for OFETs developed for other applications. Similarly, the design of the device should follow the same rules. However, there are some unique aspects that are intrinsic to sensors and deserve special attention. For instance, the sensing area of the device: this is the area where the external stimulus (chemical or physical) must be applied to the device without affecting its integrity and/or robustness. The first example of an ISOFET reported in the literature [13] was fabricated on silicon and the only organic component of it was the semiconductor employed as the active layer. In order to develop this structure, it is necessary to etch the highly doped silicon from the back side of the device in order to have free access to the gate oxide thin film. This procedure is not trivial and can dramatically compromise the robustness and functionality of the final device. Furthermore, it does not appropriately exploit the unique properties of organic materials such as, for instance, mechanical flexibility. Therefore, although the basic structure seems to be interesting, a different device design would be preferable. Similarly, for pressure/strain sensors, the mechanical stimulus should be applied so as to avoid possible damage to the active layer of the device, i.e., not directly to the semiconductor, but possibly to the supporting layer, that should therefore be able to transmit the stimulus to the layer itself.

Another relevant issue for sensors is packaging. In particular, for chemical sensors designed for working in solution, it is necessary to prevent the solution from any contact with the semiconductor layer (if this is not the sensitive layer of the device). Microfluidic systems [35, 36] coupled with the sensor's active areas offer a valid solution to this problem because they allow the flow of the solution to the active area to be controlled and channeled, without compromising the semiconductor layer. For pressure/strain sensors the packaging should not compromise the mechanical flexibility of the whole structure.

Besides these unique aspects of sensing structures, materials issues also deserve consideration. It is widely recognized that organic dielectrics may be the best solution for fabricating the insulating layer of OFETs. Organic dielectrics can be solution-processed, produce transparent films on any kind of mechanical support and show far higher dielectric permittivity values than silicon dioxide (up to 10). As a result, fully flexible OFETs with good electrical performance have already been reported in the literature [37]. Furthermore, the employment of this new class of gate dielectric materials allows a higher versatility in the choice of device configurations which minimized some of the problems related to the structure of the device itself. For instance, it is possible to construct top gate structures, reported in Fig. 6.19. High-performance devices, such as ambipolar OFETs and light-emitting transistors, have already been reported using this configuration [38]. In this way, it is also possible to obtain a free gate dielectric surface without the limits imposed by a bottom gate configuration. Moreover, by coating the organic semiconductor with an insulating film, the dielectric layer can also be used to protect the surface to avoid aging problems generally due to polluting species present in the surrounding environment. A possible drawback is a constraint on the

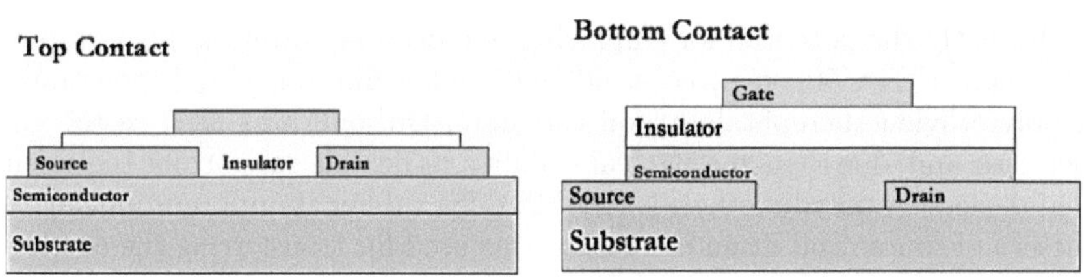

Fig. 6.19. OFETs in top gate configurations

compatibility between the solvent used for depositing the insulating layer and the underlying semiconductor.

Recent studies [33, 39, 40] have shown that it is possible to use flexible organic dielectrics such as poly(ethyleneterephthalate) (PET) and poly(dimethylsiloxane) (PDMS) to produce transparent, free-standing films. By using free-standing films, a (usually stiff and opaque) support is no longer required for the mechanical stability of the device. Another advantage of free-standing devices, which is exploitable in sensors like ISOFETs, is that they are characterized by two working surfaces. It can therefore be possible to pattern the active part of the device (source and drain electrodes and the organic semiconductor) on one side, leaving the opposite side of the free-standing foil free for the contact with the surrounding environment without compromising the overall robustness and functionality of the assembled device [9], and also with no constraints for materials deposition, as in top gated structures. Furthermore, the transparency of the dielectric layer can be particularly attractive because this structure offers the possibility of optical access to the semiconductor layer.

From an industrial perspective, a fundamental issue in device fabrication concerns the availability of suitable materials, not only for the active semiconductor layer but also for contacts that, so far, have been mainly fabricated with metals. Metals have several problems: first, though deposited in very thin layers, they are usually not mechanically flexible, thus compromising the overall robustness of devices; second, organic semiconductors offer the possibility of employing very simple and low-cost device assembly techniques some of which cannot be applied to metals. Several device fabrication techniques have been reported, allowing the easy and low-cost all-polymer device patterning such as ink jet printing, soft lithography, electro-polymerization, laser patterning, and dry printing [41–43]. From an industrial perspective, it would also be desirable to employ such techniques to obtain each layer of the device. Printing contacts with conducting polymers [44, 45] is one possibility, but it has several limitations, such as, for instance, the spatial resolution of the printed pattern, and the compatibility between the employed "ink" and the printing hardware [46]. Soft lithography [47] represents a step forward to easily achieving structures with small dimensions at a low cost and with good reproducibility suitable for mass production with roll-to-roll processes. It has also been successfully applied to organic devices, with interesting results concerning the lamination [48] or the surface chemistry mediated transfer of metal contacts [49].

Recently the potential for employing microcontact printing for patterning all-organic FETs [50] on a free-standing dielectric film was also demonstrated. A poly(ethylenetherephtalate) foil (Mylar®, DuPont) was used as the gate dielectric and, due to its mechanical stability, as flexible support for the assembled devices. A polydimethylsiloxane (PDMS) rubber stamp, reproducing the pattern of source and drain electrodes, was used for transferring the conducting polymer poly(ethylene dioxithiophene) doped with polystyrene sulfonate (PEDOT:PSS) pattern on the mechanical support in order to obtain the final

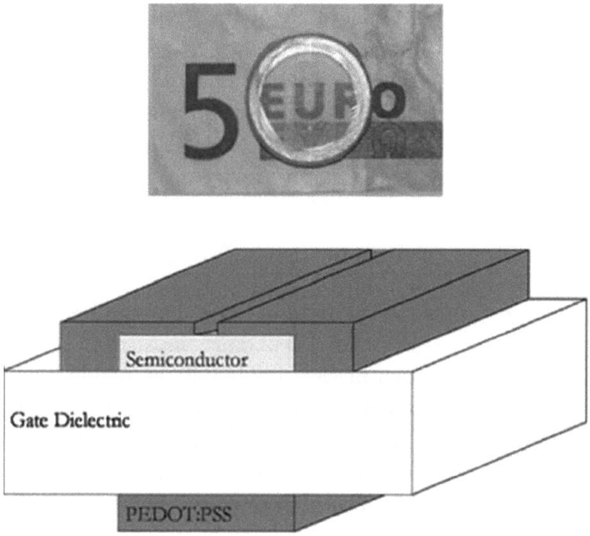

Fig. 6.20. Schematic of the all-organic FET structure; a picture of the final assembled device is provided, demonstrating the transparency of the resulting structure

structure. In this way, it was possible to obtain fully flexible and transparent all-organic FETs (see Fig. 6.20) with channel mobility and ON/OFF current ratios up to $0.1\,cm^2\,V\,s^{-1}$ and 10^5, respectively, [51].

All these findings contribute to new opportunities for the fabrication of flexible and efficient devices, with small dimensions, that may have potential applications for the fabrication of all-plastic sensing devices [34]. In particular, for strain/pressure sensors, the use of conducting polymers instead of metals allows to apply a pressure on the device without risks of irreversible damage to the contacts.

6.6 Applications for Organic Field-Effect Sensors

The possibility of obtaining low-cost, sensitive, selective, plastic chemo-, bio-, and physical sensors is attractive for a number of fields of research. In the field of biomedicine, a large amount of interest has been generated for the use of biosensors, in part due to the large variety of molecules, complexes, and also of physical parameters that are necessary to detect. Many systems have been developed in the past years to fill this need [52]. Organic field-effect sensors are still in their embryonic phase. As far as we know, no commercial application has been developed as yet. Despite this, many groups are working toward innovative solutions that make use of this technology. Here we will focus on two cases that seem particularly interesting: the first is the development of an artificial skin for robots, while the second concerns e-textiles. Both deal with a common need, i.e., the conformability of the final product to a 3D shape, a robot in the first case, the human body in the second. This requirement is fully satisfied by devices realized on flexible mechanical supports that are able to adapt their shape. Furthermore, large area is another desirable characteristic. In the following sections, an overview of these applications is given with a special focus on requirements for future devices.

6.6.1 Artificial Sense of Touch

The skin is the largest organ of the human body. For the average adult human, the skin has a surface area between 1.5–2.0 m^2, most of it is between 2–3 mm thick. The average square inch of skin holds 650 sweat glands, 20 blood vessels, 60,000 melanocytes, and more than a thousand nerve endings. Skin is composed of three primary layers: the epidermis, which provides waterproofing and serves as a barrier to infection; the dermis, which serves as a location for the appendages of skin; and the hypodermis, which is called the basement membrane (see Fig. 6.21). The dermis is tightly connected to the epidermis by a basement membrane. It also contains many nerve endings that provide the sense of touch and heat (amongst others).

Somatic sensation is given by the signals provided by several sensory receptors that trigger the experiences labeled as touch or pressure, temperature (warm or cold), pain (including itch and tickle), and the sensations of muscle movement and joint position including posture, movement, and facial expression. Touch is considered one of the five human senses; however, tactile sensation is given by an unique combination of different feelings: the perception of pressure (hence shape, softness, texture, vibration, etc.), relative temperature, and sometimes pain. The high degree of dexterity which

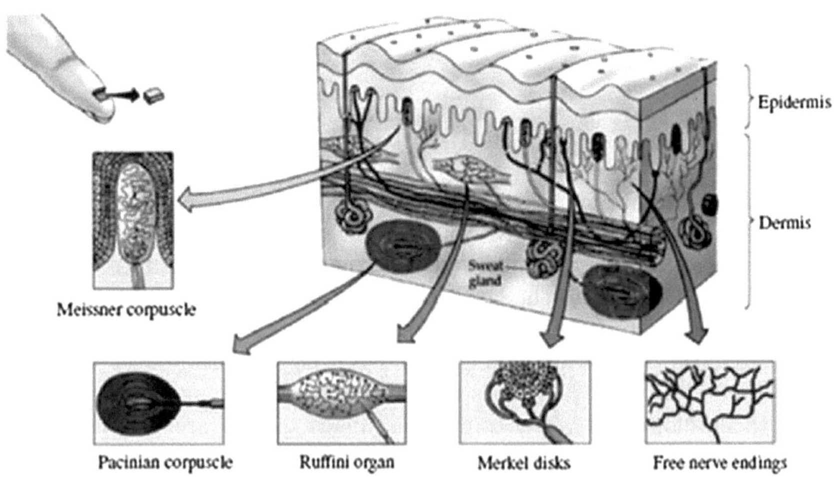

Fig. 6.21. Cross section of human skin [53]

characterizes grasping and manipulative functions in humans and the sophisticated capability of recognizing the features of an object are the results of a powerful sensory-motor integration which fully exploits the wealth of information provided by the cutaneous and kinaesthetic neural afferent systems [53]. A very accurate description of tactile units is available [53], where a classification of these units according to receptive fields and response time is given.

Obviously, reproducing the human sense of touch with an artificial system is a very challenging task, primarily, because the term "touch" is actually the combined term for several senses. Research and development on tactile sensing has experienced an impressive growth in the last three decades, and several kinds of materials and techniques have been proposed. The tentative specifications for tactile sensors have been defined [54] as:

1. The sensor surface or its covering should be both robust and durable.
2. The sensor should provide stable and repeatable output signals. Loading and unloading hysteresis should be minimal.
3. Linearity is important, while a monotonic response is absolutely necessary. Some degree of nonlinearity can be corrected through signal processing.
4. The sensor transduction bandwidth should not be less then 100 Hz, intended as tactile image frame frequency. Individual sensing units should accordingly possess faster responses to allow multiplexing.
5. Spatial resolution should be at least of the order of 1–2 mm, as a reasonable compromise between gross grasping and fine manipulation tasks.

The development of tactile sensors is one of the most challenging aspects of robotics research. Many technologies have been explored, including carbon-loaded elastomers, piezoelectric materials, and micro-electromechanical systems. Artificial skin examples, able to detect pressure, already exist, but these are difficult to manufacture in large enough quantities to cover a robot body, and, moreover, they do not stretch. The most promising examples of "electronic skin-like" systems with large areas are based on organic semiconductors and have been reported by Takao Someya's group at the University of Tokyo [55]. They have developed conformable, flexible, wide-area networks of thermal and pressure sensors, in which measurements of temperature and pressure mapping were performed simultaneously. The device structure is shown in Fig. 6.22.

The above-mentioned skin is stretchable, and remains as sensitive to pressure and temperature when it is at full stretch and also when it is relaxed. In the presented design, both sensor networks contain their own organic transistor active matrices for data read-out. This arrangement means that each network is self-contained and electrically independent and that organic transistors are only used to address sensitive (pressure or thermal) elements. The arrays are less sensitive than human skin, but already mark an improvement on previous efforts, while sensing temperatures in the range of 30–80°C. Moreover the structure is flexible enough to be rolled or bent around a 2 mm bending cylinder. Someya estimates that his E-skin will be commercially

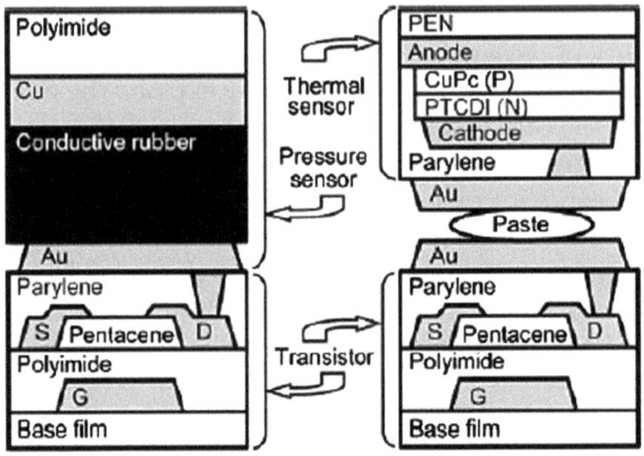

Fig. 6.22. Schematic of the device structure reported in [55]. A cross-sectional illustration of the pressure (*Left*) and the thermal (*Right*) sensor cells with organic transistors is shown. Copyright 2005 National Academy of Sciences, U.S.A.

available within four years and in the near future it will be possible to make an electronic skin that has function that human skin lacks by integrating various sensors not only for pressure and temperature, but also for light, humidity, strain, sound, or ultrasound.

Moreover, it may also be possible in the next few years to develop electronic skin made entirely out of organic transistors. In particular the possibility of developing strain and pressure sensors that can simultaneously act as switches and as sensors, without the need of any further sensing element, will be an interesting possibility. Furthermore, flexible chemosensitive transistors, biosensors, and temperature sensors could be developed using the same technologies allowing new features for this application.

6.6.2 E-Textiles

There is an increasing interest in the emerging area of E-textiles, by which the idea of endowing garments and fabrics with new electronic functions is meant, in particular aimed at monitoring physiological parameters in patients [56] and in subjects exposed to particular risks or external harsh conditions [57].

Strain and pressure sensors and also biochemosensors for measuring characteristics of the human skin are particularly interesting for this kind of applications because they could measure a wide set of parameters such as posture, breathing activity, body fluids composition, etc. in a totally nonintrusive way. This characteristic is in fact very interesting for practical applications. For instance, it would allow doctors to monitor the patient status in real time, 24 h a day; additionally, it would afford a better quality of life to patients for whom they would be perceived as noninvasive monitoring systems.

Basic specifications for this application are rather similar to those listed for the electronic skin. In addition, these systems, being in contact with (or

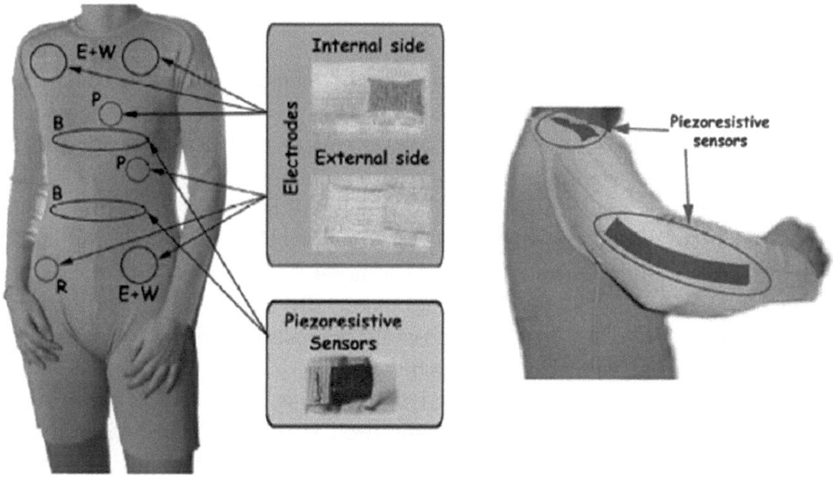

Fig. 6.23. Examples of sensorized garments for recording body signals [63]. Copyright 2005 IEEE

in close proximity to) the human body, must be subject to strict safety standards. Organic field-effect sensors developed on plastic flexible films are good candidates to accomplish this function as they can be assembled in arrays on flexible mechanical supports to be inserted in the fabric itself.

At present, the first attempts to develop strain sensors on garments are being made with piezoresistive stripes deposited on the garments [58–63]. The detection is made through piezoresistive tracks running on the fabric along, for example, a sleeve or parallel to the chest in a T-shirt. In this way, the movement results in a deformation of the track and in a variation of its resistance. No spatial resolution is achievable with such a strategy that is based on the measurement of the resistance of the whole track. The employment of OFETs could allow arrays and matrices to be built, so as to enable full spatial resolution (Fig. 6.23).

A potentially interesting development is the incorporation of these functions directly on yarns [64–66]. In this way, it could be possible to build a whole fabric made with sensitive elements that will as a result be distributed over a large area. Not only the distributed function in the fabric is a truly novel element, but also the possibility of leveraging an existing industrial technology like textiles for creating a new technological approach, though challenging, could really have a huge impact. Nevertheless, high-operating voltages of devices and chemical safety of materials are still open questions and must be carefully considered in designing such an application.

6.7 Conclusions

In this chapter we presented an overview of recent progress in the field of organic field-effect based sensors. In particular, we focused our attention on two kinds of devices: (bio)chemosensors for charged species detection in solution and sensors for strain/pressure measurements. A large amount of activity is currently taking place in these fields, which will hopefully contribute to the development of novel applications by harnessing the interesting and useful properties of organic semiconductors. Biomedicine, robotics, and the rapidly growing field of smart/E-textiles could especially benefit from the introduction of new, flexible, low cost, and possibly disposable sensors for detecting a variety of chemical and physical parameters.

Many issues remain to be solved for focusing on realistic applications: among them, the high voltages still required to drive these devices. Nevertheless, chemical tailoring ability and the possibility of distributing different sensing functions on large and flexible areas are the main strengths of this technology, which has a real potential of being beneficial in future applications for robotics, biomedicine, and biotechnology.

Acknowledgments

The authors acknowledge European Commission for funding the Integrated Project PROETEX and the Italian Research Ministry for funding the Project PRIN 2006 "Plastic BIO-FET Sensors"

References

1. Special Issue on Flexible Electronics Technology, Part I: systems and Applications, IEEE Trans. Electron. Dev. **93**(7) (2005)
2. Special Issue on Flexible Electronics Technology, Part II: materials and devices, IEEE Transaction on Electron Devices, 93, 8 (2005)
3. P.K. Weimer, Proc. IRE **50**, 1462 (1962)
4. C.D. Dimitrakopoulos, D.J. Mascaro, IBM J. Res. Dev. **45**(1), 11 (2001)
5. L. Torsi, A. Dodabalapur, L. Sabbatini, P.G. Zambonin, Sens. Act. B **67**, 312 (2000)
6. L. Torsi, A. Dodabalapur, Anal. Chem. **77**, 380 (2005)
7. M. Barbaro, A. Bonfiglio, L. Raffo, IEEE Trans. Elec. Dev. **53**(1), 158 (2006)
8. M. Barbaro, A. Bonfiglio, L. Raffo, A. Alessandrini, P. Facci, I. Barak, IEEE Electron. Devices Lett., **27**, 595 (2006)
9. A. Loi, I. Manunza, A. Bonfiglio, Appl. Phys. Lett. **86**, 103512 (2005)
10. T. Someya, T. Sakurai, El. Dev. Meet. IEEE **8**(4), 1 (2003)
11. P. Bergveld, IEEE Trans. Biom. Eng. **17**, 70 (1970)
12. P. Bergveld, Sens. Act. B, **88**, 1 (2003)
13. C. Bartic, A. Campitelli, K. Baert, J. Suls, IEDM Tech. Dig., 411 (2000)
14. C. Bartic, B. Palan, A. Campitelli, S. Borghi, IEEE Transducers 01 (2001)

15. A. Campitelli, C. Bartic, J.-M. Friedt, K. De Keersmaecker, W. Laureyn, L. Francis, F. Frederix, G. Reekmans, A. Angelova, J. Suls, K. Bonroy, R. De Palma, S. Cheng, G. Borghs, *in Proceedings of Custom Integrated Circuit Conference IEEE*, 2003
16. C. Bartic, G. Borghs, Anal. Bioanal. Chem. **384**, 354 (2006)
17. C. Bartic, A. Campitelli, S. Borghs, Appli. Phys. Lett, **82**, 3 (2003)
18. C. Gao, X. Zhu, J.-W. Choi, C.H. Ahn, IEEE Transducers 03 (2003)
19. J. Heitz, T. Gumpenberger, H. Kahr, C. Romanin, Biotech. Appl. Biochem. **39**, 59 (2004)
20. P. Mougenot, M. Koch, I. Dupont, Y.-J. Schneider, J. Marchand-Brynaert, J. Colloid Interface Sci. **177**, 162–170 (1996)
21. J. Hu, C. Yin, H.-Q. Mao, K. Tamana, W. Knoll, Adv. Funct. Mater. **13**(9), 692 (2003)
22. R. Jain, A.F. Von Recum, Wiley Periodicals, (2003)
23. B. Crone, A. Dodabalapur, Y.-Y. Lin, R.W. Filas, Z. Bao, A. LaDuca, R. Sarpershkar, H.E. katz, W. Li, Nature **403**, 521 (2000)
24. T. Someya, T. Sakitani, S. Iba, Y. Kato, H. Kawaguchi, T. Sakurai, PNAS **101**, 27 (2004)
25. Y. Noguchi, T. Sekitani, T. Someya, Appl. Phys. Lett **89**, 253507 (2006)
26. Z. Rang, M.I. Nathan, P.P. Ruden, R. Chesterfield, C.D. Frisbie, Appl. Phys. Lett. **85**(23), 5760 (2004)
27. G. Darlinski, U. Böttger, R. Waser, H. Klauk, M. Halik, U. Zschieschang, G. Schmid, C. Dehm, Appl. Phys. Lett. **97**, 093708 (2005)
28. S. Jung, T. Jackson, Dev. Res. Conf. Dig. IEEE **1**, 149 (2005)
29. S. Jung, T. Ji, V.K. Varadan, Smart Mater. Struct. **15**, 1872 (2006)
30. T. Sekitani, Y. Kato, S. Iba, H. Shinaoka, T. Someya. T. Sakurai, S. Takagi, Appl. Phys. Lett. **86**, 073511 (2005)
31. V. Ambegaokar, B.I. Halperin, J.S. Langer, Phys. Rev. B. **4**, 2612 (1971)
32. I. Manunza, A. Sulis, A. Bonfiglio, Appl. Phys. Lett. **89**, 143502 (2006)
33. A. Bonfiglio, F. Mameli, O. Sanna, Appl. Phys. Lett. **82**, 3550 (2003)
34. I. Manunza, A. Bonfiglio, Biosens. Bioelectron., in press
35. D.C. Duffy, J.C. McDonald, O.J.A. Schueller. G.M. Whitesides, Anal. Chem. **70**, 4974 (1998)
36. G.M. Whitesides, Nature **442**(27), 368 (2006)
37. R.H. Reuss, B.R. Chalamala, A. Moussessian, M.G. Kane, A. Kumar, D.C. Zhang, J.A. Rogers, M. Hatalis, D. Temple, G. Moddel, B.J. Eliasson, M.J. Estes, J. Kunze, E.S. Handy, E.S. Harmon, D.B. Salzman, J.M. Woodall, M. Ashraf Alam, J.Y. Murthy, S.C. Jacobsen, M. Olivier, D. Markus, P.M. Campbell, E. Snow, Proc. IEEE Flexible Electron. Technol., Part 1: Syst. Appl. **93** (2005)
38. J. Zaumseil, C.L. Donley, J.S. Kim, R.H. Friend, H. Sirringhaus, Adv. Mater. **18**, 2708 (2006)
39. P. Cosseddu, F. Mameli, I. Manunza, O. Sanna, A. Bonfiglio, Proc. SPIE Photonics Europe, in *Organic Optoelectronics and Photonics*, vol. CDS122-5464, No. 74, ed. by P. Heremans, M. Muccini, H. Hofstraat (2004)
40. E. Orgiu, I. Manunza, M. Sanna, P. Cosseddu, A. Bonfiglio, Thin Solid Films, in press
41. T. Kawase, T. Shimoda, C. Newsome, H. Sirringhaus, R.H. Friend, Thin Solid Films **438**, 279 (2003)

42. G.B. Blanchet, Y.-L. Loo, J.A. Rogers, F. Gao, C.R. Fincher, Appl. Phys. Lett. **82**, 463 (2003)
43. E. Becker, R. Parashkov, G. Ginev, D. Schnider, S. Hartmann, F. Brunetti, T. Dobbertin, D. Metzdorf, T. Riedl, H.H. Johannes, W. Kowalsky, Appl. Phys. Lett. **83**, 4044 (2003)
44. G. Blanchet, J. Rogers, J. Imag. Sci. and Technol. **47**, 303 (2003)
45. M. Lefenfeld, G. Blanchet, J. Rogers, Adv. Mat. **15**, 1188 (2003)
46. P. Calvert, Chem. Mater. **13**, 3299 (2001)
47. Y. Xia, G.M. Whitesides, Annu. Rev. Mater. Sci. **28**, 153 (1998)
48. V.C. Sundar, J. Zaumseil, V. Podzorov, E. Menard, R.L. Willett, T. Someya, M.E. Gershenson, J.A. Rogers, Science **303**, 1644 (2004)
49. Y.-L. Loo, R.W. Willett, K. Baldwin, J.A. Rogers, Appl. Phys. Lett. **81**, 562 (2002)
50. P. Cosseddu, A. Bonfiglio, Appl. Phys. Lett. **88**, 023506 (2006)
51. P. Cosseddu, A. Bonfiglio, Thin Solid Films, in press
52. D. De Rossi, E.P. Sciligno, in *Encyclopaedia of Sensors*, vol. X, ed. by C.A. Grimes, E.C. Dickey, M.V. Pishko (2006), pp. 1–22
53. D. De Rossi, Meas. Sci. Tech. **2**, 1003 (1991)
54. G. Harsanyi (ed.), *Sensors in Biomedical Applications: Fundamentals, Technology and Applications* (CRC Press, Boca Raton, 2000)
55. T. Someya, Y. Kato, T. Sekitani, S. Iba, Y. Noguchi, Y. Murase, H. Kawaguchi, T. Sakurai, PNAS **102**, 35 (2005)
56. E.P. Scilingo, A. Gemignani, R. Paradiso, N. Taccini, B. Ghelarducci, D. De Rossi, IEEE Trans. Inf. Technol. Biomed. **9**(3), 345 (2005)
57. A. Bonfiglio, N. Carbonaro, C. Chuzel, D. Curone, G. Dudnik, F. Germagnoli, D. Hatherall, J.M. Koller, T. Lanier, G. Loriga, J. Luprano, G. Magenes, R. Paradiso, A. Tognetti, G. Voirin, R. Waite, Proc. Mobile Response 2007, in press
58. F. Lorussi, W. Rocchia, E.P. Scilingo, A. Tognetti, D. De Rossi:, IEEE Sens. J., **4**(6), 807 (2004)
59. F. Lorussi, E. Scilingo, M. Tesconi, A. Tognetti, D. De Rossi, IEEE Trans. Inf. Technol. Biomed. **9**(3), 372 (2005)
60. A. Tognetti, N. Carbonaro, G. Zupone, D. De Rossi, 28th Annual International Conference of the IEEE Engineering in Medicine and Biology Society, New York, USA, September 2006
61. L.E. Dunne, S. Brady, B. Smyth, D. Diamond, J. Neuroengineering Rehabil. **2**(8), (2005)
62. P.T. Gibbs, H.H. Asada, J. NeuroEngineering and Rehabil. **2**, 7 (2005)
63. R. Paradiso, G. Loriga, N. Taccini, IEEE Trans. Inf. Technol. Biomed. **9**, 337 (2005)
64. J.B. Lee, V. Subramanian, IEEE Trans. Electron. Dev. **52**, 269 (2005)
65. A. Bonfiglio, D. De Rossi, T. Kirstein, I.R. Locher, F. Mameli, R. Paradiso, G. Vozzi, IEEE Trans. Inf. Technol. Biomed. **9**, 319 (2005)
66. M. Maccioni, E. Orgiu, P. Cosseddu, S. Locci, A. Bonfiglio, Appl. Phys. Lett. **89**, 143515 (2006)

7

Performance Requirements and Mechanistic Analysis of Organic Transistor-Based Phosphonate Gas Sensors

K. See, J. Huang, A. Becknell, and H. Katz

7.1 Overview of Electronic Sensors for Chemical Vapors and Warfare Agents

7.1.1 Introduction and Response Targets

There are a myriad of applications for chemical sensors, and one important area is the detection of toxic gases and vapors. Toxic industrial chemicals and materials (referred to as TICs and TIMs) represent one class of materials for which chemical sensors are designed and applied. Chemical warfare agents (CWAs) are another set of materials for which chemical sensors are used, and they represent one of the most challenging groups of analytes due to their extreme toxicity, which translates to a very low required detection limit. Detection of explosives additionally requires high sensitivities due to the low vapor pressures of the explosive materials (order of magnitude 6×10^{-6} Torr for TNT and 5×10^{-9} Torr for RDX).

The most toxic of the traditional CWAs are the organophosphonate nerve agents such as the "G" series, comprising GA, GB, GD, and GF, and the "V" series, of which VX is a typical example. The structures of the materials are shown in Fig. 7.1 below. Key structural features include the phosphorous–oxygen double bond, and the presence of a good leaving group such as F^-, CN^- (in the case of GA), or $iPr_2NCH_2CH_2S^-$ (in the case of VX).

There are many factors that compound the difficulty of detection of the organophosphonate agents, including the fact that the agent may be delivered as vapor, liquid aerosol, or as a liquid absorbed onto a dusty carrier (such as silica) to form a dry aerosol referred to as a dusty agent. Another factor contributing to detection difficulty is that due to their polarity, the molecules tend to be "sticky," and will adsorb onto many surfaces including the internal plumbing of a chemical sensor system, including filters, cold spots, etc., which can preclude the agents actually making it to the detector. A further difficulty that will be dealt with in a later section is that of selectivity and the propensity of most chemical sensors to false alarm with disturbing frequency.

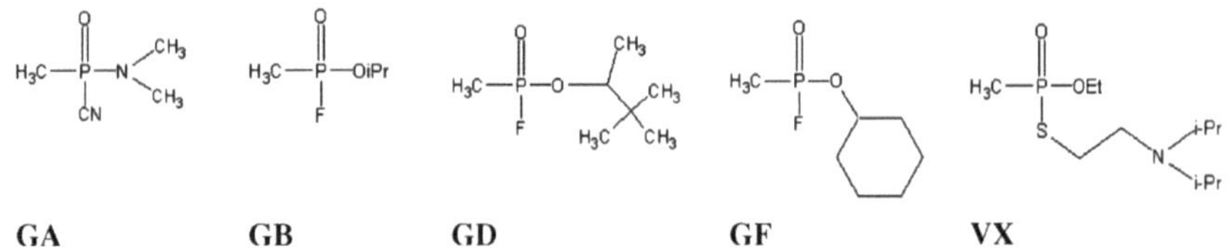

Fig. 7.1. Chemical warfare agent (CWA) structures

Table 7.1. Detection sensitivities from DTRA (http://www.bt.cdc.gov/agent/nerve/)

Agent(s)	Concentration	Time for detection
GA, GB, GD, GF	$0.1\,\text{mg m}^{-3}$	30 s
VX	$0.04\,\text{mg m}^{-3}$	30 s

As mentioned above, the organophosphonate nerve agents are some of the most toxic vapors known to man, and as such they require extraordinary sensitivities from chemical sensors. As an example, the Defense Threat Reduction Agency (DTRA) set the targets shown in Table 7.1.

As aggressive as these numbers appear, they represent concentrations of agents that are potentially lethal, as opposed to a "safe" level. Thus, perhaps the most aggressive target for materials of this type is to detect at the allowable exposure level for VX, termed the threshold limit value (TLV), which is given as $0.00001\,\text{mg m}^{-3}$ ($\sim$1 ppt).

7.1.2 Selectivity

A critical attribute of a successful chemical sensor is its ability to reject false alarms from background interferent vapors. Sensors that false alarm frequently, whether it be daily, weekly, or even monthly, tend to be ignored with time as the personnel monitoring the devices assume that each new alarm is a false alarm. Sensors which false alarm less frequently (perhaps only once a year or less) tend to be taken more seriously.

There are a number of sources of interferent vapors that can generate false alarms. For example, in an indoor environment one sees interferent vapors from cleaning, maintenance, air freshening, painting, welding, degreasing, construction, new materials emissions (carpets, furniture, etc.). In an outdoor environment, one sees interferent vapors from engine exhausts, paints, natural vapors/aerosols, firefighting materials, smokes, etc.

Although each particular environment must be separately assessed, as an example of the challenge facing a chemical sensor a "worst-case" scenario will be presented. Assume that a large quantity of interferent vapor is generated through an operation such as painting, floor stripping, etc., and that a concentration of 1% of the interferent's volatility is presented to the chemical sensor.

Table 7.2. Volatilities for two potential inteferent vapors

Chemical name	Vapor pressure at 20°C	Volatility	1% of volatility
Toluene	22 Torr	110,000 mg m^{-3}	1,100 mg m^{-3}
2-Butoxyethanol	0.7 Torr	4,400 mg m^{-3}	44 mg m^{-3}

In Table 7.2 are listed concentrations for two representative interferent materials. The ratio of concentrations for the interferent to the analyte of interest is quite high. For example, the sensor could need to detect GB at 0.1 mg m^{-3} while rejecting toluene at 1,100 mg m^{-3}, a four orders-of-magnitude difference. Similarly, the sensor could need to detect VX at 0.04 mg m^{-3} while rejecting 2-butoxyethanol at 44 mg m^{-3}, a three orders-of-magnitude difference. While these inteferent concentrations may be "worst-case," they give the reader some indication of the challenge facing the design of chemical sensors (and this is not even considering the potential need to detect VX at 0.00001 mg m^{-3}!). Clearly, selectivity is every bit as much of a challenge as sensitivity.

A caveat to the above discussion is that some sensor system implementations utilize low cost, low selectivity sensors deployed at a high rate across the area of interest. Higher cost, high-selectivity instruments are deployed to less locations across the area of interest. In the event of an alarm by one of the higher selectivity instruments, the lower selectivity instruments can be used to pinpoint the location of the vapor cloud and perhaps its origin. Thus, it may be that not every chemical sensor needs perfect selectivity, but clearly a high rate of false alarms can rarely be tolerated.

7.1.3 Stability

A successful chemical sensor must be able to tolerate a reasonable operational temperature range, probably at minimum 10–40°C for indoor applications where the sensor may be placed in utility rooms with HVAC equipment. Considering a wider scope of potential applications for both indoor and outdoor environments with various potential placements, the temperature range could be as wide as −25–65°C. Across whatever temperature range is employed, the sensor should be able to tolerate the humidities expected, for example 20% RH to 70% RH for less challenging indoor applications and 5% RH to 100% RH for the complete range of conditions across all climates.

Ideally the sensor should "warm up" rapidly after power-on, achieving its baseline condition fairly quickly. However, a number of current commercial chemical sensors can take up to 24 h to achieve a stable baseline after long periods of inactivity. Although handheld sensors are often used for short durations, it is preferable to have a sensor that is capable of continuous operation.

The output of the chemical sensor element, in whatever form it takes (electrical, optical, etc.), should be as stable as possible with time. It should not be adversely affected by interferent vapors or environmental conditions. It is strongly preferred that the sensor recover rapidly from a real detection event with little hysteresis. However, it is conceivable that if the sensor element is inexpensive enough it could be viewed as "replaceable" after detection of toxic chemicals such as the organophosphonate nerve agents. Recovery from detections should follow as closely as possible the actual concentration in the air stream, with no propensity for "capturing" the analyte for long periods which can prevent a rapid reset.

7.1.4 Response Time

Obviously it would be desired for toxic chemical detection to occur in as little time as possible, but practically speaking a sensor that responds in a matter of seconds, and definitely less than a minute, is adequate. The respiration rate of an adult is on the order of 12–20 breaths per minute, and if a detection is made before too many breaths are taken then lives can be saved. The required response time for toxic gases is also related to concentration – higher concentrations require faster detections. Fortunately, this is the way most sensors respond anyway.

7.1.5 Power Consumption and Form Factor

Existing chemical sensors span a wide range of size and power consumptions. The largest, most advanced systems employing mass spectrometric detection can occupy many cubic feet, weigh roughly 50–100 pounds, and require several thousand watts for operation. On the other end of the spectrum are sensors such as the LCD from Smiths Detection, which is little bigger than a large paperback novel and can operate on 4 AA batteries. A recent DTRA BAA requirement was for a sensor occupying less than 40 in.^{3} of space and weighing less than 2 lb, which includes whatever power source (battery) is needed. Power requirements for fixed point operation are typically less stringent, but limiting consumption to a few hundred watts or less is beneficial so that units can be plugged into existing receptacles without loading down the lines.

7.2 Organic Semiconductor Transistor Sensors

7.2.1 Organic Electronics and Chemical Sensing

The field known as "organic" or "plastic" electronics is centered on organic field-effect transistor (OFET)-based circuits mounted on large-area and/or flexible mechanical supports. Work on OFETs has been extensively reviewed, and readers are referred to *Materials Today, Chemistry of Materials*, and *Journal of Materials Research* special issues [1–4]. A general goal of OFET studies is to provide some degree of silicon-like functionality at a small fraction of

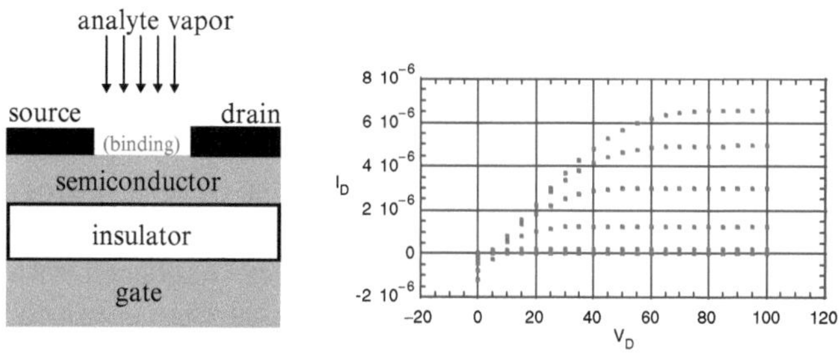

Fig. 7.2. OFET and drain current–drain voltage plot, current in amperes, voltage (negative) in volts, gate voltages 0–100 V in 20 V increments, for a diphenylbithiophene oligomer on Si/SiO_2. The depiction of the vapor binding is for the intended use of the OFET as a sensor

the cost, with the cost savings arising from the extensive use of printing processes. Anticipated applications are in circuits of moderate complexity, such as display drivers, radio frequency identification tags, or pressure mapping elements [5–7]. An additional and perhaps greater opportunity is to utilize capabilities of organic semiconductors (OSCs) that are not as readily available in silicon. A compelling application of organic electronics is chemical sensing, especially for analytes in the vapor phase, because of the ability to covalently attach receptors for compounds of interest to the molecules that make up the semiconductor, in locations where analyte-receptor binding will strongly influence the current flowing across a transistor channel. Figure 7.2 shows an OFET schematic, consisting of an organic semiconductor-dielectric bilayer addressed with source, gate, and drain electrodes. The semiconductor, through which a lateral current is modulated, is a molecular solid film of pi-conjugated cores that can be as thin as 10 nm, and to which electronically, chemically, and morphologically influential functional groups may be conveniently attached. Also shown are typical plots of drain current vs. drain voltage for a set of gate voltages. Responses to vapors are simply noted as changes in the output source drain current for a given set of input drain voltages and gate voltages as the vapor adsorbs onto the OFET.

The device is generally operated in the accumulation mode, where only one majority carrier contributes to conduction. There are two regimes of current flow, the linear region at low drain voltage V_{D} where the drain current I_{D} changes linearly with V_{D}, and the saturation region which occurs when V_{D} is approximately equal to the gate voltage V_{G} and I_{D} becomes independent of V_{D}. In the linear region, the drain current is given by:

$$I_{\mathrm{D}} = \frac{W}{L} C_i \mu \left[\left(V_{\mathrm{G}} - V_{\mathrm{TH}} \right) V_{\mathrm{D}} - \frac{V_{\mathrm{D}}^2}{2} \right], \tag{7.1}$$

where W and L are the transistor channel width and length respectively, C_i is the capacitance per unit area of the gate dielectric, μ is the field-effect mobility, and V_{TH} is the threshold voltage. In the saturation region, the drain current is given by

$$I_{\mathrm{D,sat}} = \frac{W}{2L} C_i \mu (V_{\mathrm{G}} - V_{\mathrm{TH}})^2. \tag{7.2}$$

Various semiconductors including phthalocyanines and naphthalenetetracarboxylic dianhydride are sensitive to gases such as oxygen, nitrogen dioxide, ammonia, carbon monoxide, and hydrogen sulfide [8, 9]. In a comparison of assorted unfunctionalized OSCs in OFET devices challenged with vapors of different molecular polarities, a range of responses were obtained, including both current increases and decreases of factors of 2–3 [10]. Analyte concentrations of 10–100 ppm in the vapor phase were detected. A second study where similar data were recorded on polymers with moderate conductivity was recently carried out [11].

7.2.2 Electronic Transduction Mechanism

While an extended discussion of OFET sensing mechanisms will be reserved for later in the chapter, we present the general idea of how an OFET could respond to vapors at this point. Complexation of an analyte can perturb the channel (drain) current of an OFET in two ways: formation of a local trap and alteration of mobile charge carrier density [12–14]. We begin with a simple electrostatic model for the electrical potential imposed by an adsorbed dipole having charges (q) separated by length (L) centered at point (O), where the potential varies inversely as the square of the distance from the dipole [15] and is maximized for regions nearly collinear with the dipole (small angle θ) (Fig. 7.3). For a single molecule physisorbed onto a nonpolar film, the perturbation in the electric potential is on the order of 0.05–0.1 V at a point X on the order of 10 Å from the molecule, and collinear with the dipole. Binding multiple molecules or larger dipoles increases the change in potential to a few tenths of volts and increases the volume in which the potential is substantially altered. For example, binding water to the nonpolar OSC film α-sexithiophene results in binding energy shifts of 0.1–0.2 eV [16].

This local carrier energy change results in significant OFET current changes, as shown, for example, in pentacene [17,18] and poly(hexylthiophene) [19] OFETs. While the voltage applied to most OFETs in published accounts

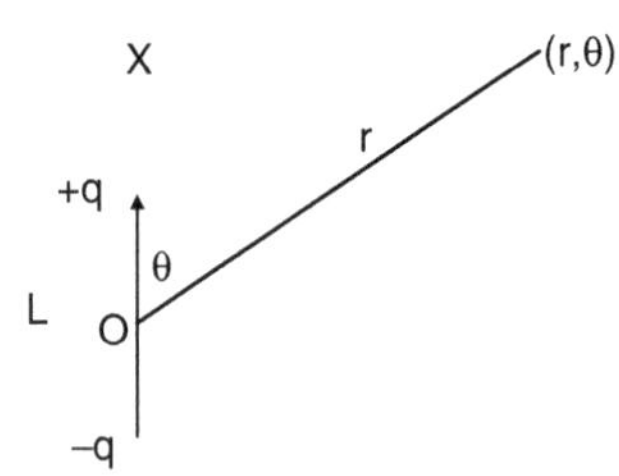

Fig. 7.3. Electrostatics of a simple electric dipole

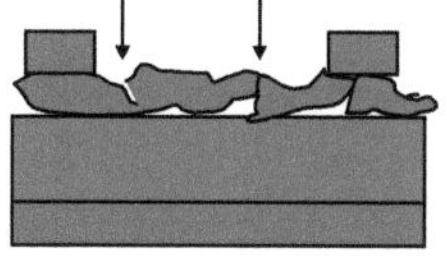

Fig. 7.4. Schematic of polycrystalline OFET with binding sites at grain boundaries

is often on the order of tens of a volts, the voltage that actually confines the gate-induced charge carriers to the channel is some fraction of that, and depends sensitively and in a complex way on the intrinsic conductivity and dielectric constant of the semiconductor [20–23]. It is conceivable that a modest dipole moment associated with an *assembly* of simple functional groups, which imposes a change in potential of 0.2 V, could have an impact on the mobile charge carrier density equivalent to tens of volts of applied gate voltage, altering the conductivity by orders of magnitude [24]. Our results on dimethyl methylphosphonate (DMMP) (see below) demonstrate this as well.

Another way to view the effects of dipole fields is to define dipoles as trap sites. The activation energy for releasing carriers from traps in OFETs is often on the order of 10–100 meV [21, 25]. The formation of traps at similar energies to those already known to markedly influence OFET currents provides a plausible means of transduction. For example, an analyte-induced trap with a depth of 100–200 meV (0.1–0.2 V) could have activation energy 2–10 times that of the activation energy of traps otherwise in the material. The current would thereby decrease by about an order of magnitude at ambient temperature as a result of the analyte binding. This would hold true especially if the binding site traps were on paths that would have to be traversed by carriers, such as grain boundaries, schematically shown in Fig. 7.4.

7.3 Testing Environments for Prototype Sensing Elements

7.3.1 Test Chambers

Robust testing of chemical sensors is not an easy task. A stable vapor source is required that can generate a known concentration of vapor. A valving arrangement must be devised so that the sensor can alternately be exposed to an ambient ("no vapor") gas stream and then the vapor of interest. If the sensor is placed in a chamber, it must be considered how quickly the concentration in the chamber can be changed given the inlet flow rate and flow pattern in the chamber. Test results where the sensor has simply been exposed to an open container of liquid in a chamber (Fig. 7.5a) are never as convincing as the "alternating exposure" technique (Fig. 7.5b), since the former can behave

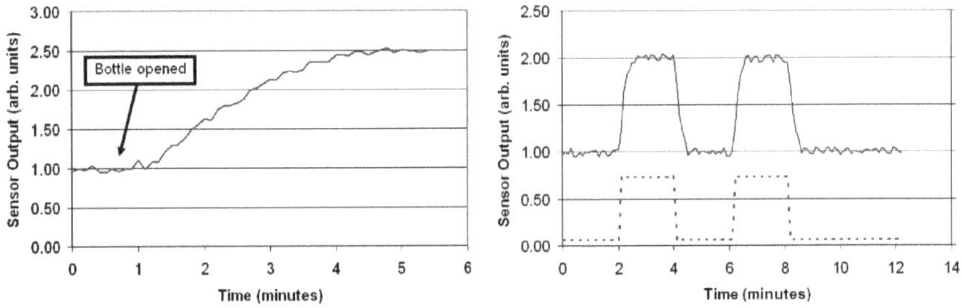

Fig. 7.5. (**a**) Hypothetical sensor response to a bottle opened inside a chamber. The sensor response is difficult to differentiate from a drifting baseline. (**b**) Hypothetical sensor response (*solid line*) to a vapor stream (*dashed line*) that is alternately switched on and off (to pure air). The sensor response clearly tracks the vapor concentration

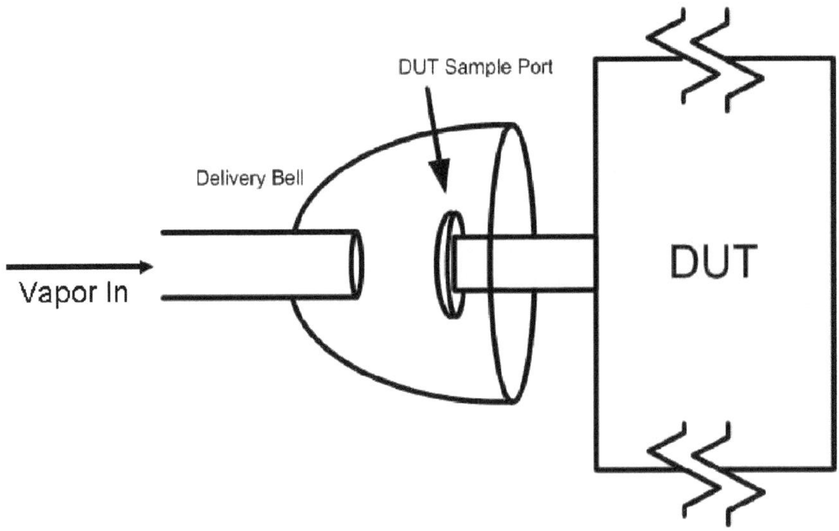

Fig. 7.6. Sample Bell arrangement for feeding vapor to a device under test (DUT)

similarly to a drifting baseline, while the latter provides fairly convincing demonstration of sensor response to the vapor.

Two general types of vapor sources have been used for sensor testing with good success. In the first type, liquid material is volatilized through some means into a pressure-driven gas flow. The advantage of this type is that the vapor is "pushed" past the sensor head, and something as simple as a sample bell (Fig. 7.6) can be used. It is not necessary to have a gas tight system around the sensor inlet, but rather the flow rates are maintained at a sufficiently high rate to feed both the sensor inlet and provide enough surplus flow to fill the bell and the area around the inlet of the device under test (DUT). The disadvantage of a pressure-driven system, as will be described

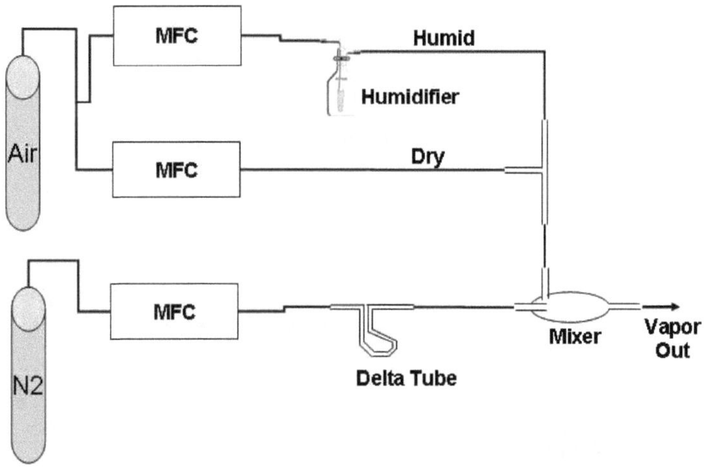

Fig. 7.7. Vapor generation system comprising three legs: (1) humidified air, (2) dry air, and (3) vapor from delta tube or similar type evaporator

below, is that the vapor concentration is not generally known exactly since it depends on the intricacies of the volatilization, and it must therefore be measured during each run.

Vapor generation in pressure-driven systems has received some attention in the literature, and an excellent source of information is the book "Detection Technologies for CWAs and Toxic Vapors" by Sun and Ong [26]. A schematic of the type of generation system advocated by Sun and Ong is shown in Fig. 7.7. Three air flows are brought together. The first is humidified air, the second is dry air, and the third is nominally saturated headspace vapor from the liquid of interest. Blending of the three flows using any desired type of flow meters (mass flow controllers (MFCs) are shown in the diagram) allows a range of vapor concentrations and humidities to be achieved. The vapor is created by an evaporator such as the delta tube (shown), saturator cell [27], or other type. Typically the dry and humid air flows are in the liters-per-minute range, while the flow through the vapor leg is on the order of 1–20 mL min^{-1} for a delta tube or higher for saturator cells [27]. Each type of vaporization tube (delta tube, saturator cell, etc.) will operate at a different efficiency that is related to the geometry and gas flow and it is therefore almost impossible to accurately predict the vapor concentration from first principles in this type of vapor generator. Thus, the vapor concentration must be determined by some calibrated method – for example gas chromatography using thermal desorption tubes.

A second topology for vapor delivery to sensor devices entails preparing the desired air/vapor mixture in a flexible gas sample bag and then drawing the gas past the sensor head by a vacuum pump placed downstream of the sensor (Fig. 7.8). The air/vapor mixture is simply prepared by adding a

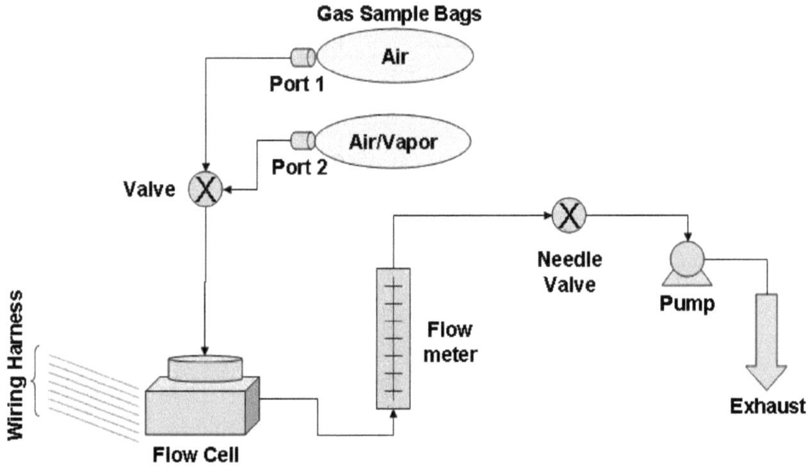

Fig. 7.8. Flow cell arrangement for challenging sensor devices with air/vapor mixtures prepared in Tedlar® gas sample bags

measured amount of liquid to a known volume of air in a bag and allowing it to evaporate. Tedlar® gas sample bagsare convenient to use in this application and available in sizes up to 100 liters. An advantage of this approach is that the concentration of vapor is fairly precisely known from the outset (minus any variation due to adsorption of the vapor on the interior bag surface). A disadvantage is that the sensor element must be placed in a relatively gas tight enclosure in order to for the downstream sample pump to be effective at drawing the sample past the sensor. Ideally the enclosure is a small volume flow cell with all required electronics placed outside of the cell in order to minimize volume.

The chamber volume in many experimental setups dramatically affects the response time of the system. As a rule of thumb, one may assume that 3–5 chamber volumes must be swept through before the concentration in the chamber approaches the concentration of the inlet. This is illustrated in Fig. 7.9 using a simplistic model.

In summary, much thought must be given to the testing arrangement for evaluating chemical sensors, and experiments should be carefully designed to produce the information desired. Pressure-driven vapor generators are often more convenient for testing of developmental sensors since they can use a sample bell arrangement around the sensor device and do not require a gas tight chamber. However, their disadvantage is the need for routine quantification of the vapor concentration. Vacuum-driven "pull" type systems with gas sample bags remove much of the need for routine quantification, however they necessitate a relatively gas tight seal in the chamber around the device. With any type of vapor generation, one must be concerned with loss of vapor due to adsorption on the walls of tubing, fittings, flow meters, etc., used in

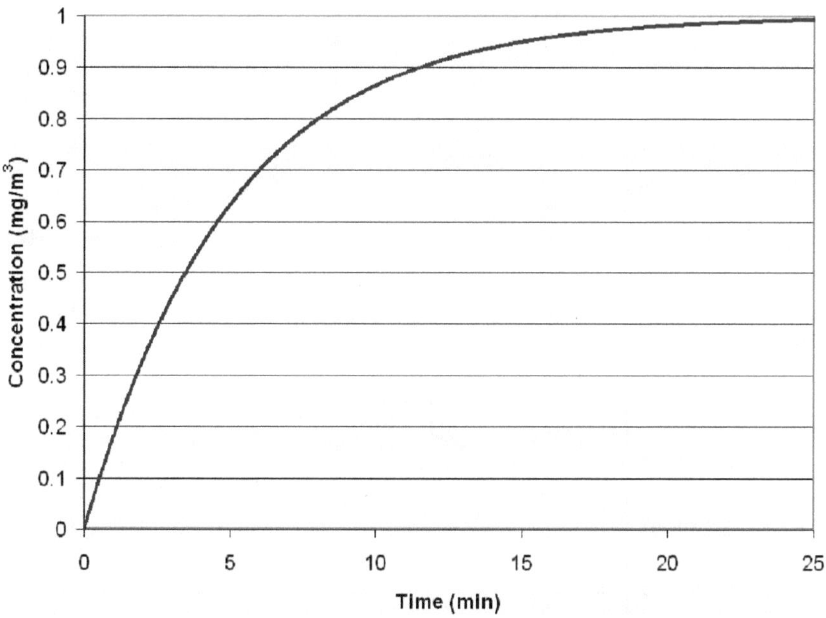

Fig. 7.9. Simplistic finite element model of concentration change in a 10 liters chamber with a 2 liters min^{-1} inlet flow at 1.0 mg m^{-3} inlet concentration. The chamber exchange time is 5 min, and three to five exchanges (15–25 min) are required for the concentration to reach 95% of the inlet concentration

fabricating the system. It is strongly recommended that fluoropolymer tubing, fittings, valves, etc. be used wherever possible and that the performance of the system be carefully verified. Lastly, chamber volume should be carefully considered as it affects evaluation of the response time for the sensors due to the need to exchange several chamber volumes worth of gas before approaching the target concentration.

Data presented in the following sections of this chapter were obtained using a variation of the "pressure-driven" method. Instead of an open "bell" surrounding the DUT, the device is analyzed in a closed chamber fitted with an inlet for the volatilized analyte mixture and an exhaust port to maintain ambient pressure. The analyte of interest was the phosphonate nerve gas simulant, DMMP. The chamber was of the order of 30 liters in capacity, and was equipped with glove ports and wire throughputs to electrical probes. The analyte source was a bubbler containing a liquid sample, through which nitrogen was passed on the way into the chamber.

In addition, we have begun more rigorous testing of the OSC sensors using the vacuum-driven "pull" method. A preliminary experiment is described here. A gas-tight fluoropolymer flow cell was constructed with approximately 300 ml internal volume as shown in Fig. 7.10. Tedlar gas sample bags were used to prepare vapor concentrations of interest and the air/vapor mixtures were

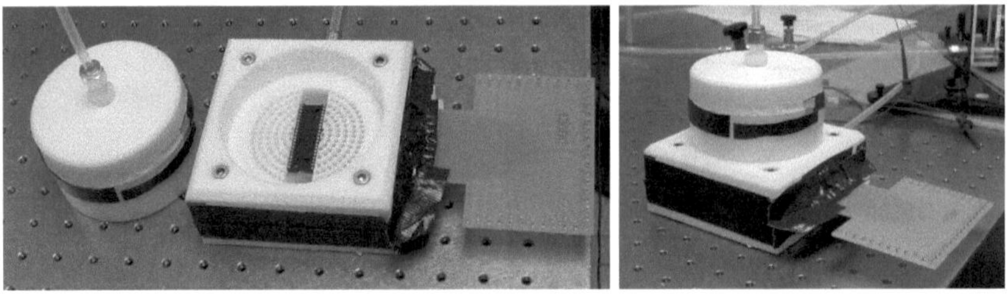

Fig. 7.10. (**a**) Flow cell (*open*) showing external breakout board for electrical connections and internal socket for mounting packaged sensor chip. (**b**) Flow cell (*closed*)

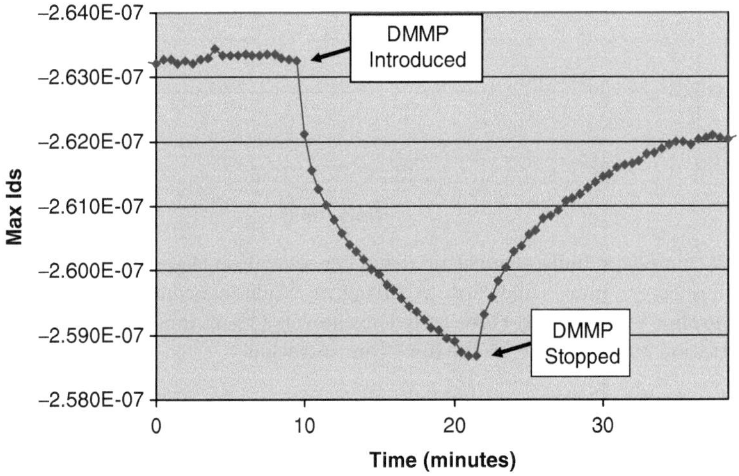

Fig. 7.11. Response of a pentacene chip to 100 mg m^{-3} DMMP vapor

pulled through the flow cell at 300 ml min^{-1} with a vacuum pump. Response to changes in inlet concentration is fairly rapid as the chamber volume is being nominally exchanged every minute.

Testing of pentacene, an unfunctionalized OSC material, showed response to dimethyl DMMP at the 100 mg m^{-3} level (Fig. 7.11), albeit with some evidence of hysteresis and somewhat slow recovery. Based on the strength of the response, it is estimated that as little as 10 mg m^{-3} could have been detected. The functionalized OSC materials are designed to have higher affinity for the phosphonate materials and should increase sensitivity even further toward the target of 0.1 mg m^{-3}. Other increases should be seen from the work on grain structure, alloys, bilayers, etc. And finally, even more sensitivity will be available with an integrated front end absorber/desorber.

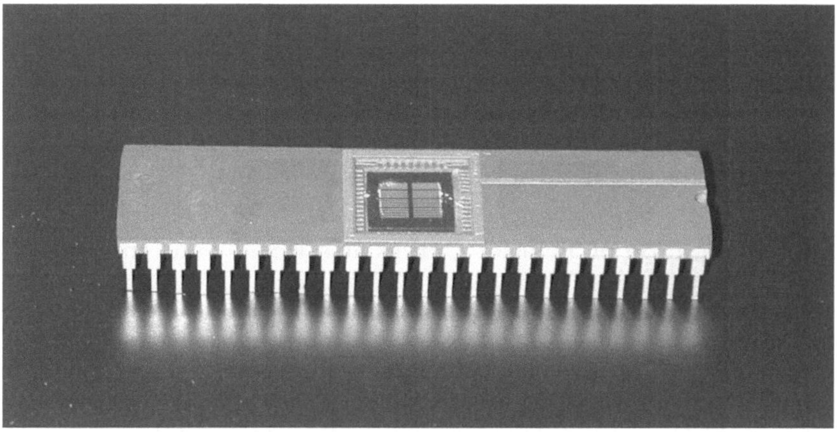

Fig. 7.12. OSC chip mounted in a ceramic DIP chip package

7.3.2 Device Packaging

A fabrication and packaging process has been developed for the OSC sensor devices. The OSC material is vacuum deposited on oxide-coated silicon wafers to which gold electrodes are then added. The electrodes extend off the side of the OSC film area onto the bare oxide where a thin titanium layer is used to provide sufficient adhesion for wire bonding. The chips are mounted in ceramic DIP packages using a room temperature cure adhesive and then wire bonded to the chip lead frame. A picture of a typical chip is shown in Fig. 7.12.

7.4 Electrical Test Procedures

There are a number of ways in which to operate an OFET-based vapor sensor. In general, any approach involves monitoring changes in device characteristics (signal output) during exposure to a vapor. Methods can be divided into two major categories (1) discrete sampling of OFET characteristics at some time interval and (2) continuous monitoring of the drain current at some fixed gate voltage. Here we will discuss the advantages and drawbacks for approaches in each category.

7.4.1 Generation of Saturation Curves at a Fixed Time Interval

This approach involves sweeping the drain voltage while measuring the resulting drain current at a series of applied gate voltages. Monitoring sensor output requires a full set of these sweeps to be performed at a desired time interval. This allows the monitoring of drain currents at a variety of gate voltages during sensor operation. The advantages of this approach are that the transfer characteristics can be extracted giving three possible quantities to monitor, the field-effect mobility μ, threshold voltage V_{TH}, in addition to the absolute drain current. The drawbacks of this approach are that at each sampling time, numerous sweeps must be conducted, which decreases the potential sampling resolution as well as exposing the device to repeated gate bias stress [28, 29] which can artificially decrease the drain current readout.

7.4.2 Generation of Transfer Curves at a Fixed Time Interval

While the method described above can provide the vital information regarding the device performance upon exposure to vapor, it is somewhat of a brute force approach. The same information can be extracted by simply conducting a single sweep of the transfer characteristics at each sampling time, as shown in Fig. 7.13. This involves fixing the drain voltage in either the linear or saturation regime (determined by an initial set of saturation curves generated as described above) and sweeping the gate voltage while recording the resulting drain current. In this way, the drain current can be monitored at many more gate voltages while at the same time being more efficient. If conducted in the saturation regime, fast sweeps can be conducted with small sampling times giving information regarding the drain current, mobility, and threshold voltage.

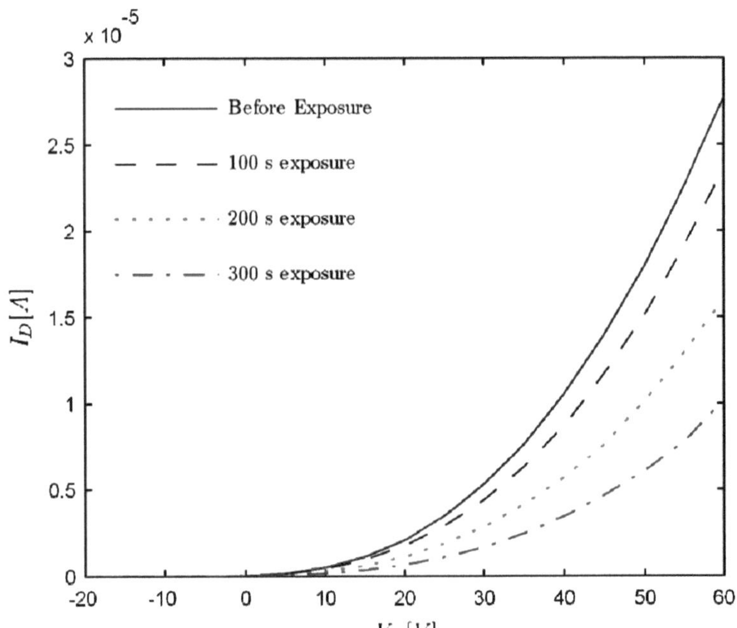

Fig. 7.13. Changes in current vs. gate voltage sweep with time of organic semiconductor exposure to DMMP

If conducted in the linear regime, the details of both the gate dependent mobility and contact resistance can be deduced in addition to the aforementioned parameters. This will be discussed later.

7.4.3 Pulsed Vs. Nonpulsed Measurements

The method of recording transfer curves would appear to be the most efficient way to maximize the information acquired during vapor exposure, however there are further details which must be considered in order to obtain a reliable baseline readout to ensure that changes in properties are truly due to interaction of the device with a vapor, as opposed to intrinsic hysteresis in the device itself. The foremost issue which has been identified in both polymer and small molecule-based organic transistors is the effects of bias stress. Bias stress refers to the deleterious effects of prolonged application of a gate bias in the accumulation mode (positive gate voltages for n-channel materials and negative gate voltages for p-channel materials). The degradation of device behavior is manifested in a decrease in drain current entirely due to a shift in threshold voltage [30–32]. This shift has been attributed to both formation of bipolarons in p-channel polymer semiconductors [31, 32] as well as trapping at the semiconductor/insulator interface in pentacene [30]. Street et al. have showed that in polymer-based devices, that reversal of the effect can occur at room temperature at vastly different time scales depending on the material used for the active layer [32]. In some cases, application of the gate voltage in a pulsed manner can prevent trapping of charges by removal of the applied gate in between each desired gate voltage applied during the sweep. The stress response of a specific materials system should be investigated to determine whether pulsed measurements are required to maintain consistent baseline performance of the OFET sensor.

7.4.4 Erasing Electrical History

Over the course of testing as well as during exposure to the ambient atmosphere the threshold voltage can shift. In the authors' experience, this shift is positive for an n-channel material, and appears to be reversible and controllable in either direction by application of the gate bias for prolonged periods. Where the application of a forward bias (positive for an n-channel) causes the previously described bias stress effect, a reverse bias can shift V_{TH} back negatively, as depicted in Fig. 7.14. The stability of this effect has been shown in nitrogen, however further studies are needed to analyze the effects of air exposure. The mechanism is currently under study, two possibilities being either a charging of the dielectric, or possibly some type of detrapping of mobile charges. If the conditions for stable shifts in V_{TH} are determined, one can imagine a device with a V_{TH} shifted far enough where even at $V_{\mathrm{G}} = 0\,\mathrm{V}$ a suitable level of current can be measured allowing for a sensor with very low power requirements due to the absence of the applied gate field during operation.

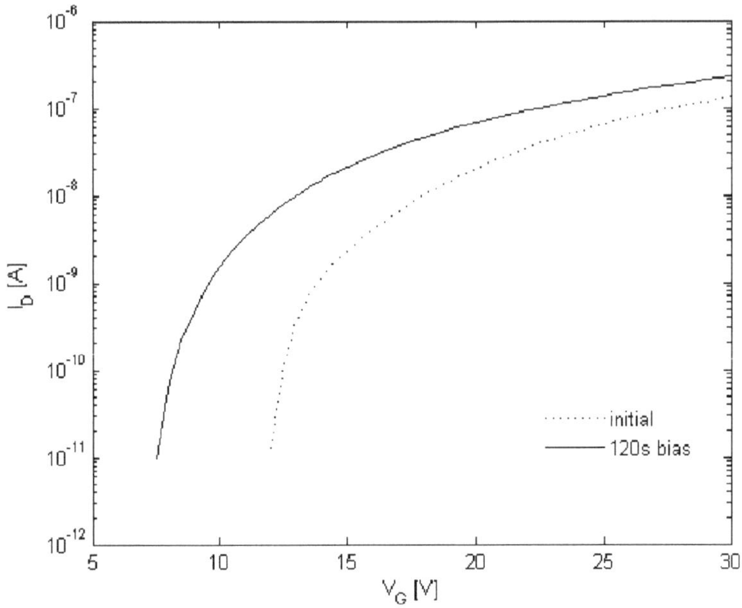

Fig. 7.14. Transfer characteristics (linear regime, $V_D = 2\,V$) in air of a top contact (Au) F15-NTCDI transistor deposited at 70°C before and after −40 V applied gate bias

7.5 Responses of Functionalized Organic Semiconductors to DMMP

We now present highlights of two experimental studies of OFET response to DMMP. The first study employed two different hole-transporting end-substituted 5, 5′-diphenyl-2, 2′-bithoiphenes, with and without terminal OH groups (and oxygens linking the side chains to the cores). The OH-substituted compound was incorporated as one component in a blend layer that was deposited over a base layer of the unfunctionalized semiconductor alone. The second study used a pair of electron-transporting naphthalenetetracarboxylic diimides, again with and without OH groups. In this case, the OH group was phenolic, and the functionalized compound was applied as a "film" with island morphology, over the unfunctionalized base layer. The base layers ensured adequate current density for easily distinguishable responses, while the OH groups increased the magnitude, reproducibility, selectivity, and speed of the current changes. We believe that the hydrogen bonding between the OH groups and the PO functionality of DMMP is responsible for the enhanced responses.

7.5.1 Responses of Functionalized Hole-Transporting Oligomers, Including Blends and Surface Modifications

Figure 7.15a is a schematic illustration of the field-effect transistor sensor with semiconductor 5, 5′-bis (4-*n*-hexyl-phenyl)-2, 2′-bithiophene (6PTTP6) film serving as the active layer. This unfunctionalized sensor shows some inherent sensitivity to DMMP. Figure 7.16 compares the change of the saturation source–drain current of 6PTTP6 sensor before and after exposure to saturated DMMP vapor, with source–drain voltage V_D set to be −100 V.

A functionalized sensor with structure as shown in Fig. 7.15b had higher sensitivity and response speed. In this functionalized OFET sensor, 15 nm of 6PTTP6 and HO6OPT semiconductor blend, serving as a functionalized receptor layer, was coevaporated on top of a 35 nm 6PTTP6 film.

An obvious comparison shows the enhanced performance from the functionalized two-layer OSC film. Figure 7.17 compares the response of two sensors as a function of time during DMMP exposure, with V_G set to be −40 V. The blank vertical line indicates the time when the vial with DMMP

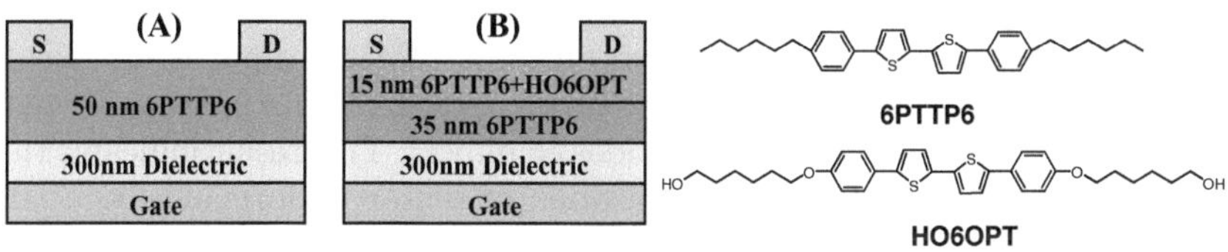

Fig. 7.15. Schematic illustration of OFET sensors and semiconductor molecular structures: (**a**) 6PTTP6 sensor; (**b**) two-layer blend sensor

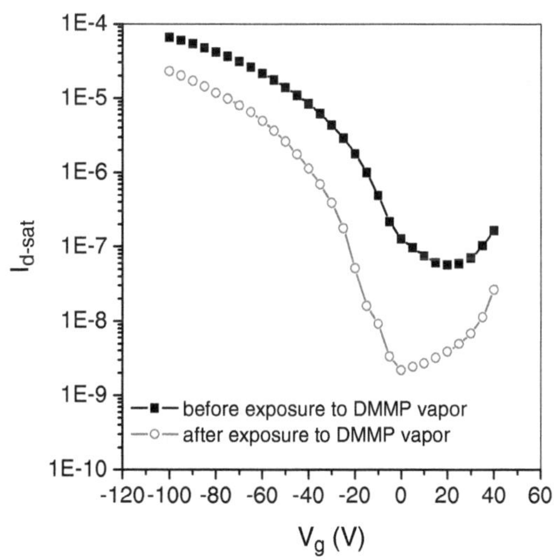

Fig. 7.16. The saturation source–drain current (in Amperes) of 6PTTP6 FET sensor before and after exposure to saturated DMMP vapor

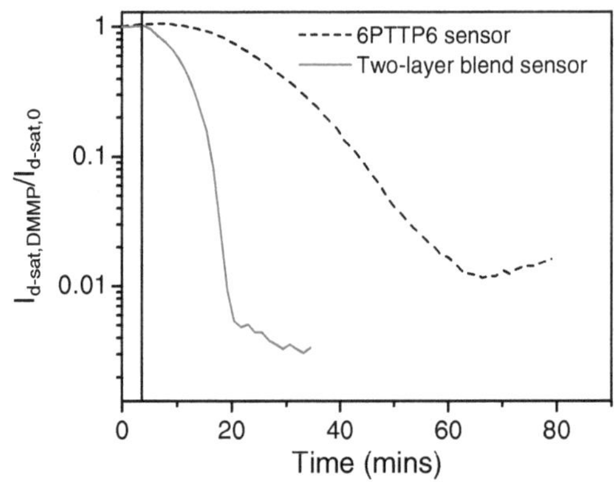

Fig. 7.17. Sensing responses of the two-layer blend sensor and the 6PTTP6 sensor to saturated vapor of DMMP

liquid was placed into the test chamber. The functionalized two-layer blend sensor exhibited much stronger and faster response than the unfunctionalized sensor especially during the first period of the evaporation of DMMP vapor from the vial, when the concentration of DMMP vapor is low due to the slow evaporation rate of DMMP (corresponding to ~1.8 ppm min^{-1} mole concentration increase in the 1 cubic foot chamber). This result indicates that receptors are essential for the optimization of sensor performance especially at low analyte concentration.

The performance of the two-layer blend FET sensor upon exposure to dilute DMMP vapor was further studied in a test chamber connected to a constant flow rate of gas, which was switched between N_2 and DMMP vapors. A bubbler containing DMMP liquid was used as the source of analyte vapor. Nitrogen gas was passed above the liquid surface and carried the analyte vapor to the test chamber. The mass of a bubbler was measured before and after testing to determine the total mass of analyte delivered by a specific volume of carrier gas, thereby allowing calculation of the average concentration of analyte vapor in the test chamber.

Figure 7.18 shows the response of the two-layer blend sensor upon exposure to 150 ppm DMMP vapor. The chamber was vented with N_2 for 2 min, and then 150 ppm DMMP was delivered for about 6 min, followed by 3 min of N_2 venting. The saturation drain current $I_{\text{D,sat}}$ decreased significantly when the DMMP vapor was injected into the chamber. The response time $t_{0.9}$ (time to reach 90% of maximum response) is about 2 min, which is a value comparable to many reports [33–35]. It should be noted that, as has been discussed in Sect. 7.3.1, the chamber volume in the experimental setup can affect the response time. In this experiment, a chamber as large as 1 ft^3 was used, and hence the apparent response time is longer than what would be obtained with smaller test chamber. The sensor recovers slowly under a gentle nitrogen

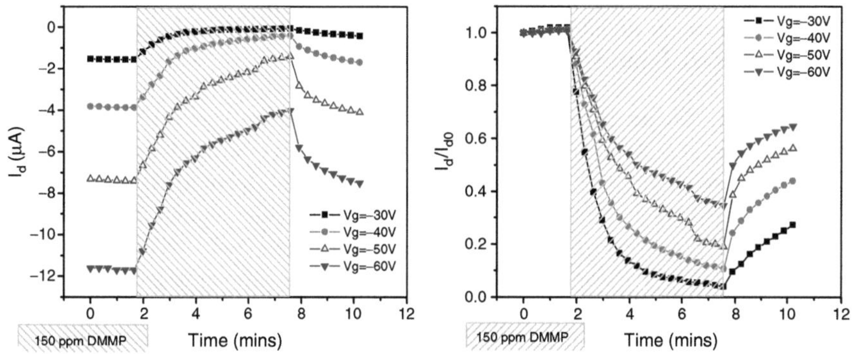

Fig. 7.18. Two-layer blend (6PTTP6+HO6OPT) FET responses to 150 ppm DMMP

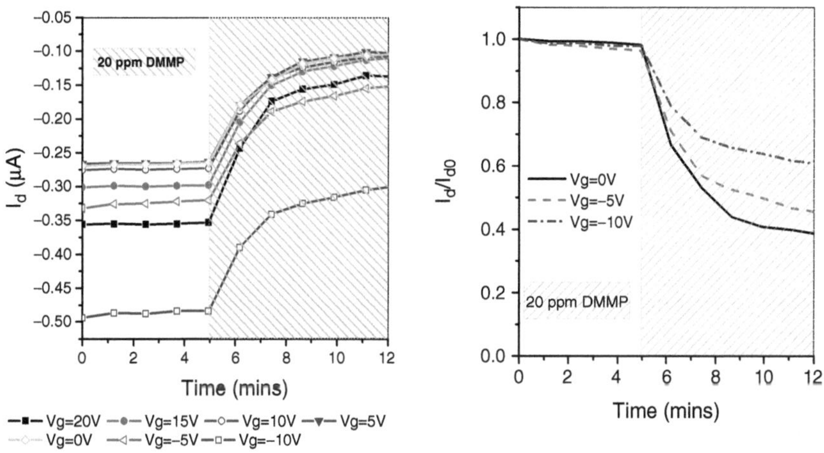

Fig. 7.19. Two-Layer blend sensor with $BuSnCl_3$ treatment response to 20 ppm DMMP

stream, but heating at 45°C for 2 min greatly accelerates the return of the sensor to its original state.

Another type of receptor material that can further improve the sensor sensitivity is an organometallic Lewis acid such as butyltin tricholoride. A two-layer blend FET sensor with adsorbed tin from butyltin trichloride shows an enhanced response, as shown in Fig. 7.19. Here the functionalized 6PTTP6-HO6OPT blend layer acts both as receptor for the analyte molecules and as a ligand for organometallic receptors. The saturation drain current of the sensor decreased dramatically upon exposure to 20 ppm DMMP vapor. At a gate voltage of −5 V, the current reduced to 60% of its initial value after exposure to 20 ppm DMMP vapor for 100 s, which suggests that sub-ppm sensing is achievable.

7.5.2 Responses of Electron-Transporting Films, Including Hydroxylated Island Overlayers

We will now discuss the response of sensors based on the *n*-type family of naphthalene-1,4,5,8-tetracarboxylic diimide (NTCDI) derivatives. The molecular structure of DMMP includes a dipole moment, principally due to the phosphorous–oxygen double bond. Previous work has shown the effectiveness of polar hydroxyl groups in forming a hydrogen bond with the DMMP molecule [36–39]. We therefore opted to introduce a phenol group onto the NTCDI core molecule (PHOH-NTCDI). Synthesis was conducted as reported previously [40]. The receptor compound was utilized in a bilayer film device structure along with a previously studied semiconductor N, N'-bis(1H,1H-perfluorooctyl)naphthalene-1,4,5,8-tetracarboxylic diimide (F15-NTCDI) [40].

Sensor layers were fabricated by either deposition of 35 nm of F15-NTCDI for the single layer (SL) case, or 25 nm F15 followed by 10 nm (nominal) of the PHOH-NTCDI for the bilayer (BL) case, onto a heavily doped $Si/Si0_2$ wafer which served as the gate electrode/oxide. For all depositions, including bilayers, the $Si/Si0_2$ wafer temperature was maintained at 70°C. Gate leakage was limited by patterning the organic into patches approximately 2×1 mm using custom designed shadow masks [41]. Device yield was high, due to the reliability of the F15 layer deposited at elevated temperatures. Molecular structures of the F15 and PHOH-NTCDI molecules are shown in Fig. 7.20.

Two sets of devices were tested, a single layer device (SL) with F15 as the active layer and the bilayer (BL) device deposited as described above. Vapor was introduced by flowing nitrogen over the analyte liquid into a chamber which contained the device. Figure 7.21a shows a semilog plot of the bilayer device transfer characteristics before and after DMMP exposure. Figure 7.21b shows a comparison between a single layer and a bilayer device exposed to DMMP. Bilayer response to potential inteferents is shown as well. The results show clearly that the bilayer device is more sensitive to DMMP than the singe layer device. Additionally, the single layer was shown to have unpredictable behavior, in some cases resulting in an *increase* in current upon

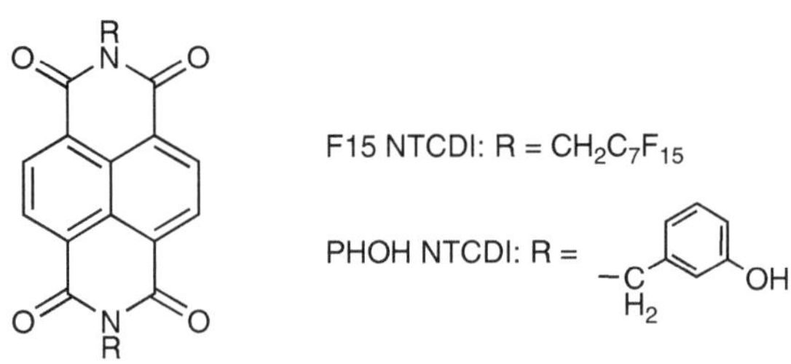

Fig. 7.20. Structures of NTCDI compounds used in this study

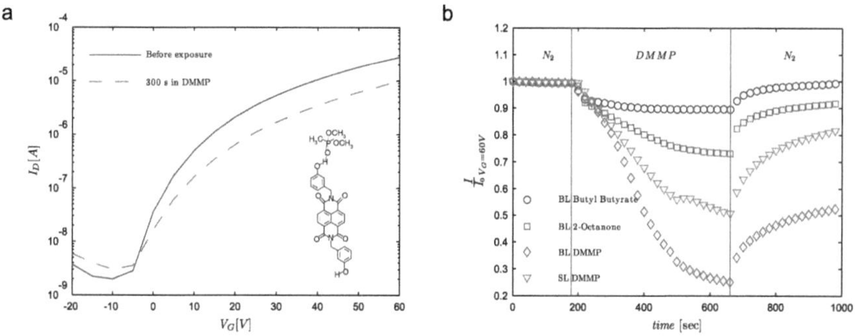

Fig. 7.21. Responses of NTCDI devices to DMMP. (**a**) semilog transfer curves before and after exposure. (**b**) linear plots of current vs. time, before, during, and after exposure

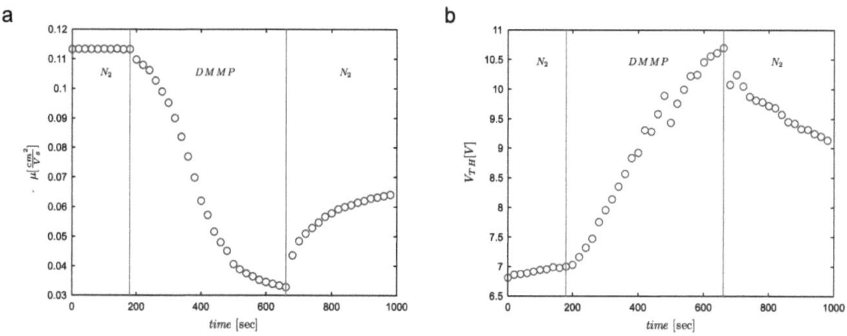

Fig. 7.22. Changes in (**a**) mobility and (**b**) threshold voltage corresponding to Fig 7.21b

DMMP exposure. The bilayer devices always showed the same behavior, the only variation being the ultimate sensitivity of the device, which is likely due to a distribution of layer thicknesses due to position in the chamber during evaporation. An added feature of the bilayer device is that it shows greater sensitivity to DMMP than to either of the inteferents.

Figure 7.22a, b show the corresponding mobility and threshold voltage changes for the bilayer device during DMMP exposure. The contribution of changes in both the mobility and threshold voltage of the device point to two distinct mechanisms of current reduction, which will be discussed later in Sect. 7.7. The results prove that the receptor compound is indeed able to enhance the performance of the OFET sensor.

7.6 Data Analysis

7.6.1 Sensitivity of an OFET Sensor: Gate Voltage Dependence and Contributions of Mobility and Threshold Voltage Changes

If we assume for now that the mobility is independent of gate voltage, the saturation drain current $I_{D,sat}$ of the unexposed OFET is given by (7.2) [32]. Accordingly, the field-effect mobility and threshold voltage can be estimated by plotting the square root of saturation drain current vs. the gate voltage (V_G).

It is common practice to utilize the relative changes in current of the transistor at some gate voltage as the transduction signal. Here we will define the relative changes in current to be the value of the current at some time during vapor exposure divided by the current at some initial time zero (I_j/I_0). Over the course of the vapor exposure we will assume that at a given time there will be change in the value of both the mobility and threshold voltage. Equation 7.2 implies that the change in relative current at some point during vapor exposure is a function of two important factors (1) the value of the *initial* mobility μ_0 and (2) the term $V_G - V_{TH_0}$. Figure 7.23 plots the value of

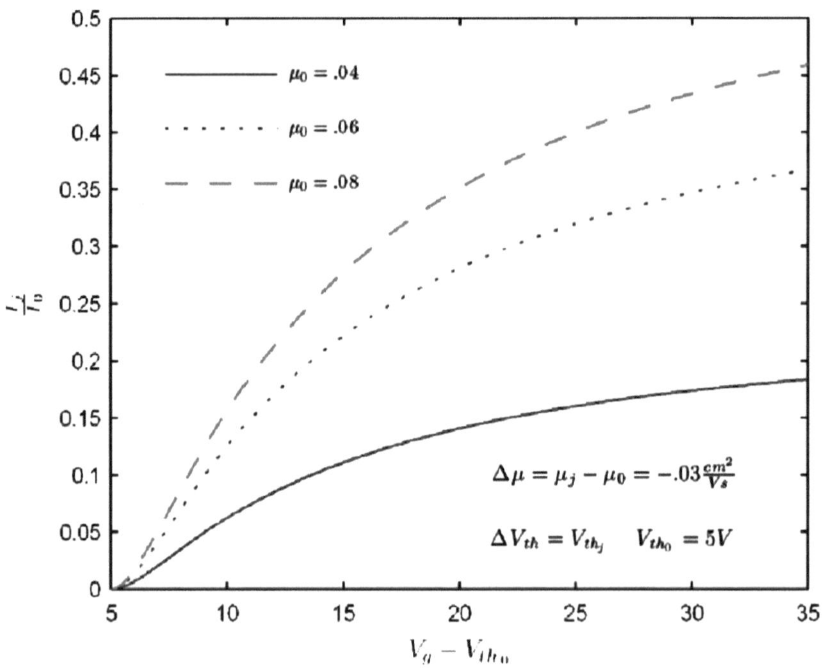

Fig. 7.23. Relative current as a function of $V_G - V_{TH_0}$ simulated using (7.4). Values of $\Delta\mu$ and ΔV_{TH_0} are representative of realistic values from experiment for an n-channel semiconductor

the relative current at some point (j) during the vapor exposure, as a function of the term $V_G - V_{TH_0}$.

This analysis provides a guideline for identifying at what gate voltage the device can be expected to display the greatest sensitivity. Broadly, it also suggests that simply comparing relative values of current at some gate voltage for different systems is not an adequate figure of merit. In fact it is the difference between the gate voltage and the initial threshold voltage which will determine the maximum change in relative current for a given experiment. As a rule, selection of a gate voltage as close as possible to the threshold voltage (but still giving adequate current for data acquisition) should result in the greatest change in relative current. Additionally, it is evident that the lower the initial mobility, the greater the sensitivity at a given gate voltage. This is an obvious result and factors into the design of OFET sensors, as fabrication of interdigitated source/drain electrodes could allow for sufficient current even with low mobility devices, allowing for enhanced sensing response.

The significance of this analysis lies in the clarification of what information is needed to interpret the changes in device current as a reliable figure of merit. It also allows the forecasting of the sensor response at unmeasured gate voltages. Equally important is understanding how best to determine the maximum sensitivity of a device, independent of the materials system utilized. The insight provided from this analysis can help in the design and operation of OFET based vapor sensors.

The field-effect mobility is a function of the potential barrier [14] between grains, and the threshold voltage can shift with the change of the density of trapped charges in the semiconductor active layer [42]. The presence of analyte molecules with large dipole moments in the semiconductor conduction channel induces a strong electric field, which not only causes a portion of the mobile charges to be trapped and lose activity, but also effectively slows down the remaining mobile charges [13]. Figure 7.24 shows the variations in the field-effect mobility and threshold voltage of the two-layer blend FET during the analyte vapor exposure. Mobility and threshold voltage can also be used as output signals of the OFET chemical sensors. Again, in this experiment, a large chamber ($1\,ft^3$) was used, which increased the observed sensor response time.

7.6.2 Self-Consistent Equation Based on Simple Saturation Current

The sensing responses of OFETs show obvious gate voltage dependence, as shown in Figs. 7.16–7.19 and 7.21. The absolute current change is larger at higher gate voltage, while the relative current change is larger at relatively lower gate voltage. Figure 7.25b plots the ratio of $I_{D,sat(DMMP)}$ to $I_{D,sat(0)}$ at various gate voltages, here the $I_{D,sat(DMMP)}$ is the saturation drain current of a 6PTTP6 FET during exposure to DMMP vapor, and $I_{D,sat(0)}$ is its original value before the exposure.

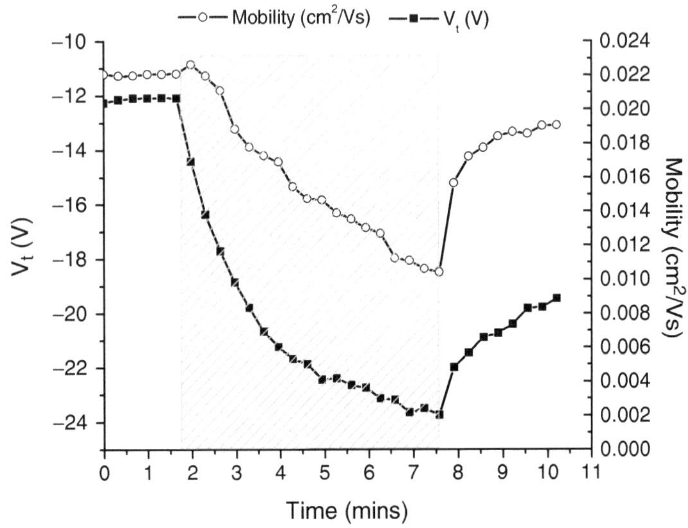

Fig. 7.24. Changes of the μ and V_{TH} of the two-layer blend FET during exposure to 150 ppm DMMP vapor

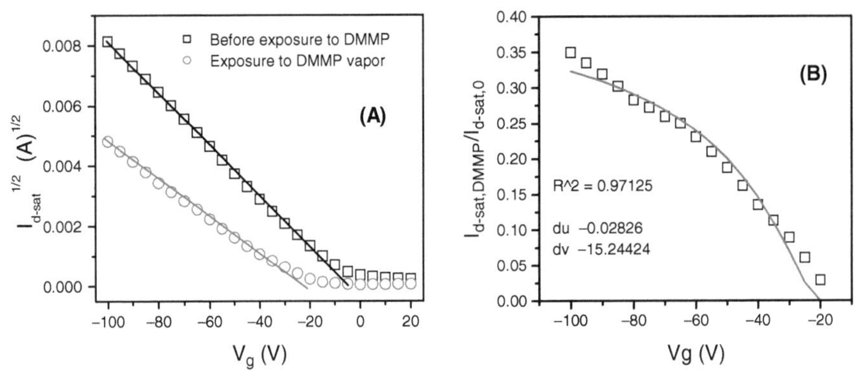

Fig. 7.25. (**a**) The square root of the saturation drain current of a 6PTTP6 FET vs. gate voltage before and during DMMP vapor exposure, with $V_{\mathrm{D}} = -100\,\mathrm{V}$. (**b**) The *open squares* correspond to the ratio of saturation source drain current $I_{\mathrm{D,sat(DMMP)}}/I_{\mathrm{D,sat(0)}}$ obtained from experiment; the *solid line* is the least squares fitted curve according to (7.3)

When analyte (DMMP for this case) molecules are diffused into the semiconductor active layer, dipoles of the molecules induce changes in μ and V_{TH}, noted as $\Delta\mu$ and ΔV_{TH}, so that the saturation drain current $I_{\mathrm{D,sat(analyte)}}$ of the OFET in analyte vapor becomes (7.3):

$$I^{1/2}_{\mathrm{D,sat(analyte)}} = \left[\frac{W}{2L}C(\mu + \Delta\mu)\right]^{1/2} |[V_{\mathrm{G}} - (V_{\mathrm{TH}} + \Delta V_{\mathrm{TH}})]|. \tag{7.3}$$

Combining (7.2) and (7.3), we can get the relationship between $I_{D,sat(analyte)}$ and $I_{D,sat(0)}$, shown as (7.4) and (7.5):

$$I_{D,sat(analyte)}/I_{D,sat(0)} = \left(1+\frac{\Delta\mu}{\mu}\right)\left[1-\left(\frac{\Delta V_{TH}}{V_G - V_{TH}}\right)\right]^2, \tag{7.4}$$

$$\Delta I_{D,sat} = I_{D,sat(analyte)} - I_{D,sat(0)} = I_{D,sat(0)}\left(\frac{I_{D,sat(analyte)}}{I_{D,sat(0)}} - 1\right). \tag{7.5}$$

Equation 7.4 is only valid for the case of $|[V_G-(V_{TH}+\Delta V_{TH})] < V_G - V_{TH}||$, but it allows the prediction of sensitivities vs. gate voltage over a useful range without measuring at each individual value. For smaller values of $|V_G|$, such as 40 V, the $[1 - \Delta V_{TH}/(V_G - V_{TH})]$ term is smaller, which results in a smaller value of the ratio of the saturation source drain currents $I_{D,sat(analyte)}/I_{D,sat(0)}$, corresponding to more noticeable sensor response. On the contrary, if $|V_G|$ is larger, such as 100 V, the term $[1-\Delta V_{TH}/(V_G - V_{TH})]$ is larger than the previous case, which results in a relatively larger value of $I_{D,sat(analyte)}/I_{D,sat(0)}$ and smaller sensor response. In (7.5), the second term

$$\left(\frac{I_{D,sat(analyte)}}{I_{D,sat(0)}} - 1\right)$$

is smaller at larger $|V_G|$, but the first term $I_{D,sat(0)}$ is larger and is the dominant term, so that the absolute current change $\Delta I_{D,sat}$ is larger at higher $|V_G|$ (Table 7.3).

This model is consistent with the responses of the OFET-based sensors on which we are basing our analyses. Figure 7.25a plots the square root of the saturation drain current of a 6PTTP6 FET, vs. gate voltages, before and during analyte vapor exposure. We can see that at lower $|V_G|$, the sensor relative response is indeed higher than at large values of $|V_G|$. The field-effect mobility μ and threshold voltage V_{TH} of the device before and after exposure to analyte vapor can then be derived by data fitting procedures. In Fig. 7.7b, the open squares correspond to the ratio of saturation source drain current $I_{D,sat(analyte)}/I_{D,sat(0)}$ obtained from experiment. A least squares fit of those data, together with μ and V_{TH} obtained from Fig. 7.7a, to (7.3), provides a value for ΔV_{TH} and $\Delta\mu$. The estimated $\Delta V_{TH} = -15.3$ V, and the estimated $\Delta\mu = -0.0283\,\text{cm}^2\,\text{V}^{-1}\text{s}^{-1}$ are very close to those obtained from experimental data, -16 V and $-0.0277\,\text{cm}^2$ V s, respectively. This estimation

Table 7.3. Field-effect mobility and threshold voltage of 6PTTP6 FET before and after exposure to DMMP vapor

	V_{TH}(V)	μ(cm^{2}V^{-1}S^{-1})	ΔV_{TH}(V)	$\Delta\mu$(cm^{2}V^{-1}S^{-1})
No DMMP	−5	0.0522	−16	−0.0277
In DMMP	−21	0.0245		

is qualitative since the field-effect mobility μ is actually also a function of V_G [43], particularly when examining the device near the threshold voltage, contrary to the assumption with which we began this discussion. When $|V_G|$ is larger than 90 V, the contact resistances cause some inconsistency between experimental and estimated data, as does the gate dependence of the mobility.

7.6.3 Contributions of Gate Dependent Mobility and Contact Resistance

Until this point in the discussion, and generally in the literature, the field-effect mobility of the device is presumed to be independent of gate voltage which enables both the mobility and threshold voltage to be simply extracted using a linear fit of the square root of the saturation current (7.2). In many cases however, this is a crude approximation and may hold true for certain values of gate voltage but rarely over the entire range of the gate voltages sampled. Horowitz et al. have outlined a procedure using the transfer line method (TLM) which allows for the extraction of the gate dependent field-effect mobility and contact resistance, as well as the threshold voltage [44]. To our knowledge, until now there have been no attempts to apply this strategy to OFET sensor analysis, even though it could provide a deeper level of understanding of OFET vapor responses.

It is assumed that the mobility follows a power law of the form:

$$\mu = \kappa \left(V_G - V_{TH} \right)^{\alpha} \tag{7.6}$$

where κ and α are empirical parameters.

The method requires measurement of transfer curves in the linear regime of the transistor for different channel lengths at each discrete signal measurement. As an example, instrumentation can be setup to acquire transfer curves for three devices at each data acquisition time, each with different channel lengths. In the linear regime, the total resistance R_{tot} can be calculated by dividing the drain voltage by the drain current. If we assume that the total resistance is composed of a contact resistance R_s and a channel resistance then R_{tot} is given by

$$R_{tot} = R_S + \frac{L}{WC_i\mu \left(V_G - V_{TH} \right)}, \tag{7.7}$$

where W and L are the channel width and length, and C_i is the capacitance per unit area of the dielectric. Then for each time, we can plot the width normalized total resistance (WR_{tot}) vs. channel length (L) for each gate voltage. Now, the y-intercept of a linear fit at each gate voltage gives us the gate voltage dependence of the width normalized contact resistance (WR_s) and the inverse slope gives the normalized channel conductance

$$\left(\frac{\Delta R_{tot} W}{\Delta L} \right)^{-1} = C_i\mu \left(V_G - V_{TH} \right). \tag{7.8}$$

By plotting the inverse slope and then fitting the data using (7.6) and (7.8), we can extract both the threshold voltage as well as the gate dependent mobility. With this method, the details of how the mobile charges in the device with different mobilities are affected by the exposure of the device to a particular vapor, emerge. This could provide unique information regarding the particulars of how specific OFET materials systems respond to different vapors in addition to considering the oftentimes neglected issue of contact resistance.

7.7 Sensing Mechanisms and OFET Models

As shown in Fig. 7.26, when the sensor is exposed to vapor, individual molecules can diffuse into the semiconductor thin film and be adsorbed mostly at the grain boundaries [13]. If the adsorbed analytes have large dipole moment, such as H_2O (~2 debye) and DMMP (~3 debye), the adsorption of those analyte molecules at the grain boundaries close to or at the semiconductor-dielectric interface can locally perturb the electrical profile around the conduction channel, and hence change the trap density in the active layer. We can interpret the trapping effects by a simple electrostatic model discussed briefly in Sect. 7.2. The electric field induced by a dipole with dipole moment of $\vec{p}$ (magnitude qL in Fig. 7.4) is:

$$\vec{E} = \frac{3K\vec{r}(\vec{p}\cdot\vec{r})}{r^5} - \frac{K\vec{p}}{r^3}, \tag{7.9}$$

where $K = \frac{1}{4\pi\varepsilon_0} = 9 \times 10^9 N \cdot m^2/C^2$ is the Coulomb's constant, and $\vec{r}$ is the vector from the position of the dipole to the position where the field is being measured. For an analyte molecule with dipole moment of approximately 3 debye, such as DMMP, the electric field induced by the dipole can be very strong in the space close enough to the dipole.

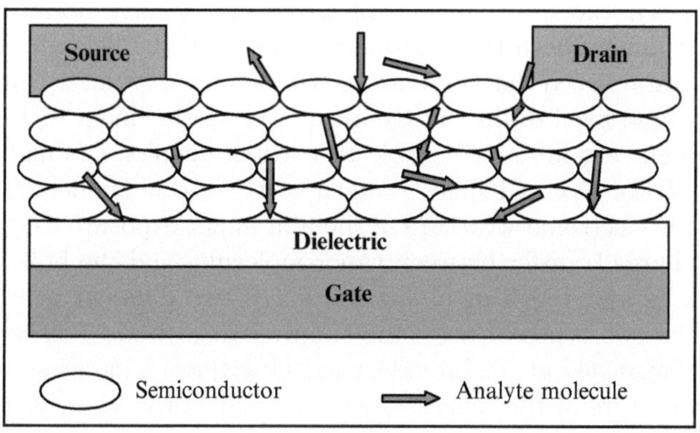

Fig. 7.26. Model of analyte diffusion through an OFET with a polycrystalline OSC

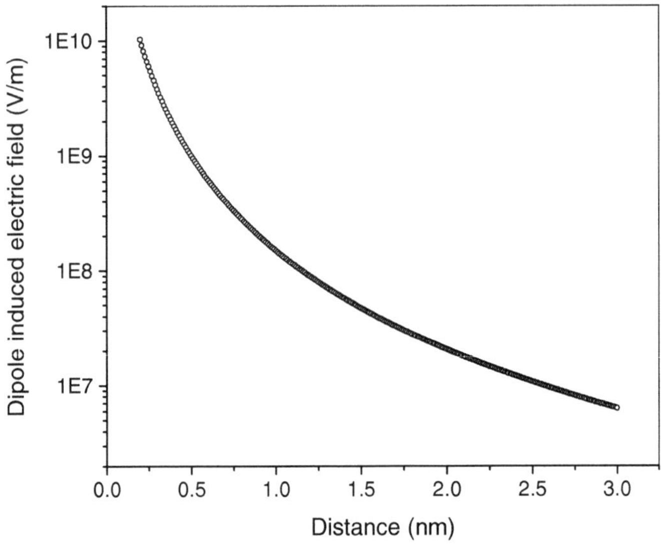

Fig. 7.27. Dipole-induced electric field at various distances in the direction parallel to the direction of the dipole moment

Figure 7.27 plots the electric field induced by a DMMP molecule in directions parallel to the direction of the dipole moment, as a function of the distance between the molecule and the point of concern.

The transverse source–gate field E_{SG} used to induce mobile charges is in the order of 10^8 V m^{-1}; and the longitudinal source–drain field E_{SD} used to drive current through the channel is usually in the order of $10^6 \sim 10^8$ V m^{-1}, depending on the device configuration. Figure 7.27 shows that when a DMMP molecule is attached to semiconductor grain boundaries, the induced electric field is in the order of $10^6 \sim 10^8$ V m^{-1}, which is comparable to E_{SG} and E_{SD}. This means that the dipoles of DMMP molecules are capable of trapping mobile charges in the semiconductor film, and significantly diminishing OFET current by changing both the threshold voltage and the field-effect mobility.

We have established that for a device with a known geometry and dielectric capacitance (and fixed drain voltage in the case of the linear regime) changes in the drain current are attributable to two parameters, the mobility μ and threshold voltage V_{TH}. Changes in each of these can further be related to changes in the electronic structure of the film under exposure to vapor. In the absence of charge transfer between vapor molecules and the bulk of the film, variations in μ and V_{TH} can be attributed to two different types of charge trapping. The two types of traps are simply categorized as either "shallow" or "deep". Sworavski et al. provide lucid descriptions in terms of both the electronic structure of the semiconductor as well as the kinetics of an OFET measurement [45]. Electronically, shallow traps are located above the Fermi level but below the LUMO of the semiconductor (for an n-channel OSC) and nearly all empty. Deep traps are located below the Fermi level and are nearly completely filled. Kinetically, shallow traps are temporary and allow multiple release of carriers over the course of a measurement, while deep traps hold charges for a time longer than the scale of the experiment.

In light of this description, threshold voltage shifts can be attributed to variations in deep trapping. In the case of an n-channel semiconductor, if the vapor induces a positive shift in the threshold voltage than an increase in deep trapping is responsible. These traps are far from the LUMO in the energy gap of the material and charges are trapped in these states for the duration of the measurement. This is reasonable if we assume a dipolar nature for the vapor molecule which produces an electric field capable of strongly binding mobile charges. As the vapor induces these deep traps, then more charges will have to be injected into the active layer in order to turn the transistor on, which for an n-channel case means a positive shift in threshold voltage.

Whereas threshold voltage shifts imply a change in the number of mobile charges present in the device, changes in mobility are related to hindrance of those mobile charges under applied field. Reduction in mobility can be related to a reduction in the velocity per unit applied (source/drain) field (hence the units $cm^2\ V^{-1}\ s^{-1}$), of charges traveling between the source and the drain. The interpretation of changes in the mobility depends on the mechanism of charge transport in the OSC layer; here we will consider two accepted theories adapted from inorganic systems.

In the first case, we consider charge transport limited by the trapping and thermal release of carriers. Here mobility is assumed to be limited by the trapping and thermal release of induced charge carriers. This model was originally developed for amorphous silicon transistors and was later adapted for OSCs [46]. In this model, the electronic structure of the semiconductor is assumed to consist of localized states in the band gap near the edge of the delocalized conduction band (for an n-channel compound). Charges are thought to drift in the extended states near the conduction band edge, where they interact with the localized states by trapping and thermal release [47]. Kinetically, the measured field-effect mobility is then limited by the transit time of trapped and released carriers in these localized states [48]. The effective mobility of the film is defined by [46]:

$$\mu = \mu_0 \frac{\sigma_F}{\sigma}, \tag{7.10}$$

Where μ_0 is the trap free mobility and σ and σ_F are the total surface charge density and free surface charge density respectively.

In the context of the multiple trapping and release (MTR) model, the decrease in transistor current upon introduction of vapor results from an increase in the density of traps due to the electrostatic properties of the analyte molecules, which introduce additional trap states in the material. As a result, the ratio of mobile charges to total charges decreases, causing a decrease in the measured mobility. Physically, the analyte molecule acts as an electrostatic trap in which charges are only temporarily trapped then thermally released over the course of the voltage sweep. Increasing the number of these shallow traps increases the transit time of the charges, decreasing the measured field-effect mobility of the charges. On removal of the analyte, traps are removed resulting in increased mobility.

Alternatively, a model which considers granular film morphology defines the mobility as

$$\mu_{\mathrm{b}} = \frac{q\bar{\nu}l}{8kT}\exp\left(-\frac{E_{\mathrm{b}}}{kT}\right), \tag{7.11}$$

where $\bar{\nu}$ is the electron mean velocity and l is the grain size [49]. This model is proposed to be applicable when the grain size exceeds the Debye length, which has been estimated at around 10 nm in previous studies [50]. In the case that grains are smaller than the Debye length then charges see the traps as being uniformly distributed and will follow the MTR model. In the grain boundary limited model the traps are localized at the grain boundaries and E_{B} is the energy barrier at the boundaries caused by the presence of defect states which induce intergrain band bending. If we accept this model, the presence of vapor molecules results in additional trapped charges in the grain boundaries, which raises the energy barrier E_{B} seen by mobile charges. The end result is a reduced field-effect mobility for the thin film device.

It is important to note that at room temperature there is little difference between the two models [46]. The grain boundary model was applied in order to explain a low-temperature saturation in the mobility, which we are not concerned with in a room temperature sensor application. Either can be used to explain how the mobility of the film is affected by the presence of an analyte vapor. In the MTR model, adsorbed vapor molecules create additional traps in the material which increase the transit time of charges and lower the mobility of the film. In the grain boundary model, the vapor results in additional traps located at the grain boundaries of the material and result in a larger energy barrier for mobile carriers traveling between grains, also lowering the mobility of the film.

7.8 Summary and Outlook

OFET responses to an analyte vapor of wide interest can be enhanced by the appending of simple receptor functionality to the OSC molecules, and by the use of OSC heterostructures. Particular protocols for addressing the devices have also been developed. Thus, progress is demonstrated toward the speed, sensitivity, stability, and selectivity requirements of the national security research community. More importantly, careful consideration of many of the electronic and chemical mechanisms that could lead to sensor response allow for performance evaluation at a range of voltages, and predictions of optimal operational conditions. With the many research groups aiming to use both organic- and inorganic-based field-effect devices in chemical sensing, it is hoped that the benchmarks and analyses presented here will stimulate further advances in chemical sensing for security, medical, and industrial applications.

Acknowledgment

This work was supported by the NSF SENSORS program in the Engineering directorate, the JHU-APL Partnership program, and the JHU Department of Materials Science and Engineering, to whom we are grateful. We thank Dr. Joseph Miragliotta for his support of this program, and Professor Andreas Andreou for valuable discussions.

References

1. H.E. Katz, Chem. Mater. **16**, 4748 (2004)
2. H.E. Katz, C. Kloc, Z. Bao, J. Zaumseil, V. Sundar, J. Mater. Res. **19**, 1995 (2004)
3. A. Facchetti, J. Letizia, M.H. Yoon, M. Mushrush, H.E. Katz, T.J. Marks, Chem. Mater. **16**, 4715 (2004)
4. A. Facchetti, Mater. Today **10**, 28 (2007)
5. T.W. Kelley, P.F. Baude, C. Gerlach, D.E. Ender, D. Muyres, M.A. Haase, D. Vogel, S.D. Theiss, Chem. Mater. **16**, 4413 (2004)
6. R. Parashkov, E. Becker, T. Riedel, H.-H. Johannes, W. Kowalsky, Proc. The IEEE **93**, 1321 (2005)
7. T. Someya, Y. Kato, T. Sekitani, S. Iba, Y. Noguchi, Y. Murase, H. Kawaguchi, T. Sakurai, Proc. Natl. Acad Sci. USA **102**, 12321 (2005)
8. G. Guillaud, J. Simon, J.P. Germain, Coord. Chem. Rev. **180**, 1433 (1998)
9. L. Torsi, A. Dodabalapur, N. Cioffi, L. Sabbatini, P.G. Zambonin, Sens. Actuators B-Chem. **77**, 7 (2001)
10. B. Crone, A. Dodabalapur, A. Gelperin, L. Torsi, H.E. Katz, A.J. Lovinger, Z. Bao, Appl. Phys. Lett. **78**, 2229 (2001)
11. F. Liao, C. Chen, V. Subramanian, Sens. Actuators B-Chem. **107**, 849 (2005)
12. M.C. Tanese, D. Fine, A. Dodabalapur, L. Torsi, Microelectron. J. **37**, 837 (2006)
13. M.C. Tanese, D. Fine, A. Dodabalapur, L. Torsi, Biosens. Bioelectron. **21**, 782 (2005)
14. L. Torsi, A. Dodabalapur, L. Sabbatini, P.G. Zambonin, Sens. Actuators B-Chem. **67**, 312 (2000)
15. D. Halliday, R. Resnick, *Physics (Part II)* (Wiley, New York, 1962)
16. C. Kendrick and S. Semancik, J. Vac. Sci. Technol. A **16**, 3068 (1998)
17. D. Li, E.-J. Borkent, R. Nortrup, H. Moon, H.E. Katz, Z. Bao, Appl. Phys. Lett. **86**, 042105 (2005)
18. Z.-T. Zhu, J.T. Mason, R. Dieckmann, G.G. Malliaras, Appl. Phys. Lett. **81**, 4643 (2002)
19. H. Hoshino, M. Yoshida, S. Uemura, T. Kodzasa, N. Takada, T. Kamata, K. Yase, J. Appl. Phys. **95**, 5088 (2004)
20. G. Horowitz, J. Mater. Res. **19**, 1946 (2004)

21. G. Horowitz, M.E. Hajlaoui, R. Hajlaoui, J. Appl. Phys. **87**, 4456 (2000)
22. M.C.J.M. Vissenberg and M. Matters, Phys. Rev. B **57**, 12964 (1998)
23. N. Tessler, Y. Roichman, Appl. Phys. Lett. **79**, 2987 (2001)
24. K.P. Pernstich, S. Haas, D. Oberhoff, C. Goldmann, D.J. Gundlach, B. Batlogg, A.N. Rashid, G. Schitter, J. Appl. Phys. **96**, 6431 (2004)
25. R.J. Chesterfield, J.C. McKeen, C.R. Newman, P.C. Ewbank, D. Filho, J.-L. Bredas, L.L. Miller, K. Mann, C.D. Frisbie, J. Phys. Chem. B **108**, 19281 (2004)
26. Y. Sun, K.Y. Ong, *Detection technologies for chemical warfare agents and toxic vapors* (CRC Press, Boca Raton, FL, 2005)
27. W.T. Muse, S. Thomson, C. Crouse, K. Matson, Inhal. Toxicol. **18**, 1101 (2006)
28. A. Salleo, F. Endicott, R.A. Street, Appl. Phys. Lett. **86** (2005)
29. A. Salleo, R.A. Street, J. Appl. Phys. **94**, 471 (2003)
30. D. Kawakami, Y. Yasutake, H. Nishizawa, Y. Majima, Jpn. J. Appl. Phys. Part 2-Lett. Exp. Lett. **45**, L1127 (2006)
31. A. Salleo, R.A. Street, Phys. Rev. B **70** (2004)
32. R.A. Street, A. Salleo, M.L. Chabinyc, Phys. Rev. B **68** (2003)
33. E.S. Snow, F.K. Perkins, E.J. Houser, S.C. Badescu, T.L. Reinecke, Science **307**, 1942 (2005)
34. A.A. Tomchenko, G.P. Harmer, B.T. Marquis, Sens. Actuators B-Chem. **108**, 41 (2005)
35. S.V. Patel, T.E. Mlsna, B. Fruhberger, E. Klaassen, S. Cemalovic, D.R. Baselt, Sens. Actuators B-Chem. **96**, 541 (2003)
36. L. Bertilsson, I. Engquist, B. Liedberg, J. Phys. Chem. B **101**, 6021 (1997)
37. L. Bertilsson, K. PotjeKamloth, H.D. Liess, Thin Solid Films **285**, 882 (1996)
38. L. Bertilsson, K. Potje-Kamloth, H.D. Liess, I. Engquist, B. Liedberg, J. Phys. Chem. B **102**, 1260 (1998)
39. L. Bertilsson, K. Potje-Kamloth, H.D. Liess, B. Liedberg, Langmuir **15**, 1128 (1999)
40. H.E. Katz, J. Johnson, A.J. Lovinger, W.J. Li, J. Am. Chem. Soc. **122**, 7787 (2000)
41. H.P. Jia, G.K. Pant, E.K. Gross, R.M. Wallace, B.E. Gnade, Org. Electron. **7**, 16 (2006)
42. L. Torsi, A. Dodabalapur, H.E. Katz, J. Appl. Phys. **78**, 1088 (1995)
43. G. Horowitz, R. Hajlaoui, R. Bourguiga, and M. Hajlaoui, Syn. Metals **101**, 401 (1999)
44. G. Horowitz, P. Lang, M. Mottaghi, H. Aubin, Adv. Funct. Mater. **14**, 1069 (2004)
45. J. Sworakowski, K. Janus, S. Nespurek, M. Vala, IEEE Trans. Dielectrics And Electr. Insulation **13**, 1001 (2006)
46. G. Horowitz, M.E. Hajlaoui, R. Hajlaoui, J. Appl. Phys. **87**, 4456 (2000)
47. W.E. Spear and P.G. Le Comber, J. Non-Cryst. Solids **8–10**, 727 (1972)
48. P.G. Lecomber, W.E. Spear, Phys. Rev. Lett. **25**, 509 (1970)
49. G. Horowitz. M.E. Hajlaoui, Adv. Mater. **12**, 1046 (2000)
50. R. Bourguiga, G. Horowitz, F. Garnier, R. Hajlaoui, S. Jemai, H. Bouchriha, Eur. Phys. J. Appl. Phys. **19**, 117 (2002)

8

Electrochemical Transistors for Applications in Chemical and Biological Sensing

A. Kumar and J. Sinha

8.1 Introduction

Organic electrochemical transistors (OECTs) are also three-electrode (source–drain–gate) devices similar to organic field-effect transistor devices however the source–drain current in OECTs is modulated by application of a gate potential across an electrolyte solution (Fig. 8.1). Since the application of the gate potential in OECTs results in electrochemical doping or dedoping of the organic semiconductor, OECTs operate at much lower voltages compared to field-effect transistors. Furthermore, because the kinetics of the doping or dedoping processes is governed by the diffusion of the ions from the adjacent electrolyte solution, OECTs in general show slower response times. In OECTs, doping levels of the active conjugated polymer layer change with change in gate voltage as opposed to field-effect transistors where only the field in the gate region changes. Upon exposure to an analyte, OECTs exhibit either an increase or a decrease in resistance, when operated in the doped or dedoped configurations, respectively. Therefore, the performance of an OECT device, for a given analyte, can be optimized by studying the response as a function of the doping of the organic semiconductor. Since the nature of the semiconductor in OECTs can vary significantly by a slight variation in applied potentials, a reference electrode is generally used as a gate electrode, and the potentials are controlled by a bi-potentiostat (Fig. 8.1). In a typical OECT, a slight voltage offset (tens of millivolts) is applied between the source and drain with respect to the gate (reference) electrode. The drain current (current flow from source to drain) is then measured as a function of analyte concentration. For example, if the OECT is operating at a gate potential of +0.5 V, then the source could be kept at +0.5 V and the drain at +0.48 V, resulting in an offset of 20 mV. The drain current is then measured in the absence of and then in the presence of analyte. The resulting change in drain current is related to the concentration of the analyte used. The device is then studied at various gate potentials (corresponding to the state of the semiconductor going from completely dedoped to doped) in order to find out

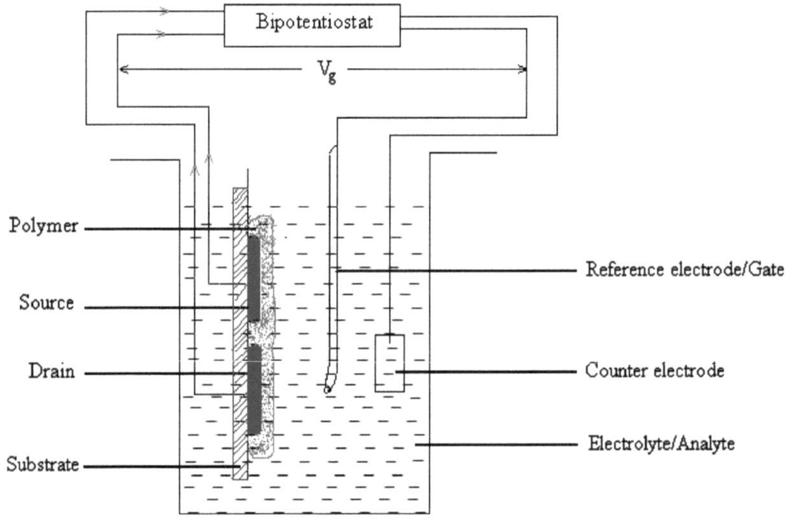

Fig. 8.1. Schematic representation of an OECT device

the optimum gate potential resulting in the maximum change in drain current for a given analyte concentration. Once the optimum gate potential is obtained, then the device response at this optimum gate potential is studied as a function of analyte concentration. In general, one starts with a very low concentration of analyte and measures the resulting device response. This is then followed by gradual addition of more analyte until the device response saturates. Generally one observes a sigmoidal response, i.e., the device is not sensitive below a certain concentration and saturates above a certain concentration. The effective detection range of the device, is therefore assumed to be in between those two concentrations, where the device shows a linear response to changes in analyte concentration. Since each device is made individually, the device response is normalized with respect to the device response in absence of analyte in order to remove device-to-device variation. This can be achieved by dividing the change in drain current (difference of drain current in the presence and absence of analyte) by the original drain current (in the absence of analyte).

In this chapter, we review OECTs using conjugated polymers as the semiconducting layer. Conjugated polymers provide an ideal platform for the design and fabrication of OECT-based sensors because they undergo large and reversible physical (conformational) or chemical (redox) changes under the influence of external physical (temperature, light, electric, magnetic, or mechanical) or chemical (pH, chemical agent) stimuli. Therefore, a measure of the change in source–drain current can be used to detect the presence of the analytes which provide these stimuli. Other advantages of using conjugated polymers as transducers for OECTs include ease of fine tuning of the redox properties, introduction of specificity by easy incorporation (covalent or physical) of the receptor for a particular analyte, low cost and finally ease of fabrication. In fact the use of OECTs developed with conjugated polymers is projected to be the future of single use disposable sensors.

8.2 Sensors Based on Electrochemical Transistors

OECTs have developed significantly from the first OECT, based on polypyrrole, reported by Wrighton and coworkers in 1984 (Reproduced with permission from [1]. Copyright 1984 American Chemical Society.) Interestingly, the first OECT employed an additional gate microelectrode between the source and drain microelectrodes (Fig. 8.2). The presence of this additional gate microelectrode resulted in slower response time and poor sensitivity, because of the increase in the distance between the source and drain microelectrodes. Later, in 1985, the same group reported the application of OECTs to chemical sensing [2–4]. The earlier OECTs employing conjugated polymers as the transducer required reasonably sophisticated fabrication facilities, not commonly available to chemists, to construct microelectrode arrays. This restricted the full exploitation of the potential of OECTs for sensing applications. This problem was solved by Contractor and coworkers by using isoporous polycarbonate membranes coated with gold on both sides [5]. Electropolymerization in the

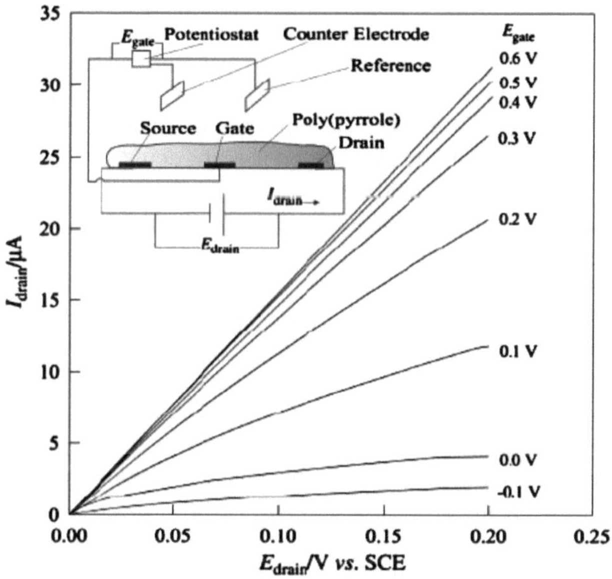

Fig. 8.2. Plots of the drain current, I_{drain} as a function of the drain voltage, E_{drain}, for a polypyrrole-based micro-eletrochemical transistor operated in MeCN containing 0.1 mol dm^{-3} Bu_4NClO_4. Each curve corresponds to a different value of the gate voltage, E_{gate}. As the gate voltage increases from −0.1 V vs. SCE to 0.6 V the resistance of the polypyrrole film decreases. The *inset* shows the arrangement used to make the measurement. Reproduced with permission from [1]. Copyright 1984 American Chemical Society

pores of these membranes resulted in the fabrication of OECTs wherein one side of the membrane was used as the source electrode and the other as the drain electrode. The thickness of the membrane was the length of the semiconducting channel, i.e., the distance between the source and drain. Since these membranes are commercially available with varying thickness and also from different materials, and additionally gold can be easily evaporated, this has resulted in increased interest for using OECTs for chemical and biological sensing applications.

8.2.1 Sensor Mechanisms

In OECT-based sensors, it is the conjugated polymer layer which acts as a transducer upon exposure to analytes. However, it is not essential that the conjugated polymer layer be the one which responds directly to the presence of the analyte. Therefore, OECTs can be broadly divided into four types based on their mechanism of detection of the analyte. These sensor mechanism types are listed in Table 8.1.

The first sensor mechanism type, Type I, utilizes direct chemical sensing whereby the analyte reacts directly with the conjugated polymer layer, resulting in a change to the chemical state of the conjugated polymer. The first OECT sensor, reported by Wrighton and coworkers [2], was based on this type of mechanism where the active polyaniline layer was found to respond to small changes in pH as well as to redox reagents such as Ru(III)–Ru(II) and Fe(III)–Fe(II). Later, the same group demonstrated that OECTs based on Poly(3-methylthiophene) can be used to detect very small quantities ($<10^{-15}$ mol) of a chemical oxidant $IrCl_6{}^{2-}$ [3]. Gaponik et al. [15] reported the use of an OECT sensor for the detection of dissolved SO_2. They used the well-known concept of polyaniline catalyzed oxidation of dissolved SO_2. Later, Saxena et al. [28], used the specific recognition of Cu(II) by polycarbazole for the design of an ion-selective OECT sensor using polycarbazole as the active transducer layer. Although the device was found to be selective for Cu(II) cations, it was found to be responsive only within a very narrow concentration range of Cu(II), from 2.5 to 100 µM. In 1996, Bartlett and coworkers [20] reported the development of an OECT device based on polyaniline, which was responsive for NADH using a highly acidic polyelectrolyte counter ion. The sensing mechanism proposed was the catalytic oxidation of NADH by polyaniline itself.

The second sensory mechanism type, Type II, involves the interaction of the analyte with the electrolyte, which then induces changes in the properties of the electrolyte. The selection of electrolyte is crucial for the design of this type of OECT. In general, the electrolyte is chosen for its ability to induce ionic conductivity changes upon exposure to the analyte, resulting in changes in the source–drain current through the active layer. Wrighton and coworkers [14] were the first to exploit this concept when they used the moisture dependence of the ionic conductivity of a solid state electrolyte:

Table 8.1. Summary of reviewed OECT sensors

Active layer	Analytes	Sensor mechanism type[a]
	DNA [6]	III
	IgG antigen–antibody [7]	III
PEDOT	Glucose [8]	IV
	H_2O_2 [8]	IV
	Humidity [9]	II
	Urea [10–13]	IV
	lipid [10–13]	IV
	Hemoglobin [10]	IV
	Humidity [14]	II
	Dissolved SO_2 [15]	I
	Metal ions [16]	III
Polyaniline	Ru (III)–Ru(II) [2], Fe(III)–Fe(II) [2]	I
	H_2O_2 [17, 18]	IV
	Alkaline phosphatase [19]	IV
	pH [2]	I
	NADH [20]	I
	Glucose [10–13, 21–25]	IV
	pH [26]	IV
Polypyrrole	Pencillin [26]	IV
	NADH [27]	IV
Polycarbazole	Cu(II) [28]	I
	H_2 [4]	IV
Poly(3-methylthiophene)	pH [4]	IV
	O_2 [4]	IV
	Ir(IV) [3]	I

[a] See text for detail

poly(vinyl alcohol)/phosphoric acid. They used an OECT containing polyaniline as the active transducer layer and found that this solid state OECT responds reversibly to the change in humidity. Later on, Nilsson et al. [9] exploited the moisture sensitivity of the nafion solid state electrolyte for the design of a flexible all-plastic printable OECT-based humidity sensor using PEDOT:PSS as the gate as well as the active transducer layer (Fig. 8.3).

The third type of sensor mechanism, Type III, involves the physical deformation (change in conformation) of the active conjugated polymer layer upon binding of the analyte (guest) to the receptor (host) species. These sensors therefore can potentially sense any guest–host binding event. The host or guest can be entrapped physically within the active conjugated polymer layer or alternatively can be attached covalently to the conjugated polymer, after which it can then bind to its partner (the guest or host). Though the physical entrapment is easier, the leaching out of the host overtime represents a major challenge for these sensors. Contractor and coworkers [16] were the first to

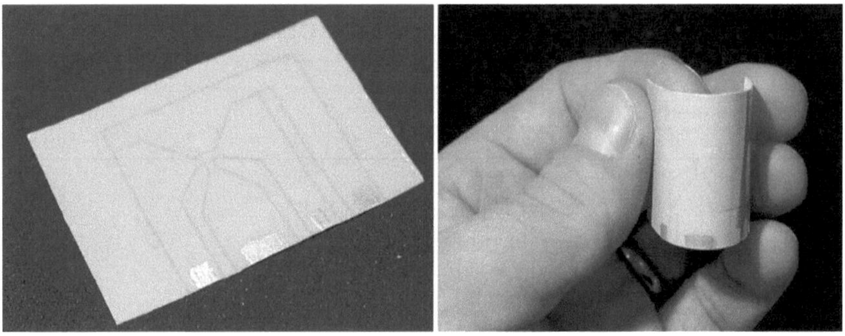

Fig. 8.3. (*Left*) Solid state OECT humidity sensor printed on polyethylene-coated paper. (*Right*) The sensor was capable of operating when bent at a radius of curvature of approximately 5 cm. Reprinted from [9] with permission from Elsevier

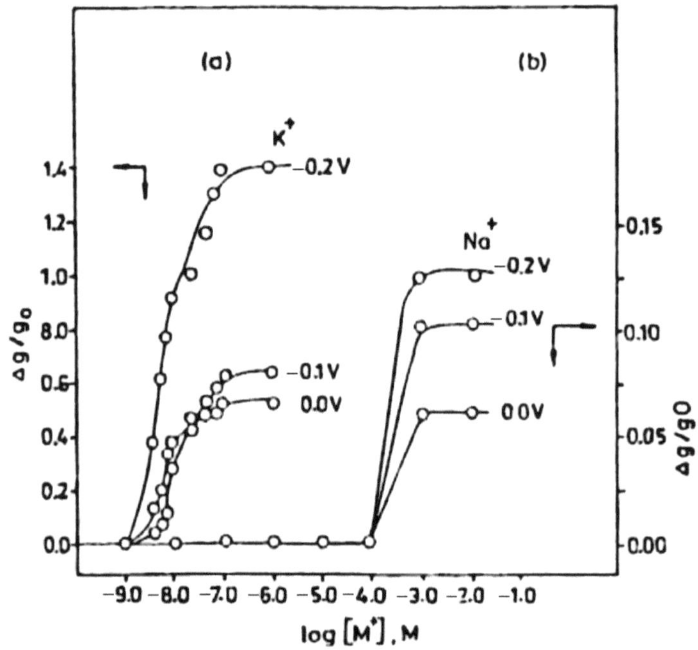

Fig. 8.4. Direct conductance of the device on exposure to various concentrations of alkali metal ions. (**a**) Set of curves recorded at three gate voltages in the presence of various concentrations of K^+ ions and (**b**) set of curves at the same voltages in the presence of various concentrations of Na^+ ions. (The counterion in both the cases is Cl^-). Reproduced with permission from [16]. Copyright 1997 American Chemical Society

exploit this concept for the design of OECT-based sensors for the detection of metal ions (K^+, Na^+, Ba^{2+}, Sr^{2+}, and Ca^{2+}). They used 18-crown-6 ether as the host dispersed in a polyaniline transducer. This allowed the presence of K^+ ion down to very low concentrations (~10^{-8} M) (Fig. 8.4). Later, they extended this concept for the design of immunosensors [7] where as Kumar and coworkers [6] used this concept for the fabrication of a DNA sensor. These development efforts are discussed in detail in Sects. 8.2.3 and 8.2.4, respectively.

The fourth sensor mechanism type, Type IV, involves a cascade of events in the presence of analyte, with the final outcome of the cascade process resulting in a change in the chemical state of the active conjugated polymer layer. The active conjugated polymer layer should be insensitive to the analyte in the absence of the reaction cascade. The first example of this type of OECT sensor was reported by Wrighton and coworkers [4], who exploited the platinum catalyzed reactions of H_2 and O_2 to design a sensor for the detection of H_2, O_2, and also pH. They used OECTs based on poly(3-methylthiophene) impregnated with platinum particles, which functioned in an aqueous solution at room temperature. This sensor was capable of detecting pH from 0–12. In 2004, Malliaras and coworkers [8] proposed a simple OECT device based on commercially available PEDOT:PSS which was fabricated by simple spin coating of the transducing layer on a glass support. They used a platinum wire as the gate electrode and observed an increase in source–drain current in presence of hydrogen peroxide. Most of the enzyme-based biological OECT sensors fall into this category and are discussed in detail in Sect. 8.2.2.

8.2.2 Enzyme-Based Sensing

Conjugated polymers provide an ideal platform for the design and fabrication of enzyme-based OECT biosensors because enzymes are very specific in their mode of action. These biosensors are based upon redox enzymes that provide chemical stimuli in the presence of an analyte by catalyzing the production or consumption of a redox active species, which in turn changes the chemical state of the active conjugated transducer layer directly or through a mediator. Since many of the biochemical processes catalyzed by redox enzymes result in changes in pH, polyaniline, which is sensitive to changes in pH, is the most extensively used conjugated polymer for OECT-based biosensors.

Matsue et al. [27] were the first to explore an enzyme-based OECT biosensor. They used Diaphorase as the entrapped enzyme in a polypyrrole transducing layer for the detection of NADH via a redox mediator (the sodium salt of anthraquinone-2-sulfonic acid). The net result was the conversion of polypyrrole from its conducting state to its insulating state in the presence of NADH. The device showed a response time of ≈15–20 min in the presence of NADH. Later Nishizawa et al. [26] exploited the pH sensitivity of the polypyrrole film for the design and fabrication of OECT sensors for pH and for pencillin. The Penicillinase enzyme was entrapped in a membrane which was coated with a polypyrrole film, in which a decrease in pH was observed in the presence of penicillin due to the hydrolysis of penicillin by Penicillinase.

Contractor and coworkers were the first to explore the pH sensitivity of polyaniline for the fabrication of an OECT biosensor for the detection of glucose, using entrapped Glucose oxidase [21]. They used the concept of "bread/butter/jam" for the fabrication of OECT wherein polyaniline (butter) was deposited on two platinum disk electrodes (bread), separated by 8 μm, acting as the source and drain, followed by another layer of Glucose oxidase entrapped polyaniline (jam). They successfully demonstrated that their device sensed the change in microenvironment pH resulting from the Glucose oxidase catalyzed oxidation of glucose, generating a linear response for up to 10 mM glucose solution. They also postulated that these devices can be used for sensing very small quantities of a range of biomolecules. Later they demonstrated that these pH responsive OECTs can be used for the detection of other biomolecules such as urea, hemoglobin, and lipids by appropriate selection of the entrapped enzyme [10]. This work introduced a new normalized way of reporting the sensor response by dividing the change in drain current (difference of drain current in the presence and absence of the analyte) by the original drain current (in the absence of analyte). This way of reporting the sensor response removed the device-to-device discrepancies introduced due to the individual manufacture of each device, and in addition allowed a comparison of the reproducibility of device fabrication and sensitivity (defined as change in response of the sensor per mM change in the concentration of the analyte) of different devices. In 1996, Contractor and coworkers [11] developed a novel way of entrapping the enzymes in the active conjugated polymer layer, using electrostatic interactions between the charged conjugated polymer surface (polyaniline in the oxidized form) and the charges on the enzymes. This allows them to design a so-called electronic tongue by fabricating closely separated multiple microelectrode pairs on a single chip. Since each microelectrode pair can be individually addressed (oxidized to form charged surface), it was possible to entrap separate enzymes on each microelectrode pair without any crossloading and also without the need for a mask. Using this method, they successfully fabricated three OECTs on a single chip and were able to quantify glucose, urea, and a lipid from a single solution containing a mixture of these three analytes (Fig. 8.5)

In 1998, Contractor and coworkers [5] demonstrated a new and simple way of making OECTs using isoporous polycarbonate membranes. This resulted in easier and cheaper device fabrication without sophisticated microfabrication facilities and provided a better alternative to the silicon-based devices. Furthermore, it was possible to study the effect of channel length (separation between the source and drain) by using membranes with different thickness. They also observed that the nature of the interface between the membrane surface and the polymerization medium plays a crucial role in the morphology of the electropolymerized conjugated polymer, which in turns affects the sensitivity of the OECT devices. When polyaniline was grown electrochemically in the pores of as-received membranes, more ordered growth was observed because of the hydrophobic nature of the membrane wall. However, when the membranes were treated with nonionic surfactant, Triton-X-100, prior to electropolymerization, the resulting polymer was found to be less ordered.

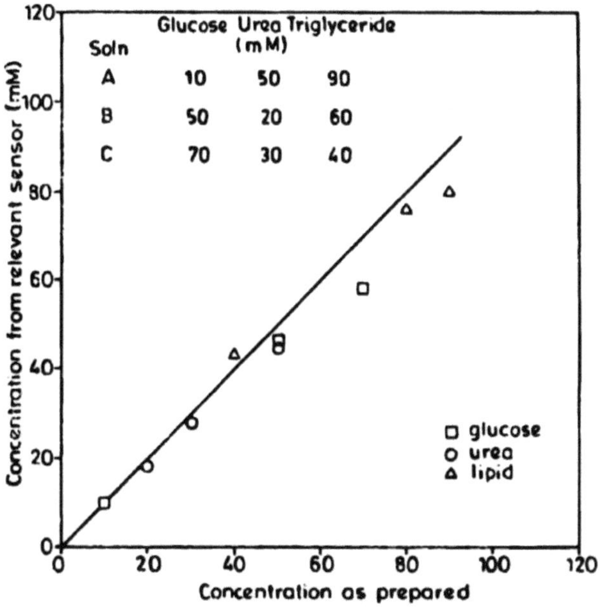

Fig. 8.5. Analysis of three different mixtures of glucose, urea, and lipid using a sensor array. Reproduced with permission from [11]. Copyright 1996 American Chemical Society

Interestingly, the OECT devices based on more disordered polyaniline exhibited larger changes in conductance between the insulating and conducting states. Furthermore, the more disordered polymer resulted in easy diffusion of the enzyme into the polymer matrix resulting in higher enzyme loading and hence higher sensitivity of OECT sensors. In 1999, after an exhaustive study on these OECT-based sensors, Contractor and coworkers [12] concluded that the sensitivity of these devices depends upon the channel length (increases with a decrease in the source–drain separation) and also on the microstructure of the polymer where the more disordered (porous) polymer exhibits better sensitivity due to higher loading of enzyme and also better availability of the entrapped enzyme for easy diffusion of the analyte. Furthermore, more disordered polymers exhibited conditions favorable for accommodation of conformation changes.

In all of the above-mentioned work of Contractor and coworkers, enzyme entrapment was carried out at higher pH after the electropolymerization of aniline at much lower pH. Enzymes could not be added along with the electropolymerizing solution because the enzymes become denaturated at such low pHs. This restricted the amount of enzyme that could be entrapped and hence limited the sensitivity of the OECT devices. In 1999, Tripathy and coworkers [29–31] demonstrated that aniline could be polymerized at pHs as high as 5.5, using a strongly acidic polyelectrolyte like poly(styrene sulfonate). Based on these reports, Contractors and coworkers fabricated OECT devices where the enzymes were added in the electropolymerizing solution containing poly(styrene sulfonate) [13]. This resulted in higher loading and better diffusion of enzyme throughout the polymer matrix. As expected, these OECT devices showed better sensitivity, higher linear ranges of detection and faster response time compared to the earlier devices, where the enzymes were entrapped postpolymerization. In fact, the sensor response in case of urea and triglyceride lipid sensors was found to improve by a factor of ≈100. Battaglini [17] and coworkers reported that it was also possible to improve the electroactivity of polyaniline at higher pH by N-alkylation using propane sultone. This resulted in partially sultonated polyaniline which was found to be electroactive at pHs as high as 7. They further demonstrated that Horseradish peroxidase entrapped OECTs based on this modified polyaniline could detect hydrogen peroxide down to 1 ppm at neutral pH.

Bartlett and coworkers have also carried out extensive studies on polyaniline-based OECTs for glucose and hydrogen peroxide sensing. For glucose sensing, they coated an insulating layer of electropolymerized 1,2-diaminobenzene, containing the enzyme Glucose oxidase over a thin film of electropolymerized polyaniline [22, 23]. This is different from the work of Contractor and coworkers who used another layer of polyaniline loaded with glucose oxidase enzyme. Bartlett and coworkers postulated that the application of a selectively permeable insulator layer, apart from producing a highly active enzyme film, prevents interference from other species present in the analyte solution. Their configuration further differs from that of Contractor and coworkers in that they have to add an additional redox mediator, tetrathiafulvalene, to couple the enzyme-mediated reaction with the transducing polyaniline layer. The switching time of the device in the presence of glucose was found to be ≈10 s. In 1998, they reported that it is possible to detect glucose down to 2 μ M by careful optimization of the fabrication of the OECT and by varying the drain voltage [24]. Interestingly, this detection limit is almost 40 times lower than the detection limit of the same device when operated as an amperometric sensor, clearly establishing the amplifying nature of the transistor configuration. These above-mentioned devices are based on the addition of a redox mediator which makes them unsuitable for in vivo detection of glucose and also for measurements in fluidic system. This problem was circumvented by Bartlett and coworkers in 2000 who covalently attached a redox mediator directly to the Glucose oxidase enzyme [25]. They used a pyridine-based osmium complex bearing either a carboxylate or an aldehyde group as redox mediator for the covalent attachment to Glucose oxidase. In 1998, Bartlett and coworkers replaced Glucose oxidase in their OECT device, based on polyaniline as a transducer, with horseradish peroxidase and showed that the device can also work as a sensor for the detection of hydrogen peroxide [18]. In 2004, Malliaras and coworkers [8] proposed a simple OECT device based on commercially available PEDOT:PSS which was fabricated by simple spin coating of the transducing layer onto a glass support. They used a platinum wire as the gate electrode and observed an

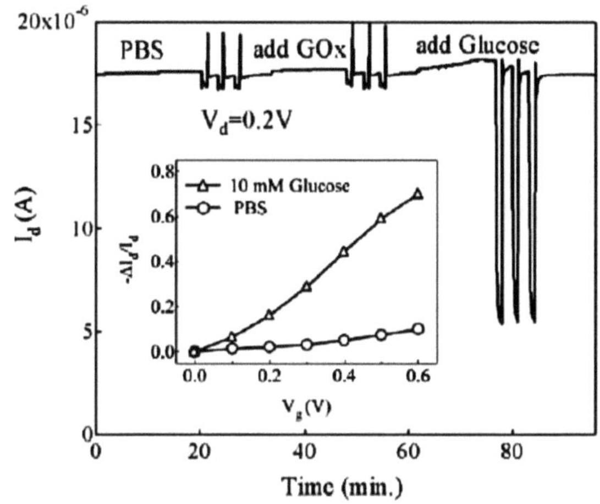

Fig. 8.6. I_{d} vs. time for the device in PBS solution, in which first GO_x, and then glucose are added. $V_{\mathrm{d}} = 0.2$ V, and V_{g} is plused to 0.6 V for 1 min. *Inset* shows the relative change of I_{d} (i.e., $-[I_{\mathrm{d}}(V_{\mathrm{g}} = 0\ \mathrm{V}) - i_{\mathrm{d}}(V_{\mathrm{g}})]/[I_{\mathrm{d}}(V_{\mathrm{g}} = 0\,\mathrm{V})]$) as a function of V_{g} for different solutions [8] – Reproduced by permission of the Royal Society of Chemistry

increase in source–drain current in the presence of both Glucose oxidase and glucose (Fig. 8.6). They postulated that the response of the sensor was not due to a change in pH caused by the generation of gluconic acid, but due to the generation of hydrogen peroxide, which gets oxidized at the platinum electrode resulting in reduction of the PEDOT:PSS layer. In principle, this design is very simple, easy to fabricate and can be used for the fabrication of OECT-based sensors for other biochemical reactions which produce hydrogen peroxide.

8.2.3 Antibody–Antigen-Based Sensing

Most of the OECTs based on enzyme detection discussed above, required an electron transfer process to switch the device between the OFF and ON states. In 1997, Contractor and coworkers demonstrated that it is also possible to trigger the switching of the OECT devices simply by changes in the conformation of the transducer layer. This change in conformation can be induced by using a guest–host system wherein one of the partners is entrapped in the transducer layer which then acts as a sensor for the detection of the other partner. This type of sensing mechanism falls into type III method of detection and, in principle, can be used to design sensor for any guest–host binding system. The specificity of these devices is governed by the specificity of the guest–host binding pair and in this case, the conjugated polymer merely acts as a transducer. One of most specific binding pairs to be found is the antigen–antibody binding and this is the basis for the design and fabrication of immunosensors. Interestingly, it has been proposed that the conformational changes upon binding of an antibody with antigen are not significant, as these are preorganized structures. They can therefore not be used for the design of type III OECT-based immunosensors [32]. However, in 2002, Contractor and coworkers [7] successfully exploited antigen–antibody binding for the design and fabrication of first reagent-less immunosensor based on OECTs. They used PEDOT as the transducing layer and goat antirabbit IgG and rabbit IgG as the biospecific binding pair. They used isoporous polycarbonate membrane (coated with gold on both sides) with a pore diameter of 1.2 μm as the working electrode for the electropolymerization of PEDOT to generate a transducing layer. Since the electroactivity of PEDOT is pH insensitive, it was possible to add goat antirabbit IgG in the electropolymerization solution itself. This resulted in higher and uniform loading of the antibody throughout the PEDOT transducing layer, resulting in a more sensitive immunosensor. They fabricated different devices by electropolymerizing 50 mM PEDOT solution in pH 7.2 phosphate buffer, using a 6:1 water:acetonitrile mixture in the presence of varying concentration of goat antirabbit IgG antibody. The sensor response was then measured at different gate potentials for given concentrations of rabbit IgG antigen and it was found that the maximum response was observed at a gate potential of −0.8 V. These devices were then exposed to different concentration of rabbit IgG antigen at a constant gate potential of −0.8 V. As expected, a sigmoidal sensor response was observed as a function of rabbit IgG concentration (Fig. 8.7). Increasing the amount of goat antirabbit IgG antibody during the electropolymerization resulted in improved sensi-

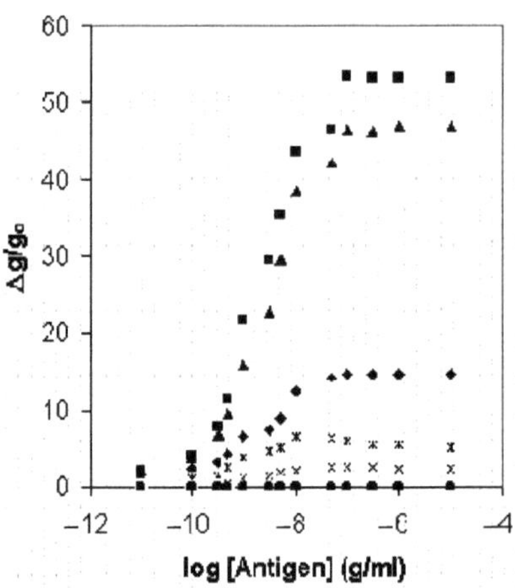

Fig. 8.7. Sensor response as a function of antigen concentration (*filled squares*, *filled triangles*, and *filled diamonds* are for 3, 5, and 1×10^{-5} g ml^{-1} of the antibody, respectively, immobilized during polymerization. *Asterisks* and *times* are for 3 and 1×10^{-5} g ml of the antibody, respectively, immobilized by physical adsorption after the polymerization. *Filled circle* is for the control experiment where no antibody was immobilized) [7] – Reproduced by permission of the Royal Society of Chemistry

tivity with a saturation of loading at 3×10^{-5} g ml^{-1}. These devices were found to exhibit a linear response from a very low rabbit IgG concentration of 0.1–100 ng ml^{-1}. The response time of the device was found to be around 3 min. Since the binding of the antibody–antigen is specific, they showed that it was possible to entrap antigen and then detect the antibody. The key reason for the success of these devices was switching over to PEDOT (a pH insensitive conjugated polymer) from polyaniline or polypyrrole (pH sensitive conjugated polymers). The former allows the incorporation of recognition elements in the electropolymerization solution as polymerization can be carried out at any pH. If polyaniline or polypyrrole are used as the transducing layer, then the antibody has to be entrapped after the polymerization of the active transducing layer which results in lower loading of the antibody and hence weaker signal upon exposure to antigen. This was clearly demonstrated by the control experiments where the antibody was entrapped within the active PEDOT layer postpolymerization, resulting in very weak sensor responses when exposed to antigen (Fig. 8.7).

8.2.4 DNA-Based Sensing

Based on the success of the reagent-less immunosensor where the change in conformation upon binding of the antibody–antigen was used to trigger the OECT based on PEDOT as the transducer, Kumar and coworkers [6] extended this concept to the fabrication of label-free DNA sensors. They exploited the specific binding of single stranded DNA (ssDNA) to a complementary DNA strand thereby forming double stranded DNA. This results to a change in the conformation of the transducing PEDOT layer, resulting in triggering of the OECT device. OECT devices for DNA detection were fabricated by electropolymerization of PEDOT in the pores of polycarbonate membranes in the presence of ssDNA probes in 0.1 M KCl dissolved in phosphate buffer, pH 8. These devices were than exposed to solutions of varying concentrations of complementary strand DNA at a gate potential of 0.8 V and the sensor response was recorded. The expected sigmoidal response was observed when the OECT devices were exposed to complementary ssDNA and, as expected, no response was observed when the OECT was exposed to noncomplementary ssDNA (Fig. 8.8).

The length of the ssDNA probe was varied from 20mer (20 nucleotide bases) to 5mer (5 nucleotide bases) in order to determine the minimum number of binding interactions necessary for the detection of target ssDNA using this configuration. It was observed that a 5mer ssDNA probe is sufficient to detect the complementary 5mer ssDNA. Furthermore, it was also observed that the linear range of detection as well as the sensitivity increased with increasing number of complementary base pairs (increasing length of the ssDNA probe). The detection range for varying base pair lengths is listed in Table 8.2. It was also demonstrated that a shorter ssDNA probe of 5mer was sufficient to detect a longer ssDNA with the complementary bases located somewhere along its

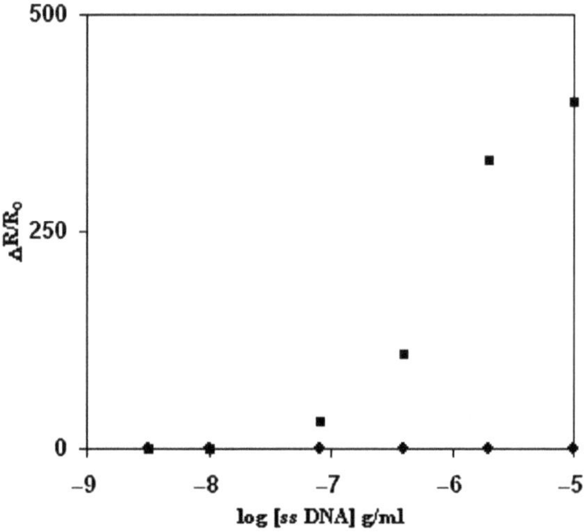

Fig. 8.8. Sensor response as a function of ssDNA concentration (A1) (*filled square*). For complementary strand; (*filled circle*) for control experiment [6] – Reproduced by permission of the Royal Society of Chemistry

Table 8.2. Detection range for various DNA sensors [6] – Reproduced by permission of the Royal Society of Chemistry

Probe ssDNA base length	Conc.[a] (μg ml^{-1})	Complementary target ssDNA base length	Detection range (g ml^{-1})
20	20	20	8×10^{-8} to 1×10^{-5}
10	20	10	6×10^{-7} to 1×10^{-5}
5	10	10	9×10^{-7} to 6×10^{-6}
10	10	5	5×10^{-7} to 3×10^{-6}

[a] Concentration in the polymerization solution

length. They postulated that their design has several advantages, such as ease of fabrication, the absence of a requirement for labeling or modification of the probe DNA. Furthermore, they used fluorescence spectroscopy to show the duplex formation between the probe ssDNA and complementary ssDNA in the OECT devices, confirming that the device response is due to the duplex formation between complementary strands.

8.3 Recent advances in Design and Fabrication of Eletrochemical Transistors

There are some interesting advances in the design and fabrication of OECTs. The most promising one is the use of microfluidic-based OECTs reported by Malliaras et al. [33]. They have successfully integrated microfluidic channels

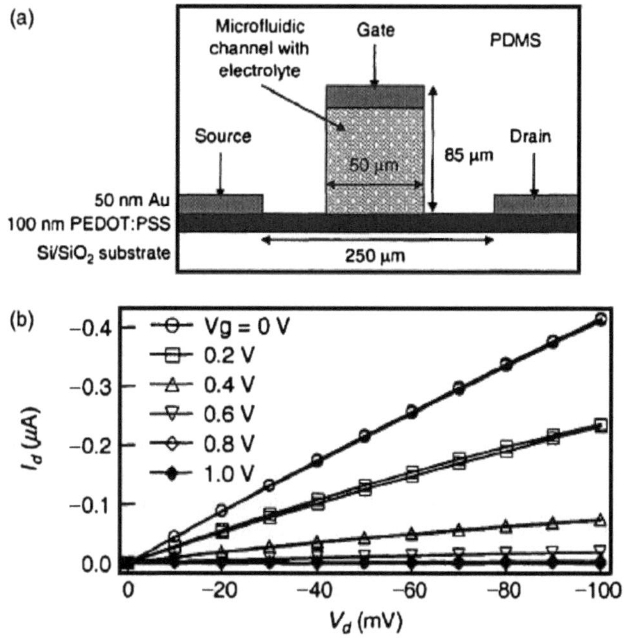

Fig. 8.9. (**a**) Schematic cross section of the device structure (not to scale) showing the gate electrode integrated into the microfluidic channel. (**b**) Typical *I–V* characteristics measured with 10 mM Tris-Cl buffer solution flowing across the PEDOT:PSS transistor at 0.1 μl min^{-1}. Reprinted with permission from [33]. Copyright 2005 American Institute of Physics

with sensor transduction elements to realize a lab-on-a-chip concept. They used PEDOT:PSS as the active transducing layer and fabricated a microchannel on top of it with a built-in gate electrode in the "ceiling" of the channel (Fig. 8.9a). Therefore, the microchannel serves the dual purposes of controlling the flow of the analyte as well as the gate potential. This OECT can be switched OFF by the application of the positive gate potential (Fig. 8.9b) and the transistor behavior was found to be dependent on the ionic strength of the analyte solution.

Bartlett and coworkers [19,34] modified the design of OECT based devices by employing a two cell configuration wherein the OECT device was located in one cell, analyte in other cell and the two cells were connected by a salt bridge (Fig. 8.10). The advantage of this design is that the device may be operated in the optimal environment (in cell 1) and also carry out the biochemical sensing reactions with the analyte in the optimal buffer conditions (in cell 2). Using this design, they were able to detect the enzyme alkaline phosphatase, in nanomolar range within 70 s, using poly(aniline)/poly(vinylsulfonate) as the active transducing layer. Since the analyte sensing is mediated by the use of enzymes, these devices fall into the type IV sensor mechanism.

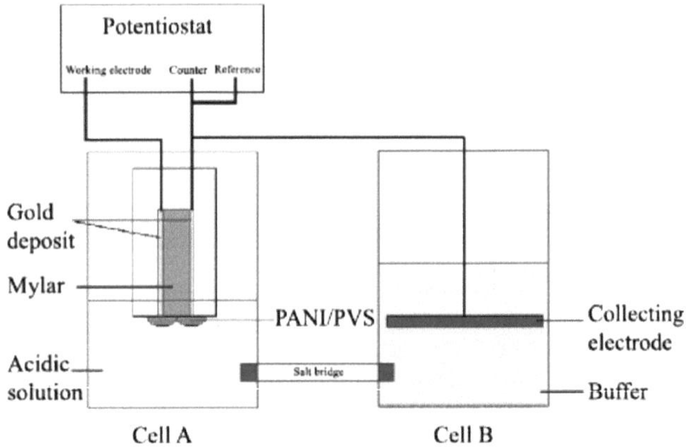

Fig. 8.10. The two compartment arrangement used in these experiments. Cell A contains acid. Cell B contains the buffer that best suits the bio-electrochemical reaction. The polymer is connected to the collecting electrode and the charge transfer is enabled between the two cells by a salt bridge. The potentiostat is used to apply a potential difference, the drain voltage, between the two electrodes of the polymer transistor and record the resulting drain current. Reprinted from [34] with permission from Elsevier

8.4 Summary and Future Directions

From the above discussion, it is clear that conjugated polymer-based OECTs provide an ideal platform for the design and fabrication of chemical and biological sensors. The sensing mechanism can consist either of direct (physical or chemical changes) or indirect (mediated by other species) interactions of the analytes with the active conjugated polymer-based transducing layers. Although it is easy to fabricate these devices, the scale-up for bulk production is still a major challenge as these devices currently require individual fabrication. Another bottleneck in this field is the inability of these devices to do in vivo detection of analytes as the device response is sensitive to the ionic strength of the analyte solution and also to the presence of other species. We feel that the OECT sensors based on the concept of changes in conformation upon guest–host binding (type III) provide excellent opportunities for the design of new chemical and biological sensors due to the plethora of known guest–host binding pairs.

The authors acknowledge the crucial discussion and inputs provided by Prof. A.Q. Contractor during the preparation of this manuscript and the Council of Scientific and Industrial Research India for providing a Junior Research Fellowship to Jasmine Sinha.

References

1. H.S. White, G.P. Kittelsen, M.S. Wrighton, J. Am. Chem. Soc. **106**, 5375 (1984)
2. E.W. Paul, A.J. Ricco, M.S. Wrighton, J. Phys. Chem. **89**, 1441 (1985)
3. J.W. Thackeray, H.S. White, M.S. Wrighton, J. Phys. Chem. **89**, 5133 (1985)
4. J.W. Thackeray, M.S. Wrighton, J. Phys. Chem. **90**, 6674 (1986)
5. S. Sukeerthi, A.Q. Contractor, Chem. Mater. **10**, 2412 (1998)
6. K. Krishnamoorthy, R.S. Gokhale, A.Q. Contractor, A. Kumar, Chem. Commun. 820 (2004)
7. M. Kanungo, D.N. Srivastava, K. Kumar, A.Q. Contractor, Chem. Commun. **7**, 680 (2002)
8. Z.T. Zhu, J.T. Mabeck, C. Zhu, N.C. Cady, C.A. Batt, G.G. Malliaras, Chem. Commun. 1556 (2004)
9. D. Nilsson, T. Kugler, P.O. Svensson, M. Berggren, Sensors and actuators B **86**, 193 (2002)
10. A.Q. Contractor, T.N.S. Kumar, R. Narayanan, S. Sukeerthi, R. Lal, R.S. Srinivas, Electrochimica Acta **39**, 1321 (1994)
11. H. Sangodkar, S. Sukeerthi, R.S. Srinivas, R. Lal, A.Q. Contractor, Anal. Chem. **68**, 779 (1996)
12. S. Sukeerthi, A.Q. Contractor, Anal. Chem. **71**, 2231 (1999)
13. M. Kanungo, A. Kumar, A.Q. Contractor, Anal. Chem. **75**, 5673 (2003)
14. S. Chao, M.S. Wrighton, J. Am. Chem. Soc. **109**, 6627 (1987)
15. N.P. Gaponik, D.G. Shchukin, A.I. Kulak, D.V. Sviridov, Mendeleev Commun. **7**, 70 (1997)
16. R.B. Dabke, G.D. Singh, A. Dhanabalan, R. Lal, A.Q. Contractor, Anal. Chem. **69**, 724 (1997)
17. D. Raffa, K.T. Leung, F. Battaglini, Anal. Chem. **75**, 4983 (2003)
18. P.N Bartlett, P.R. Birkin, J.H. Wang, F. Palmisano, G.D. Benedetto, Anal. Chem. **70**, 3685 (1998)
19. Y. Astier, P.N. Bartlett, Bioelectrochem. **64**, 53 (2004)
20. P.N. Bartlett, J.H. Wang, E.N.K. Wallace, Chem. Commun. **3**, 359 (1996)
21. D.T. Hoa, T.N.S. Kumar, N.S. Punekar, R.S. Srinivas, R. Lal, A.Q. Contractor, Anal. Chem. **64**, 2645 (1992)
22. P.N. Bartlett, P.R. Birkin, Anal. Chem. **65**, 1118 (1993)
23. P.N. Bartlett, P.R. Birkin, Anal. Chem. **66**, 1552 (1994)
24. P.N. Bartlett, J.H. Wang, W. James, Analyst **123**, 387 (1998)
25. F. Battaglini, P.N. Bartlett, J.H. Wang, Anal. Chem. **72**, 502 (2000)
26. M. Nishizawa, T. Matsue, I. Uchida, Anal. Chem. **64**, 2642 (1992)
27. T. Matsue, M. Nishizawa, T. Sawaguchi, I. Uchida, J. Chem. Soc.; Chem. Commun. 1029 (1991)
28. V. Saxen, V. Shirodkar, R. Prakash, J. Solid State Electrochem. **4**, 234 (2000)
29. W. Liu, J. Kumar, S.K. Tripathy, K.J. Senecal, L. Samuelson, J. Am. Chem. Soc. **121**, 71 (1999)
30. W. Liu, A.L. Cholli, R. Nagaranjan, J. Kumar, S. Tripathy, F.F. Bruno, L. Samuelson, J. Am. Chem. Soc. **121**, 11345 (1999)
31. R. Nagaranjan, S. Tripathy, J. Kumar, F.F. Bruno, L. Samuelson, Macromolecules **33**, 9542 (2000)
32. D.T. McQuade, A.E. Pullen, T.M. Swager, Chem. Rev. **100**, 2537 (2000)
33. J.T. Mabeck, J.A. DeFranco, D.A. Bernards, G.G. Malliaras, S. Hocde, C. Chase, J. Appl. Phys. Lett. **87**, 13503 (2005)
34. Y. Astier, P.N. Bartlett, Bioelectrochem. **64**, 15 (2004)

9

PEDOT:PSS-Based Electrochemical Transistors for Ion-to-Electron Transduction and Sensor Signal Amplification

M. Berggren, R. Forchheimer, J. Bobacka, P.-O. Svensson, D. Nilsson, O. Larsson, and A. Ivaska

9.1 The PEDOT:PSS-Based Electrochemical Organic Thin Film Transistor

9.1.1 Electrochemical Transistors: A Brief Introduction and a Short Historical Review

Conjugated polymers have been extensively studied in electrochemical cells as the working electrode. Typically, the conducting polymer film is deposited either onto a conducting supporting electrode or just onto an insulating carrier surface. This is then included in a two- or a three-electrode setup, all immersed in a common liquid electrolyte. As the oxidation state of the conjugated polymer is altered the number of free charge carriers is controlled within the polymer bulk. Upon oxidizing (reducing) the neutral polymer positive (negative) polarons start to dominate the character of the polymer, as evidenced by changes in the optical, chemical, and electronic properties. Based on this electronic control of the fundamental properties of the polymer, scientists and engineers have explored using conjugated polymers as the active material in electrochromic windows and displays [1,2], surface wettability switches [3,4], actuators [5], and so on. To enable further development of the devices, it is of utmost importance to understand the impedance properties of the electrolyte/polymer film system [6,7]. Commonly, different kinds of impedance spectroscopy techniques are employed to gain insights in how the ions couple to the electrons inside the polymer and how the electrical potential distributes in the device structures. In experiments where the polymer films exhibit high electronic conductivity, typically ion diffusion in the electrolyte is the main limiting mechanism for switching of the polymer oxidation state. Conversely, as very fast electrolytes are used and when the polymer possesses a relatively low electronic conductivity, the inherent electronic conductivity of the polymer bulk limits the switch speed. The conductivity of the polymer varies over several orders of magnitude as the oxidation level of the conjugated material is switched. In actuators and electrochromic windows this large variation of the electronic conductivity of the polymers inhibits fast updating, which becomes a severe problem while switching the polymer from the neutral to its doped state.

A transistor is a device which defines its operation principle via electric gating of the electronic charge transport along a channel. While conductance switching of the conjugated polymers is undesirable in many of the different electrochemical device it represents the key device functionality in electrochemical transistors. In 1984, Wrighton and colleagues reported electrochemical transistor action in a three-electrode device configuration including polypyrrole as the active material [8]. This work paved the way for a series of elegant studies demonstrating the successful use of different conducting polymers in electrochemical transistors. Typically, an electrochemical transistor is built-up from the combination of three electrodes, of which two are connected to the polymer channel material while the remaining third electrode serves as the gate. As this gate is addressed, the doping level (oxidation state) of the conducting polymer is controlled, thus enabling direct electrochemical control of the electronic current transport in between the two first electrodes. Besides polypyrrole many other conjugated polymers have been explored in electrochemical organic thin film transistors (OECT) [9–11], such as polythiophene [12] and polyaniline [13] derivatives.

9.1.2 The Operation Principle of the PEDOT:PSS-Based Electrochemical Organic Thin Film Transistor

In 2002, David Nilsson reported an electrochemical transistor using poly(3,4-ethylenedioxythiophene) doped with poly(styrene sulfonate) (PEDOT:PSS) as the active layer [14]. In this system, both the electronically conducting system (PEDOT) and the dopant are polymers and therefore nonvolatile as the oxidation state of the PEDOT is altered. PEDOT is pristinely doped and is therefore in its high conducting state. This allows us to define the drain, source, and gate as well as the transistor channel in PEDOT:PSS solely, see Fig. 9.1. As the gate is addressed positively, vs. the source electrode, PEDOT in the channel is reduced to its neutral state.

The principal of operation of this OECT transistor is as follows: at zero gate voltage, the transistor is in the "ON" (low impedance state) state allowing a high current to pass through the channel. Proper transistor function requires that the drain electrode be negatively biased and that the gate electrode be positively biased with respect to the source. This then allows the transistor to operate in the third quadrant. The OECT finds its analogue in the depletion mode p-type field-effect transistor. The current vs. voltage characteristics are governed by electrochemical reduction in the channel driven by two different sources. As the gate is positively addressed, cations of the electrolyte enter the PEDOT: PSS channel to react according to (9.1), where M^+ denotes the cation from the electrolyte and e^- the electron provided by the grounded source electrode:

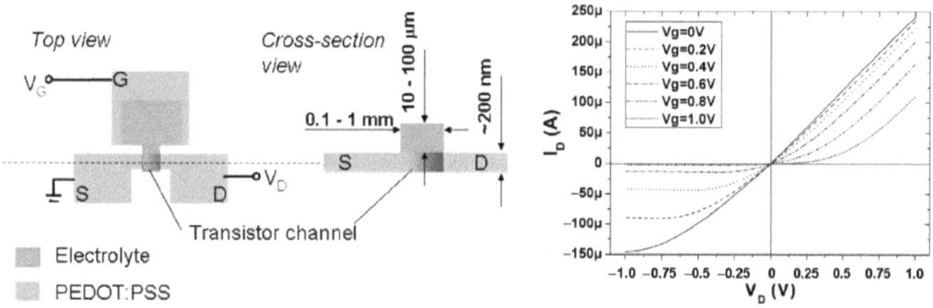

Fig. 9.1. (*Left*) the architecture of the PEDOT:PSS-based OECT is illustrated viewed from the *top* and at a cross section, respectively. The area of the PEDOT:PSS film, in between the source and drain contact, that is capped by the electrolyte, defines the transistor channel. (*Right*) the current vs. voltage characteristics of the PEDOT: PSS-based OECT. At zero gate voltage, the transistor is in the "ON" state and at $V_G = 0.8\,\mathrm{V}$ the channel current is suppressed by more than two orders of magnitude as compared to the "ON" state current

$$\mathrm{PEDOT{:}PSS} + \mathrm{M}^+ + \mathrm{e}^- \Leftrightarrow \mathrm{PEDOT} + \mathrm{M{:}PSS}. \tag{9.1}$$

At low negative drain voltages a nearly linear drain current vs. gate voltage behavior is found at all gate voltages. As the drain voltage is further decreased toward more negative values the drain current starts to saturate and a clear channel pinch-off is observed. The pinch-off effect originates from the fact that a potential difference exists along the channel. The associated electric field inside the channel is then counter-mirrored inside the electrolyte, closest to the PEDOT:PSS channel, and establishes a steady-state reduction–oxidation gradient with a concentration of reduced PEDOT closest to the negatively biased drain electrode. As the drain voltage increases further, there is a corresponding increase in the reduction of PEDOT closest to the drain electrode, i.e., to the right side of the channel in Fig. 9.2. As the gate voltage is grounded or at least is made more negative, reaction (9.1), inside the channel, goes in the reverse direction and PEDOT:PSS gets reoxidized and a higher electronic current can again be conducted between the drain and source electrodes. The operation of the OECT includes two electric current loops. First, the gate–electrolyte–channel–source represents one loop and its associated impedances are illustrated in Fig. 9.2. In this loop charges are transported as electronic (Z^{e}) and ionic (Z^{i}) species, respectively. Second, the source–channel–drain loop is entirely electronic besides a small parallel ionic current contribution, occurring in the electrolyte due to the induced electrochemistry inside the channel.

9.1.3 Design Criteria and Device Operation Parameters

The relative lateral dimensions of the different parts of the OECT transistor couple strongly to a number of the crucial performance parameters of the

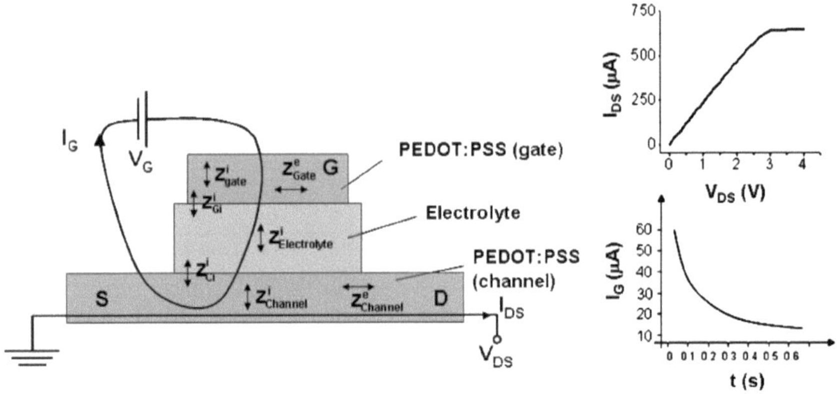

Fig. 9.2. (*Left*) the impedances of the OECT for the associated electronic (e) and ionic (i) currents. Two current loops exist; the gate electrode–electrolyte–channel–source electrode loop and the source–channel–drain. (*Right*) the steady-state drain current (I_{DS}) vs. voltage (*top*) and the gate current (I_{G}) vs. time

device. The ON/OFF ratio of the transistor and the switch speed are perhaps the most important for sensors. PEDOT:PSS is pristinely doped to a level above 80%. To enable maximum reduction of the channel and therefore also to generate the highest possible drain current ON/OFF ratio, the number of available oxidation (or charging) sites of the gate electrode must equal or exceed the number of available reduction sites in the channel. This is achieved simply, by making the volume (typically the area) of the gate electrode ten times larger than the channel, see Fig. 9.3. Also, the length and width of the transistor channel affect the device performance to a great extent. For gelled electrolytes with a high salt concentration we found that the width should equal the length of the channel in order to gain the highest ON/OFF ratio, in this case above 10^3.

In the table below the typical device parameters are given for a PEDOT:PSS-based OECT including an aqueous electrolyte with a concentration of 100 mM.

Electrolyte concentration	Channel length/width	Gate area	ON/OFF ratio	Switch speed (OFF ⇒ ON)[a]	Switch speed (ON ⇒ OFF)[a]
100 mM	0.5/0.5 mm	2.5 mm^2	2,000	2 s	0.1 s

[a] OFF-to-ON (ON-to-OFF) switch time: defined as the time it takes for drain current to increase (decrease) from 10% (90%) to 90% (10%) of the "ON" drain current preceded by a momentarily change of the gate voltage from 1 to 0 V (0 to 1 V)

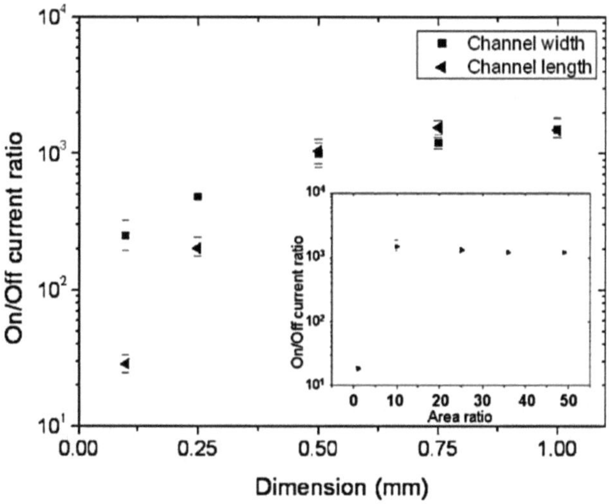

Fig. 9.3. The ON/OFF ratio vs. channel width and channel length (one of the parameters is kept at 1 mm while the other was changed). It turns out that a square-shaped channel is the optimum in reaching the highest ON/OFF ratio. (*Inset*) the ON/OFF ratio variation vs. the area ratio between the gate electrode and the channel area. When the volume (area) of the gate exceeds ten times the volume (area) of the channel the highest ON/OFF ratio is reached

9.1.4 Manufacturing Techniques

The PEDOT:PSS transistor is based on a lateral device configuration, in which the gate, transistor channel, drain, and source contacts (all defined in PEDOT:PSS) can be placed adjacent to each other in the same plane. Vertical transistor configurations have been realized, in which the gate and the channel sandwich the electrolyte medium. The vertical configuration gives a relatively faster turn-on and turn-off responses, however for electrochemical sensor transistor applications a lateral configuration is desired since it freely exposes its channel and gate to the target analyte. In its simplest form, when the analyte is within the electrolyte, the laterally defined PEDOT:PSS features fully defines the electrochemical transistors. In some cases, when proper sensing selectivity or signal transduction is not defined in the PEDOT:PSS-electrolyte system alone, an additional material is then required inside the gate electrode-to-channel loop. The PEDOT:PSS-based OECT does not require any ultrasmall features; it is a device that is driven by electric potentials and not by charge injection or via field-effect-induced charge accumulation, mechanisms that both require ultrasmall features in the form of thin films and/or narrow separations of conductors. The relaxed requirements regarding feature sizes for the OECTs, which give a very robust device structure, permit the use of ultralow cost manufacturing techniques.

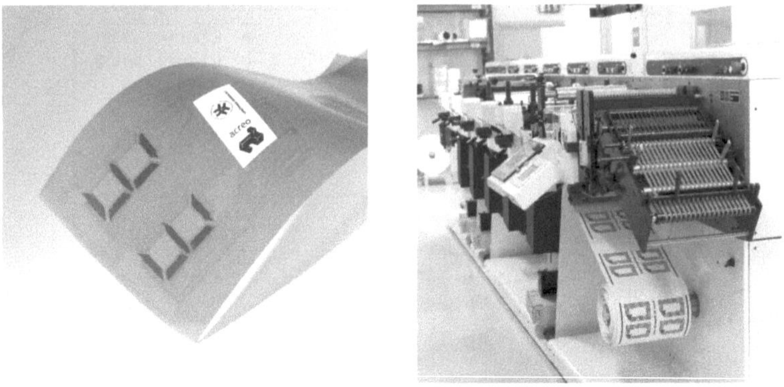

Fig. 9.4. (*Left*) a seven-segment display including electrochromic display elements based on PEDOT:PSS. (*Right*) the FA 3300/5 Nilpeter label printing press utilized to manufacture electrochemical devices based on PEDOT:PSS

At Acreo AB, in Sweden, efforts are devoted to developing an all-in-line manufacturing process for PEDOT:PSS-based electrochemical devices, such as electrochromic displays, transistors, and also sensors. The Acreo AB team has established the Electronic Paper Printing House (EPPH), a laboratory combining traditional processes and technologies for large area electronics with those of the traditional printing industry. The FA 3300/5 Nilpeter label printing press, shown in Fig. 9.4, is one of the corner stones of the EPPH lab and contains stations for rotary screen-, offset-, and flexo-printing [15]. The machine is fed with a glossy fine paper, similar to a photopaper, precoated with 200 nm of PEDOT:PSS (performed by AGFA-Gevaert NV, Belgium). To define the gate and channel etc., lines along the PEDOT:PSS coating must be deactivated. Various kinds of deactivation processes exist for conducting polymers [16] but these are typically very slow. For example, the deactivating ink must be in contact with the PEDOT:PSS polymer coating for minutes, and therefore is difficult to include in an all-in-line processes including rotary-processing running at speeds of 10–100 m min^{-1}. To speedup the deactivation process Acreo AB has developed a chemical deactivation process that is driven by electrochemistry. Now instead, nonconducting patterns are generated in the PEDOT:PSS coating in less than a second [17]. This electrochemical overoxidation patterning process has been implemented in the FA 3300/5 Nilpeter printing press and is hosted in the first printing station. At this station the rotary screen technique is utilized to generate the nonconductive patterns along the PEDOT:PSS film. A printable electrolyte gel is used as the ink and a metallic squeegee combined with a cylinder, in contact with the PEDOT:PSS coating, as the anode and cathode, respectively. At anodic overpotentials PEDOT undergoes an irreversible oxidation process which forces the material to become isolating to electronic conduction. In the second station of the press, another screen-printing process serves as the deposition and patterning tool for electrolytes. In the preceding station(s) device encapsulation and annealing processes takes place.

9.2 The PEDOT:PSS OECT as an Ion-to-Electron Transducer

9.2.1 Different Sensor Principles of the PEDOT:PSS Electrochemical Transistor

To define a sensor function in the PEDOT:PSS OECT a selective sensing mechanism must be coupled to control any of the operational parameters of the transistor. In other words, as a consequence of a sensing event the activity of any of the components in the gate/electrolyte/channel/source loop is affected, which translates into a readout signal in the drain/channel/source loop. Recently, several approaches have been explored by adding an oxidation derivative [18] and the inclusion of a lipid bilayer membrane into the electrolyte compartment [19]. In these cases the sensing mechanisms were understood to affect the $Z^{\mathrm{i}}_{\mathrm{Gi}}$ and the $Z^{\mathrm{i}}_{\mathrm{electrolyte}}$ impedances of the PEDOT: PSS OECT device, respectively. Along the same lines, work has been done on two different versions of OECT sensors via the inclusion of the sensing layer to enable coupling of the sensing mechanism to changes in the $Z^{\mathrm{i}}_{\mathrm{C}}$ and the $Z^{\mathrm{i}}_{\mathrm{electrolyte}}$ impedances.

9.2.2 Humidity Sensing

Proton conducting membranes have been developed for a number of different applications, exemplified by fuel cell applications [20]. Different groups of materials are explored for these applications and one of the most commonly studied materials is Nafion, a richly fluorinated polymer including sulfonic acid side groups. Migration of protons and other small cationic species in the Nafion membrane is understood to occur via clusters of sulfonic acid groups, covalently bonded to the polymer backbone. As cations moves along the Nafion membrane, they jump from cluster to cluster of sulfonic acid groups [21]. As the humidity level of the ambient atmosphere is controlled, i.e., the water content inside the Nafion membrane is changed, thereby causing the activation barrier for cations to jump from cluster to cluster of sulfonic acid groups, to change dramatically. Therefore, the cationic conductivity in Nafion is strongly dependent on the humidity level of the environment and can be used in sensor applications [22].

In these experiments, a 5 µl droplet of a Nafion perfluorinated ion-exchange resin was added onto a PEDOT:PSS lateral transistor configuration [14]. After drying, the Nafion membrane formed a thick film structure covering both the gate and the transistor channel. Addressing the gate yields normal EC-transistor action but the speed it takes to gain a steady drain current

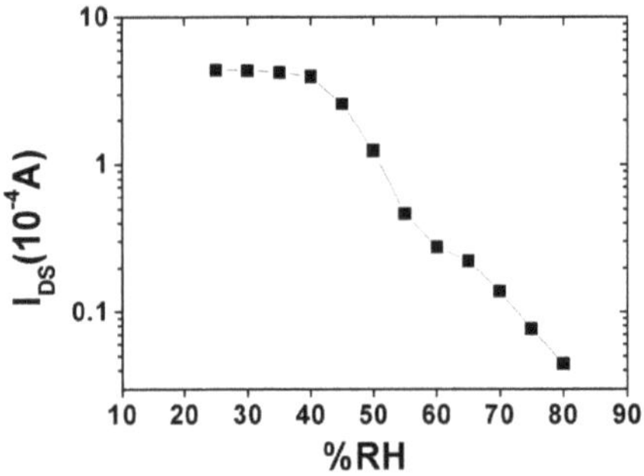

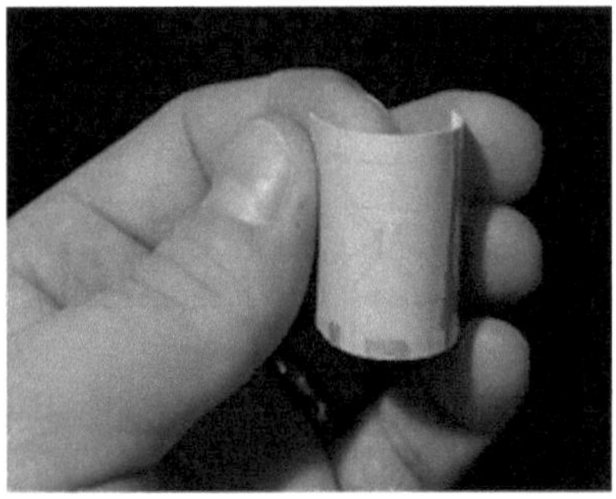

Fig. 9.5. (*Left*) the drain current after 15 s of gate biasing ($V_G = 1.2\,V$). Reproduced with permission from [23]. Copyright Wiley-VCH Verlag GmbH & Co. KGaA. (*Right*) the PEDOT:PSS/Nafion OECT sensor printed on fine paper. Reproduced with permission from [23]. Copyright Wiley-VCH Verlag GmbH & Co. KGaA

state, at a specific gate voltage, varies considerably depending on the ambient humidity level. However, the steady-state drain current level differs only slightly with respect to the different humidity levels. Therefore, the sensor can only be used in its dynamic mode, i.e., proper humidity sensing using the OECT is achieved in a mode where the drain current level is measured after a given time of gate biasing. In Fig. 9.5, the drain current is given for different relative humidity levels 15 s after which the gate voltage was applied. At room temperature, in the 40–80% RH range, we find that the drain current level (after 15 s of gate biasing) drops nearly exponentially with respect to the RH over two orders of magnitude. To prove its robustness, the sensor transistor was made onto an ordinary coated fine paper, see Fig. 9.5.

9.2.3 Ion-Selective Membranes

Ionophore-based solvent polymeric membranes are widely used as sensing membranes in ion-selective electrodes (ISEs) [24, 25]. This type of potentiometric sensor has attracted great interest in the last decade due to the extraordinary improvement in the detection limit down to picomolar (10^{-12} M) levels [26, 27]. Furthermore, solid-contact ISEs have been developed by using various conducting polymers, including PEDOT, as the ion-to-electron transducer [28–31].

In this work, an ionophore-based solvent polymeric membrane is cast on top of the PEDOT:PSS channel in order to obtain an ion-selective organic electrochemical transistor (IS-OECT). The polymeric membrane is composed of 2-nitrophenyl octyl ether (o-NPOE, 57.2% ww), poly(vinyl chloride) (PVC, 27.9% ww), N,N,N',N'-tetracyclohexyl-3-oxapentanediamide (ETH 129, 9.6% ww) and potassium tetrakis(4-chlorophenyl)borate (KTpClPB, 5.4% ww), where ETH 129 is a Ca^{2+}-selective ionophore. The channel region of the transistor that is coated by the Ca^{2+}-selective membrane is immersed in an aqueous solution containing 0.1 M KCl as a background electrolyte and

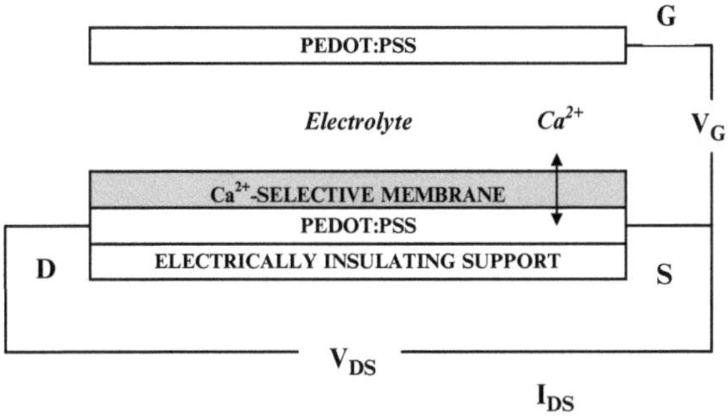

Fig. 9.6. Schematic description of the Ca^{2+}-selective organic electrochemical transistor. Modulation of the drain current (I_{DS}) by the gate potential (V_G) depends on the ion transport properties of the selective membrane (see Fig 9.7)

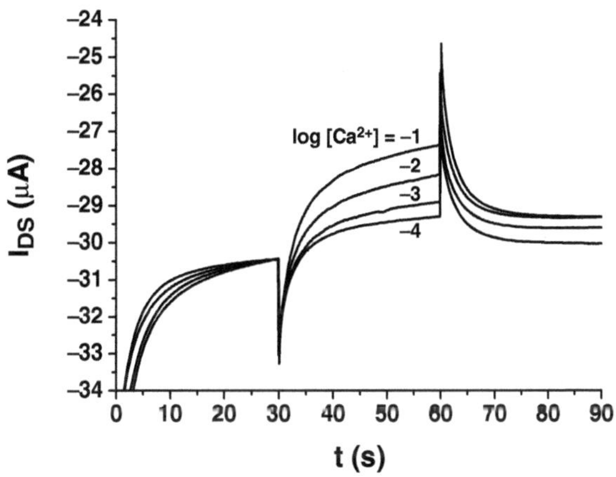

Fig. 9.7. Response curves for different concentrations of Ca^{2+} (at a constant background electrolyte of 0.1 M KCl). The gate voltage ($V_G = 0.15$ V) is applied at $t = 30$–60 s. The response curves are shifted to have the same I_{DS} just before application of V_G at $t = 30$ s

different concentrations of $CaCl_2$. The gate is immersed in the same solution completing the measuring circuit of the transistor, as shown schematically in Fig. 9.6.

The Ca^{2+}-selective OECT is characterized by measuring the drain current (I_{DS}) as a function of time at a constant drain voltage ($V_{DS} = -0.8$ V) in the absence and presence of an applied gate voltage (V_G). The response curves for different concentrations of Ca^{2+} (at a constant background electrolyte of 0.1 M KCl) are shown in Fig. 9.7. Here, the gate voltage ($V_G = 0.15$ V) is applied at $t = 30$–60 s. For an easy comparison, the response curves are shifted to have the same I_{DS} just before $t = 30$ s. The results in Fig. 9.7 show the dependence of I_{DS} on the concentration of Ca^{2+} at an applied $V_G = 0.15$ V.

9.3 The PEDOT:PSS Electrochemical transistor in logic and amplification circuits

9.3.1 Introduction to Electrochemical Circuits and Systems

SPICE [32] simulations indicate that the electrochemical transistor can be used favorably to implement a number of useful digital and analogue circuits. Naturally, the slower speed, compared to traditional electronics (by about a factor 10^8!), should be taken into consideration. The area of (silicon-based) electronic circuit theory offers numerous solutions to the implementations of circuits. However, depletion-type transistors are less favorable to use in system designs than their enhancement counterparts. For this reason, it is rare to find circuits that use depletion transistors. One has to go back to the electron tube era to find solutions that can be used as templates for circuits suitable for the electrochemical transistor.

A closer investigation of the possible modes of operation given by the transistor characteristics shown in Fig. 9.1 reveals that the third-quadrant mode is superior to the first-quadrant mode. It is not possible to achieve amplification at all in the first quadrant while this is easy to achieve in the third quadrant. Thus, circuit designs should utilize the transistor in the third-quadrant mode. However, the range of the input and output voltages do not overlap, which introduces some added complexity in system designs.

In the design of circuits there are certain basic electrical design rules for the electrochemical transistor that should be kept in mind:

- The source/drain terminal that is connected to the highest potential defines the source. The opposite terminal thus becomes the drain.
- Switching of the device (modulation) is obtained when the gate potential is higher than the source potential.
- The gate potential must never become lower than the drain potential. This is to avoid overoxidation.

The third rule can be made more specific: The gate potential should never become more negative than the most negative potential in the channel of the transistor. Usually the most negative potential is the drain potential but it is possible (if the gate voltage falls rapidly), that part of the channel may overoxidize due to dynamic effects. A more conservative rule that is always safe is to avoid decreases of the gate potential below the source potential.

The organic-based OECT, reported here, can handle remarkably higher power than its OFET relatives. As an example, an Orgacon (particular grade of PEDOT:PSS from AGFA)-based transistor with active channel size $0.5 \times 0.5\,\mathrm{mm}^2$ can handle a switching power of up to 20 mW. Above this power level breakdown occurs, most likely due to thermal stress. The transistor can also handle voltages of up to 70 V as well as currents up to 1 mA.

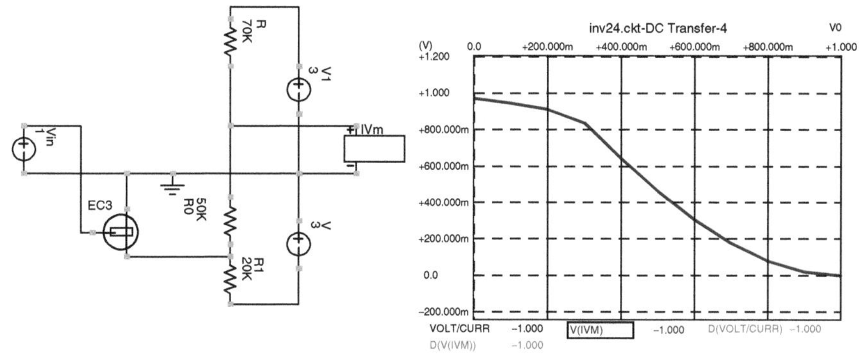

Fig. 9.8. (*Left*) the circuit of a logic inverter including a depletion mode transistor and (*right*) its associated input–output characteristics

9.3.2 Electrochemical Digital Circuits

Digital (or logic) circuits are robust toward component variations and thus become a natural starting point for "electrochemical electronics" [33]. The logical inverter shown in Fig. 9.8 illustrates the principles of depletion-based circuitry. When the gate receives an input voltage of 0 or 1 V, the transistor will turn ON or OFF, thus either shorting its drain to the ground rail or disconnecting it. In the first case the output voltage, indicated by the rectangular "voltage probe" will show a value that is defined by the resistors R and $R0$ and the positive 3 V supply. The resulting voltage will be about 1 V. In the second case, $R1$ and the negative voltage supply are also involved and it is seen that the output voltage will now be around 0 V. Figure 9.8 also shows the input–output characteristics as the input signal is changing gradually between 0 and 1 V.

The inverter requires a symmetrical +/− 3 V supply. It is possible to lower the supply voltage to +/− 2 V at the cost of one more transistor by exchanging the resistor R for a "current generator" as shown in Fig. 9.9.

Logical gates with more than one input are straightforward extensions of the inverter. Figure 9.10 shows the circuit diagram of a NAND gate and its behavior for different input signals. Similarly, a NOR gate is implemented by connecting the input transistors in series. A physical implementation based on ink-jet printing is shown in Fig. 9.11.

9.3.3 Electrochemical Analog Circuits

Provided that the circuit constraints are respected, several useful analog circuits can be designed. In particular, we will focus on amplifying circuits. Such circuits open up the possibility to implement sensor amplifiers, comparators, frequency-selective filters, oscillators, timers, feedback-control systems, etc. Figure 9.12 displays the basic one-transistor amplifier. The load resistor gives

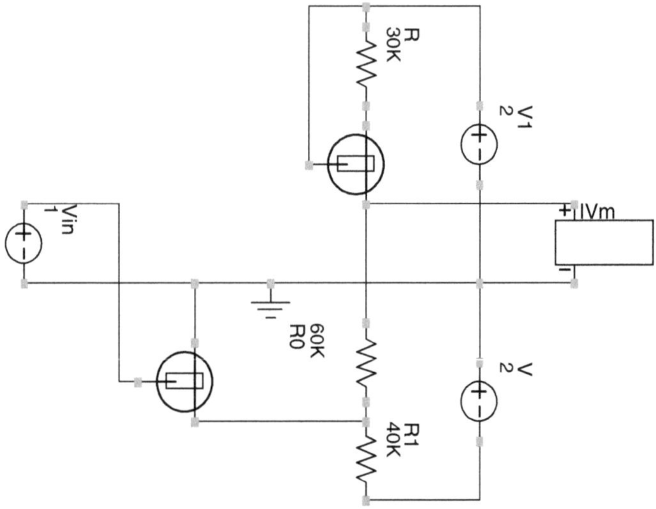

Fig. 9.9. The circuit of an inverter operating at a bias voltage of 2 V

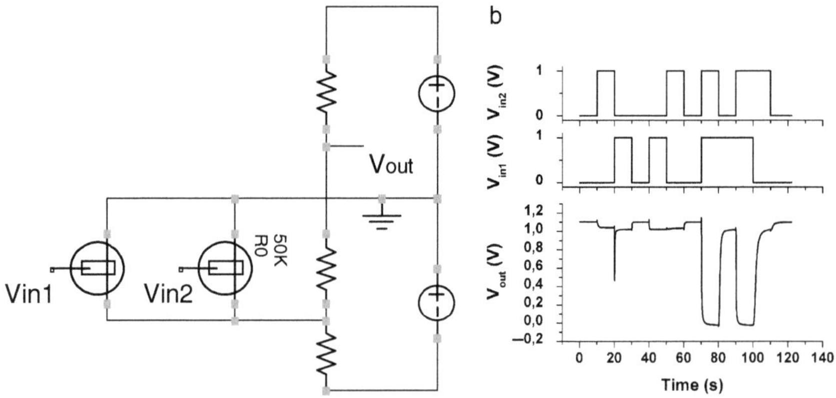

Fig. 9.10. The circuit diagram of a NAND-gate including two depletion mode transistors

the amplification. With a resistor value of 50 kΩ this yields an estimated amplification of about five times for small input signals; see Fig. 9.12, provided that the input DC-level is set so that the transistor is correctly biased.

Higher amplification can be achieved by increasing the load resistor. However, this requires a very strict control of the DC-biasing of the transistor. A well-known technique for autobiasing is shown in Fig. 9.13. Here, an additional resistor $R2$ is added in series with the source. By proper choice of the two resistors, additional DC biasing of the input signal can be avoided. The amplification is given here by the ratio of $R/R0$. However, if the circuit is only used for AC signals, then higher amplification can be achieved by decoupling

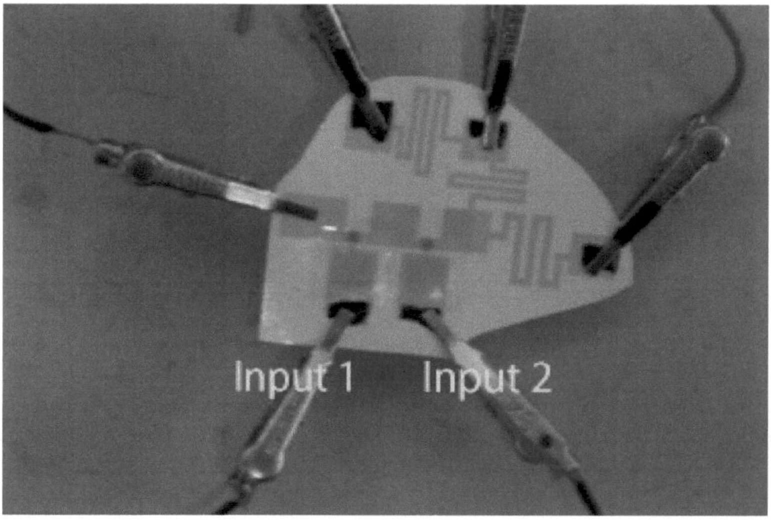

Fig. 9.11. Two ink-jet-printed PEDOT:PSS-based OECTs forming a NOR gate

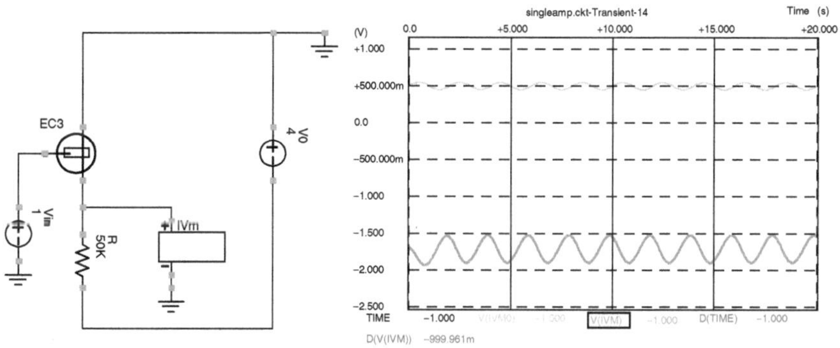

Fig. 9.12. (*Left*) the circuit of a one-transistor amplifier and, to the (*right*) the associated amplification characteristics

$R2$ with a large capacitor. This will in effect lower the impedance of the source load, giving rise to a corresponding increase in amplification. The capacitor can favorably be implemented as a symmetric 2-electrode PEDOT:PSS electrochemical capacitor.

An interesting variation of this circuit is the "cascode" amplifier shown in Fig. 9.13. Here, the drain load is exchanged for a constant-current generator. Theoretically this gives infinitely high impedance thus yielding a very high level of amplification.

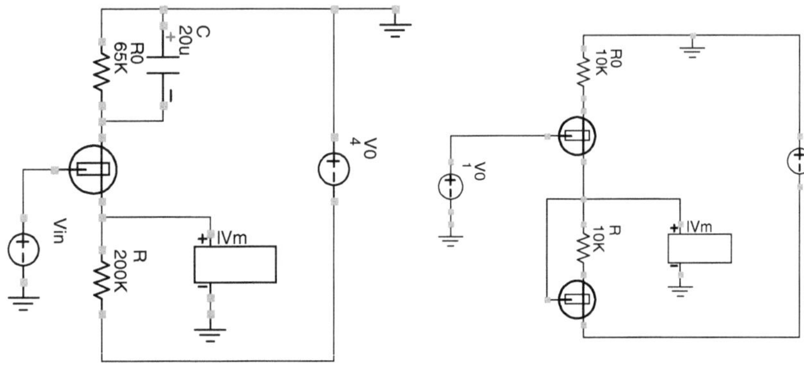

Fig. 9.13. (*Left*) the circuit outline of a one-transistor amplifyer, for which the amplification is dictated by the ratio of R and $R0$. (*Right*) the same principal circuit but R has been replaced by a constant-current generator

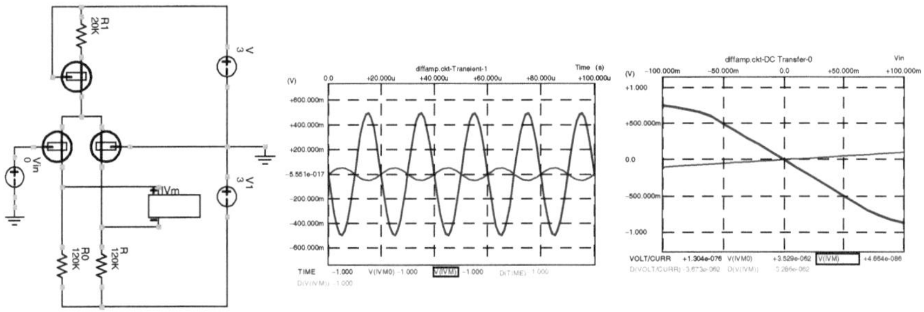

Fig. 9.14. (*Left*) the differential amplifier. (*Middle*) and (*right*) the amplification of a varying signal and the input/output relationship for the differential amplifier. As is seen, the circuit shows a good linear response in a large regime of its total dynamic range. Many variations of the basic differential amplifier are possible. In a simpler circuit, the M1 transistor can be eliminated (resistor $R1$ then needs to be enlarged). For a single polarity input, the second gate can be connected to ground

9.3.4 The Differential Amplifier

In many sensing applications the sensor signal is compared to a reference level. In other applications only the difference between two signals is significant and needs to be amplified. A suitable circuit for these situations is the differential amplifier, shown in Fig. 9.14. The two input signals are connected to each of the gates. The output signal is available in both inverted and noninverted forms on the drain side of the transistors. This also allows for a differential output with a signal level which is twice as high as that measured across the two drain terminals.

9.3.5 Zero Detector

In some cases, a signal is to be tested against a fixed reference level, for example to see whether it is positive or negative. In this instance, the following circuit, shown in Fig. 9.15, can be used. The two leftmost transistors act as a high-gain stage while the third transistor is used as a "source follower." The amplification of this stage is less than 1 but acts to restore the output to levels which are compatible with digital circuits. Also in Fig. 9.15, the SPICE simulation of the input/output relationship is shown. In practice the performance of the circuit will depend greatly on the quality of the electrochemical transistors.

9.3.6 Oscillators

Periodic signals, either sinusoidal or square-wave can be used for many purposes. Here we present a few oscillator circuits based on electrochemical amplifiers. The first is the differential amplifier-based oscillator shown in Fig. 9.16. The circuit will generate an approximately sinusoidal signal as shown in Fig. 9.16.

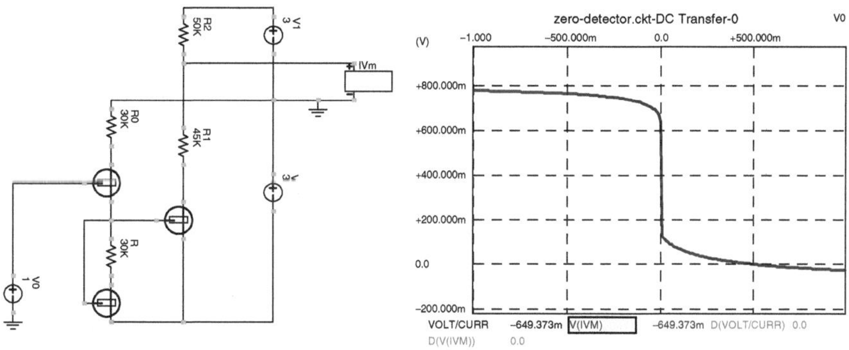

Fig. 9.15. The zero detector and its associated transfer characteristics

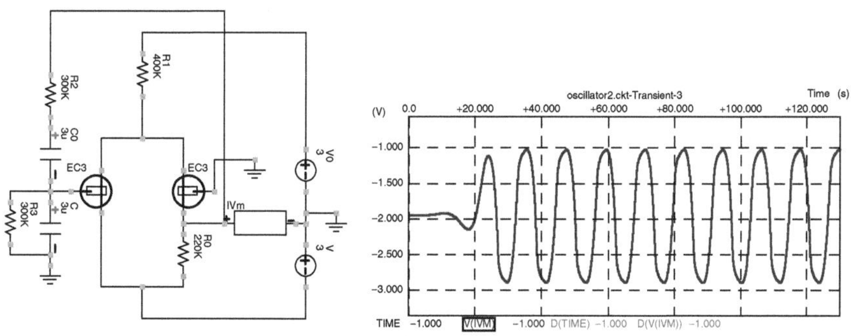

Fig. 9.16. A differential amplifier-based oscillator

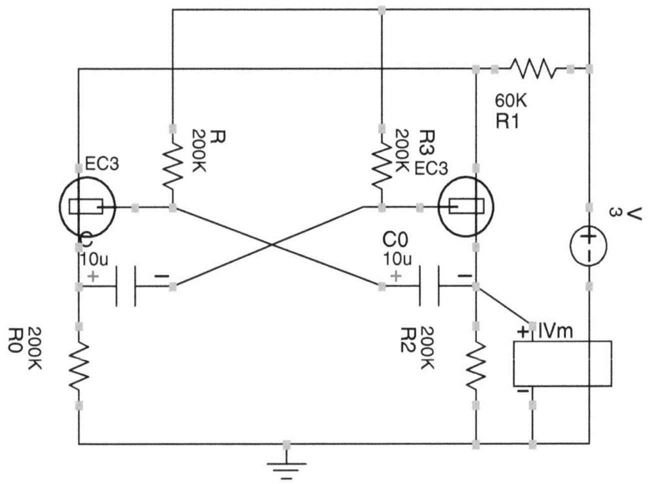

Fig. 9.17. The stable flip-flop producing a square-shape oscillation output

A more square-way shaped output is generated by the unstable flip-flop (Fig. 9.17). This can be seen to consist of two "crossconnected" amplifier stages. This circuit can act as an example where a traditional (silicon-based) design cannot be directly mapped onto an EC-design. Detailed analysis will show that, during the start-up phase, there will be an increased risk of overoxidation of the transistors. This may occur when the gate potential of the transistor swings to a low value thus passing the potential of the source. Although the gate potential may not pass the drain level, the rate of change can be too high, thus leading to overoxidation. A way to decrease this risk is to insert resistors in series with the capacitors.

Finally, the ring oscillator is a classical device to produce a periodic, square-wave signal. There are no frequency-controlling components; rather it is the inherent delays in the components themselves that set the frequency. For this reason, the ring oscillator is often used to evaluate the performance of a new technology. The ring oscillator consists of an odd number of logical inverters connected in a circular fashion (Fig. 9.18). The output signal from this device and the physical implementation are also shown in Fig. 9.18.

9.4 Outlook

The PEDOT:PSS-based electrochemical transistor balance at the boundary between amperometric and potentiometric sensor functionality. Its device architecture is simple and robust and these PEDOT:PSS sensors can easily be manufactured using standard printing technologies. In addition, its proper ion-to-electron transduction characteristics and the possibility to integrate biologically active substances promise for simple and sensitive sensors in the

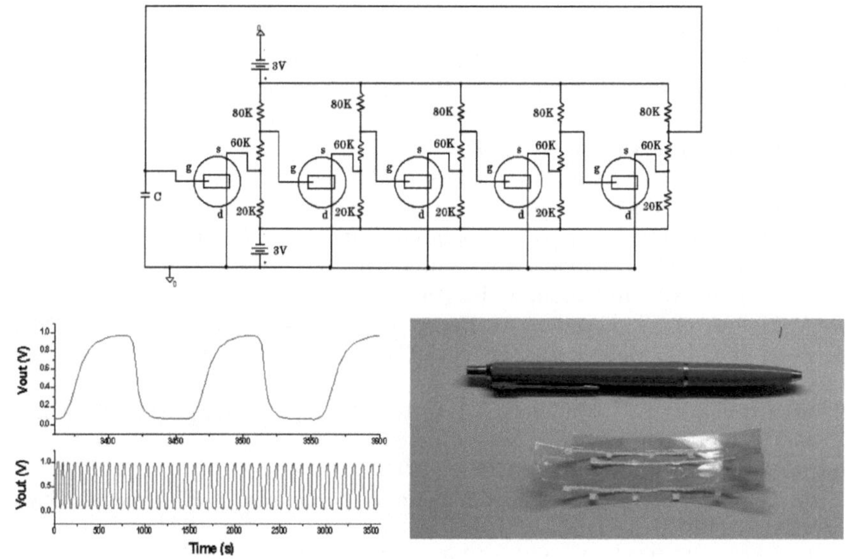

Fig. 9.18. The classical ring-oscillator circuit (*top*) and the associated output characteristics (*bottom left*) and an electrochemical transistor ring oscillator printed on a plastic foil (*bottom right*)

area of biology and chemistry. Also, display cells, logics and amplification circuits can be realized in PEDOT:PSS electrochemical devices. Together this enables all-integrated sensor device systems made out of few individual materials, that all can be manufactured using standard printers. Therefore, this technology platform promises for novel single-use sensors integrated into different kinds of carriers, such as paper and foils, enabling sensor labels and stickers possible to add onto for instance packages.

References

1. Q. Pei, G. Zuccarello, M. Ahlskog, O. Inganäs, Polymer **35**, 1347 1994
2. B.D. Reeves, B.C. Thompson, K.A. Abboud, B.E. Smart, J.R. Reynolds, Adv. Mater. **14**, 717 (2002)
3. T. Sun, G. Wang, L. Feng, B. Liu, Y. Ma, L. Jiang, D. Zhu, Angew. Chem. Int. Ed. **43**, 357 (2004)
4. J. Isaksson, C. Tengstedt, M. Fahlman, N. Robinson, M. Berggren, Adv. Mater. **16**, 316 (2004)
5. E. Smela, Adv. Mater. **15**, 481 (2003)
6. M.M. Musiani, Electrochim. Acta **35**, 1665 (1990)
7. C. Deslouis, T. El Moustafid, M.M. Musiani, B. Tribollet, Electrochim. Acta **41**, 1343 (1996)
8. H.S. White, G.P. Kittlesen, M.S. Wrighton, J. Am. Chem. Soc. **106**, 5375 (1984)
9. J. Huang, M.S. Wrighton, Anal. Chem. **65**, 2740 (1993)

10. V. Rani, K.S.V. Santhanam, J. Solid State Electrochem **2**, 99 (1998)
11. V. Saxena, V. Shirodkar, R. Prakash, J. Solid State Electrochem **4**, 231 (2000)
12. J.W. Thackeray, H.S. White, M.S. Wrighton, J. Phys. Chem. **89**, 5133 (1985)
13. D.G.S. Nikolai, P. Gaponik, A.I. Kulak, D.V. Sviridov, Mendeleev Communications Electronic Version, 47 (1997)
14. P. Andersson, D. Nilsson, P.-O. Svensson, M. Chen, A. Malmstrom, T. Remonen, T. Kugler, M. Berggren, Adv. Mater. **14**, 1460 (2002)
15. M. Berggren, D. Nilsson, N.D. Robinson, Nat. Mater. **6**, 3 (2007)
16. Y. Yoshioka, G.E. Jabbour, Adv. Mater. **18**, 1307 (2006)
17. P. Tehrani, N.D. Robinson, T. Kugler, T. Remonen, L.-O. Hennerdal, J. Hall, A. Malmstrom, L. Leenders, M. Berggren, Smart Mater. Struct. **14**, 21 (2005)
18. Z.-T. Zhu, J.T. Mabeck, C. Zhu, N.C. Cady, C.A. Batt, G.G. Malliaras, Chem. Commun. **13**, 1556 (2004)
19. D.A. Bernards, G.G. Malliaras, G.E.S. Toombes, S.M. Gruner, Appl. Phys. Lett. **89**, 53505 (2006)
20. T. Norby, Solid State Ionics **125**, 1 (1999)
21. A. Lehmani, P. Turq, M. Perie, J. Perie, J.-P. Simonin, J. Electroanal. Chem. **428**, 81 (1997)
22. F. Opekar, K. Tulík, Analytica Chimica Acta **385**, 151 (1999)
23. D. Nilsson, M. Chen, T. Kugler, T. Remonen, M. Armgarth, M. Berggren, Adv. Mater. **14**, 51 (2002)
24. E. Bakker, P. Bühlmann, E. Pretsch, Chem. Rev. **97**, 3083 (1997)
25. P. Bühlmann, E. Pretsch, E. Bakker, Chem. Rev. **98**, 1593 (1998)
26. T. Sokalski, A. Ceresa, T. Zwickl, E. Pretsch, J. Am. Chem. Soc. **119**, 11347 (1997)
27. E. Pretsch, Trends Anal. Chem. **26**, 46 (2007)
28. A. Cadogan, Z. Gao, A. Lewenstam, A. Ivaska, D. Diamond, Anal. Chem. **64**, 2496 (1992)
29. J. Bobacka, Anal. Chem. **71**, 4932 (1999)
30. J. Bobacka, A. Ivaska, A. Lewenstam, Electroanalysis **15**, 366 (2003)
31. J. Bobacka, Electroanalysis **18**, 7 (2006)
32. *SPICE: A Guide to Circuit Analysis using Spice*, P. Tuinega, 3rd edn. (Prentice-Hall, Englewood Cliffs, NJ, 1995), ISBN 0-13-158775-7
33. D. Nilsson, N. Robinson, M. Berggren, R. Forchheimer, Adv. Mater. **17**, 353 (2005)

Index

Springer Series in
MATERIALS SCIENCE

Editors: R. Hull R. M. Osgood, Jr. J. Parisi H. Warlimont

50 **High-Resolution Imaging and Spectrometry of Materials**
Editors: F. Ernst and M. Rühle

51 **Point Defects in Semiconductors and Insulators**
Determination of Atomic and Electronic Structure from Paramagnetic Hyperfine Interactions
By J.-M. Spaeth and H. Overhof

52 **Polymer Films with Embedded Metal Nanoparticles**
By A. Heilmann

53 **Nanocrystalline Ceramics**
Synthesis and Structure
By M. Winterer

54 **Electronic Structure and Magnetism of Complex Materials**
Editors: D.J. Singh and D. A. Papaconstantopoulos

55 **Quasicrystals**
An Introduction to Structure, Physical Properties and Applications
Editors: J.-B. Suck, M. Schreiber, and P. Häussler

56 **SiO_2 in Si Microdevices**
By M. Itsumi

57 **Radiation Effects in Advanced Semiconductor Materials and Devices**
By C. Claeys and E. Simoen

58 **Functional Thin Films and Functional Materials**
New Concepts and Technologies
Editor: D. Shi

59 **Dielectric Properties of Porous Media**
By S.O. Gladkov

60 **Organic Photovoltaics**
Concepts and Realization
Editors: C. Brabec, V. Dyakonov, J. Parisi and N. Sariciftci

61 **Fatigue in Ferroelectric Ceramics and Related Issues**
By D.C. Lupascu

62 **Epitaxy**
Physical Principles and Technical Implementation
By M.A. Herman, W. Richter, and H. Sitter

63 **Fundamentals of Ion-Irradiated Polymers**
By D. Fink

64 **Morphology Control of Materials and Nanoparticles**
Advanced Materials Processing and Characterization
Editors: Y. Waseda and A. Muramatsu

65 **Transport Processes in Ion-Irradiated Polymers**
By D. Fink

66 **Multiphased Ceramic Materials**
Processing and Potential
Editors: W.-H. Tuan and J.-K. Guo

67 **Nondestructive Materials Characterization**
With Applications to Aerospace Materials
Editors: N.G.H. Meyendorf, P.B. Nagy, and S.I. Rokhlin

68 **Diffraction Analysis of the Microstructure of Materials**
Editors: E.J. Mittemeijer and P. Scardi

69 **Chemical-Mechanical Planarization of Semiconductor Materials**
Editor: M.R. Oliver

70 **Applications of the Isotopic Effect in Solids**
By V.G. Plekhanov

71 **Dissipative Phenomena in Condensed Matter**
Some Applications
By S. Dattagupta and S. Puri

72 **Predictive Simulation of Semiconductor Processing**
Status and Challenges
Editors: J. Dabrowski and E.R. Weber

73 **SiC Power Materials**
Devices and Applications
Editor: Z.C. Feng

Springer Series in
MATERIALS SCIENCE

Editors: R. Hull R. M. Osgood, Jr. J. Parisi H. Warlimont

MIX
Papier aus verantwortungsvollen Quellen
Paper from responsible sources
FSC® C105338

If you have any concerns about our products,
you can contact us on
ProductSafety@springernature.com

In case Publisher is established outside the EU,
the EU authorized representative is:
Springer Nature Customer Service Center GmbH
Europaplatz 3, 69115 Heidelberg, Germany

Printed by Libri Plureos GmbH
in Hamburg, Germany